Springer Series in Optical Sciences

Volume 42

Edited by Theodor Tamir

Springer Series in Optical Sciences

Volume 42 **Principles of Phase Conjugation**
By B. Ya. Zel'dovich, N. F. Pilipetsky, and V. V. Shkunov

Volume 43 **X-Ray Microscopy**
Editors: G. Schmahl and D. Rudolph

Volume 44 **Introduction to Laser Physics**
By K. Shimoda

Volume 45 **Scanning Electron Microscopy**
By L. Reimer

Volume 46 **Holography in Deformation Analysis**
By W. Schumann, J.-P. Zürcher, and D. Cuche

Volume 47 **Tunable Solid State Lasers**
Editors: P. Hammerling, A. B. Budgor, and A. Pinto

Volumes 1–41 are listed on the back inside cover

B.Ya. Zel'dovich   N.F. Pilipetsky
V.V. Shkunov

Principles
of Phase Conjugation

With 70 Figures

Springer-Verlag
Berlin Heidelberg New York Tokyo

Dr. Boris Ya. Zel'dovich
Dr. Nikolai F. Pilipetsky
Dr. Vladimir V. Shkunov

Institute for Problems in Mechanics, Academy of Sciences of the USSR,
Prospekt Vernadskogo 101, SU-117526 Moscow, USSR

ISBN 3-540-13458-1 Springer-Verlag Berlin Heidelberg New York Tokyo
ISBN 0-387-13458-1 Springer-Verlag New York Heidelberg Berlin Tokyo

Library of Congress Cataloging in Publication Data. Zel'dovich, B. Ya. (Boris Ya.), 1944- Principles of phase conjugation. (Springer series in optical sciences ; v. 42) Bibliography: p. Includes index. 1. Optical phase conjugation. I. Pilipetsky, N. F. (Nikolai F.), 1931-. II. Shkunov, V. V. (Vladimir V.), 1953-. III. Title. IV. Series. QC446.3.067Z45   1985   535'.2   84-23546

© Springer-Verlag Berlin Heidelberg 1985
Printed in Germany

The use of registered names, trademarks, etc. in this publication does not imply, even in the absence of a specific statement, that such names are exempt from the relevant protective laws and regulations and therefore free for general use.

Offset printing: Beltz Offsetdruck, 6944 Hemsbach/Bergstr. Bookbinding: J. Schäffer OHG, 6718 Grünstadt.
2153/3130-543210

Preface

This book has been prompted by our desire to share with others our apprecia-
tion of the harmony and beauty in a particular sphere of modern optics
known as "optical phase conjugation". Practical applications of the phase-
conjugated wave are likely to be far-reaching. Optical phase conjugation
(OPC) combines in itself aesthetic and pragmatic attractiveness, a synthesis
that has made OPC a subject of general attention. The figure presents the approximate rate of publications (number of articles per year) on OPC in the
world literature for recent years, the lower curve denoting the work carried
out in the USSR. The efforts of a large unofficial international collective have
yielded an impressive result.

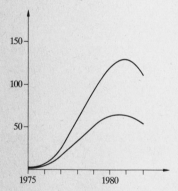

At present, the physical processes underlying various OPC methods are
quite understandable, and it is the physics of OPC to which our book is
devoted. Practical and scientific applications of phase-conjugated waves,
which are of no less interest, have been touched upon in short, as major
achievements in this sphere are a matter of the future.

Today there are two main methods of OPC: i) by backward stimulated
light scattering, ii) by four-wave mixing. Naturally, much attention is given to
these methods in our book which, after the introductory Chap. 1, can be
divided into two almost independent parts – Chaps. 2 – 5, and Chaps. 6 – 8.

We have made every endeavor to present the basic ideas and theoretical material as comprehensively as possible so that the reader does not need to refer to

the original papers. In discussing experimental work we have confined our-selves to the main results, avoiding particularities. We do not refer to theoretical work in the body of the book, which, in our opinion, makes the reading (and writing) of the book much easier. However, we do refer to original experimental work so that the reader may become acquainted with necessary details.

We express our deep gratitude to Academician N. G. Basov and Academician A. Yu. Ishlinsky for their support of our investigations into the OPC problem. We are thankful to Associate Member of the Academy of Sciences of the USSR D. M. Klimov, Professor V. B. Librovich, Professor Yu. P. Raizer, and Professor I. I. Sobelman for their help, attention, and interest in our work.

Valuable discussions and collaboration with V. V. Ragulsky played an important role in our presentation of the material. To him and many others at the Institute for Problems in Mechanics, the Academy of Sciences of the USSR, we are very much obliged, for scientific collaboration and help. In particular, we should like to mention the contributions made by V. N. Blaschuk, A. V. Sudarkin, A. V. Mamaev, O. L. Kulikov, A. L. Gyulamiryan, N. B. Baranova, M. A. Orlova, T. V. Yakovleva, V. I. Popovichev, and other co-authors of our work. We are also grateful to Yu. P. Alexeenko for help in the translation of our manuscript.

We are grateful to Dr. H. Lotsch of Springer-Verlag for encouragement and support of this book. The copy editor, Gillian Hayes, helped us enormously to improve both the scientific meaning of the text and the English, for which we are very indebted.

Moscow, December 1984

B. Ya. Zel'dovich
N. F. Pilipetsky
V. V. Shkunov

Contents

1. **Introduction to Optical Phase Conjugation** 1
   1.1 What a Phase-Conjugated Wave Is 1
   1.2 Ways of Generating a Phase-Conjugated Wave 5
       1.2.1 OPC in Static and Dynamic Holography 5
       1.2.2 Parametric Methods of OPC 7
       1.2.3 What Stimulated Scattering Is 9
       1.2.4 Discrimination Mechanism of OPC-SS 11
       1.2.5 Comparison of Various Methods 13
   1.3 What Is OPC For? 14
       1.3.1 Self-Compensation for Distortions 14
       1.3.2 Self-Targeting of Radiation 17
       1.3.3 Control of Spatial-Temporal Structure 18
       1.3.4 Scientific Applications 22
   1.4 Literature 24

2. **Physics of Stimulated Scattering** 25
   2.1 Steady-State Picture of Stimulated Brillouin Scattering 25
   2.2 Spontaneous and Stimulated Scattering;
       Quantum Representation of SS 31
   2.3 Other Types of Stimulated Scattering 35
       2.3.1 Stimulated Raman Scattering (SRS) by Molecular
             Vibrations 35
       2.3.2 Other Types of SRS 36
       2.3.3 Stimulated Rayleigh-Wing Scattering (SRWS) 37
       2.3.4 Stimulated Temperature Scattering Due to Absorption
             (STS-II) 39
       2.3.5 General Properties of Stimulated Scattering 40
   2.4 Effect of Pump Depletion in Stimulated Scattering 41
   2.5 Dynamics of Stimulated Scattering 45
       2.5.1 Spectral Width of Stimulated Scattering 46
       2.5.2 Stimulated Scattering Build-Up Wave 47
       2.5.3 Calculation of SS Dynamics 48
       2.5.4 Determination of Threshold in Nonstationary Condition . 50
       2.5.5 Above-Threshold Behavior of SS and Loop Scheme 52

2.6 Stimulated Scattering of Non-Monochromatic Radiation 53
    2.6.1 Quasi-Static Case 53
    2.6.2 Broad-Band Pump in a Short Medium 54
    2.6.3 Effect of Group Velocity Detuning 55
2.7 Excitation of Anti-Stokes Components – Four-Wave Processes 57
    2.7.1 Presentation of Nonlinear Polarization 57
    2.7.2 Excitation of Anti-Stokes Components Under
          Forward Scattering 59
    2.7.3 SBS of Counterpropagating Waves 60
2.8 Polarization Properties of Stimulated Scattering 63
2.9 Literature 64

3. **Properties of Speckle-Inhomogeneous Fields** 66
3.1 Central Limit Theorem. Gaussian Statistics of Speckle Fields .. 66
    3.1.1 Characteristic Function 67
    3.1.2 Central Limit Theorem 68
    3.1.3 Gaussian Statistics of Speckle Fields 70
3.2 Parabolic Wave Equation 72
3.3 Characteristic Dimensions of Speckle-Structure Inhomogeneities 73
3.4 Focused Speckle Field 76
3.5 Dislocations of Wavefront 79
3.6 Literature 84

4. **OPC by Backward Stimulated Scattering** 85
4.1 Experimental Discovery of OPC-SS 85
4.2 Discrimination Conditions Against Non-Conjugated Waves
    in Light Guides 87
4.3 The Specklon 91
4.4 Structure of Uncorrelated Waves 94
4.5 Experimental Investigation of Discrimination in a Light Guide . 96
4.6 OPC-SS in Focused Beams 98
    4.6.1 Speckle Beam 98
    4.6.2 Ideal Beam 101
    4.6.3 Weakly Distorted Beam 102
4.7 Registration Methods and Quality Estimation of OPC 103
4.8 Literature 107

5. **Specific Features of OPC-SS** 108
5.1 Theory of the Specklon 108
    5.1.1 "Serpentine" Distortions 109
    5.1.2 Spectral and Angular Distortions 113
    5.1.3 Specklon Envelope Equation 117

5.2 Specklon Phase Fluctuations 119
    5.2.1 Temporal Phase Fluctuations 119
    5.2.2 Transverse Coherentization 120
5.3 Theory of OPC-SS in Focused Speckle Beams 121
5.4 OPC of Depolarized Radiation 124
    5.4.1 Theory of OPC-SS of Depolarized Radiation 125
    5.4.2 Experimental Results 127
5.5 Nonlinear Selection of Non-Monochromatic Radiation 130
5.6 Effect of Saturation in OPC-SS 133
5.7 OPC-SS with Reference Wave 136
    5.7.1 SS of Radiation with Incomplete Speckle Modulation ... 136
    5.7.2 SS with Reference Wave Through Focusing with a Lens .. 138
    5.7.3 OPC-SS of Under-Threshold Signals 140
    5.7.4 OPC-SS of Large-Diameter Beams 141
5.8 OPC by Other Types of SS 142
5.9 Literature 143

6. OPC in Four-Wave Mixing 144
6.1 Principles of OPC-FWM 144
6.2 Selective Properties of OPC-FWM 146
6.3 Polarization Properties of OPC-FWM 150
6.4 Effect of Absorption on OPC-FWM Efficiency 152
6.5 Theory of Coupled Waves 153
6.6 Nonstationary Effects in OPC-FWM 158
6.7 Reference Wave Instability 161
    6.7.1 Self-Focusing of a Beam as a Whole 161
    6.7.2 Plane Wave Instability 162
    6.7.3 Instability of Counterpropagating Waves 165
    6.7.4 Effect of Wave Self-Action on Efficiency and Quality of
          OPC-FWM 168
6.8 Literature 169

7. Nonlinear Mechanisms for FWM 171
7.1 Molecular Orientation 171
7.2 Saturation Nonlinearities 173
    7.2.1 Phenomenological Approach 173
    7.2.2 Resonant Gases 175
    7.2.3 Absorbing Dyes and Thermal Effects 180
    7.2.4 Gain Saturation 182
7.3 Nonlinearities in Semiconductors 183
    7.3.1 Bound Electron Nonlinearity 183
    7.3.2 Nonparabolicity of Conduction Band 184

7.3.3 Generation of Free Carriers 185
7.3.4 Photorefractive Crystals 187
7.4 SBS Nonlinearity and Parametric SBS Generation 189
7.5 Literature 192

8. **Other Methods of OPC** 194
8.1 OPC by a Reflecting Surface (OPC-S) 194
8.1.1 Principle of OPC-S 194
8.1.2 Mechanisms of Surface Nonlinearity 196
8.2 OPC in Three-Wave Mixing (TWM) 199
8.3 Medium Parameter Modulation at Doubled Frequency 201
8.4 OPC of Sound Waves 203
8.5 Quasi-OPC: Retroreflectors 205
8.6 Exotic Methods of OPC 209
8.6.1 Photon Echo 209
8.6.2 Bragg's Three-Wave Mixing 211
8.6.3 OPC by Superluminescence 213
8.7 OPC in Forward FWM 214
8.8 Resonators with Phase-Conjugate Mirrors (PCM) 217
8.8.1 Structure of OPC-FWM Resonator Modes 217
8.8.2 Matrix Method in OPC Resonator Theory 220
8.8.3 Selection of Mode with Lowest Transverse Index 223
8.8.4 Resonators with an SBS Cell 226
8.9 Literature 227

References 229

Subject Index 249

1. Introduction to Optical Phase Conjugation

This chapter contains the basic ideas of optical phase conjugation. We have tried to present them in as popular a manner as possible and almost without the use of mathematics. We also discuss applications of optical phase conjugation.

1.1 What a Phase-Conjugated Wave Is

If you want to quickly get the idea of what a phase-conjugated wave is, you should take a summer holiday and make yourself comfortable on the bank of a lake on a clear windless day. Then, take a stone in your hand and throw it into the lake. A low splash will be accompanied by water sprays flying in different directions and on the surface of the water you will see waves running to all sides from the place where the stone was thrown. If you have a motion-picture camera with you, do not spare your film and shoot the events observed from beginning to end. When you return home, you will be eager to show the film to your friends. Naturally, they will want to see it again and you will have to rewind the film. It is in doing just this (if the projector is in operation) that you can observe a reversed (phase-conjugated) wave. From all over the lake the waves run towards the place where the stone dropped. All water sprays gather in the center with an amazing accuracy, the stone itself leaps out of the water, and a moment later you can see the glassy surface of the lake.

Using cinema we have succeeded in reversing time. To our disappointment, this almost never works in real life. Any motion is usually accompanied by dissipation, that is to say, by irreversible conversion of mechanical energy into heat. That is why the picture of "reversed motion" would violate the second law of thermodynamics. Even if energy dissipation were negligible, we would not be able to define the initial positions and velocities of all particles, which is necessary to reproduce all the details of a reversed process.

The spreading of waves and the scatter of splashes caused by a fallen stone have an analogy in coherent optics. Indeed, a highly directional laser beam is scattered, "splashed" into smaller beams of different directions (Fig. 1.1), when it passes through a medium whose refractive index has irregular in-

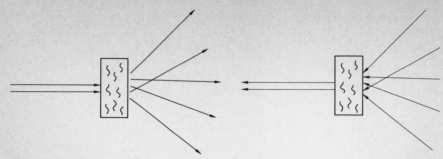

Fig. 1.1. A parallel beam of light is scattered in different directions after passing through an inhomogeneous medium

Fig. 1.2. The wave is reversed with respect to that shown in Fig. 1.1: an inhomogeneous medium corrects for the beam

homogeneities. Should we succeed, as in the case of a cine projector, in reversing time, we could observe a picture no less wonderful: a "shaggy" light beam approaches the inhomogeneous medium and becomes ideally directional after passing it (Fig. 1.2). It is interesting that in optics such a procedure, i.e., time reversal, can indeed be realized. A number of methods for generating a phase-conjugated wave have been suggested and realized in recent decades. The optical situation is favorable in two aspects.

First, the high transparency of optical media provides insignificant dissipation which could lead to time irreversibility. Thus, in both geometrical and wave optics of transparent media the principle of time reversibility holds true: the equations remain invariant under sign reversal of time. Therefore, for any solution of the wave equation (for instance, for a beam "splashed" by an inhomogeneous medium, Fig. 1.1) there exists a "reversed" solution of the same equation (Fig. 1.2).

Second, in coherent optics it is possible to define positions and directions, amplitudes and phases of elementary rays that provide a detailed reproduction of the propagation of the reversed wave. In particular, this can be achieved because a coherent laser beam possesses a relatively small number of degrees of freedom (field oscillators) whose "generalized velocities" are to be reversed. Indeed, polarized radiation of wavelength λ in a solid angle $\Delta\Theta$ and within a frequency interval Δf passing an area S within a time interval T has no more than $N = T\Delta f S \Delta\Theta/\lambda^2$ independent degrees of freedom. If the radiation is monochromatic, then according to the uncertainty relation $\Delta f \cdot T \approx 1$, and with $S = 1\,\mathrm{cm}^2$, $\lambda = 10^{-4}\,\mathrm{cm}$, $\Delta\Theta = (10^{-2}\,\mathrm{rad})^2 = 10^{-4}\,\mathrm{sterad}$, we have $N \lesssim 10^4$. Simple acoustic movements (waves) in a liquid possess, in a typical situation, even fewer degrees of freedom ($\lambda \gtrsim 10^{-1}\,\mathrm{cm}$), and, therefore, it is possible to reverse them in the absence of dissipation. The reversal of a complex molecular motion in $1\,\mathrm{cm}^3$ of a liquid would, however, require information on $N \sim 10^{23}$ degrees of freedom, i.e., of the order of Avogadro's number.

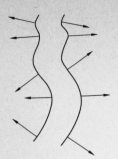

Fig. 1.3. The wavefronts of counterpropagating incident and phase-conjugated waves

The oscillations of a real monochromatic wave field $E_{\text{real}}(R, t)$ are characterized by a local amplitude $|E(R)|$ and a local phase $\phi(R)$:

$$E_{\text{real}}(R, t) = |E(R)|\cos[\omega t - \phi(R)] , \tag{1.1.1}$$

where ω is the field frequency. The wave

$$E_{\text{conj}}(R, t) = |E(R)|\cos[-\omega t - \phi(R)] \equiv |E(R)|\cos[\omega t + \phi(R)] \tag{1.1.2}$$

is then a conjugate wave relative to (1.1.1), that is to say, for monochromatic fields time reversal is equivalent to a sign reversal of the phase, $\phi(R) \to -\phi(R)$, at all points in space. For instance, to a traveling plane wave

$$E_1(R, t) = A\cos(\omega t - kz - \phi_0)$$

corresponds the phase-conjugated wave

$$E_2(R, t) = A\cos(\omega t + kz + \phi_0) ,$$

which propagates in the opposite direction. In the case of fields with a more complicated spatial structure it is convenient to illustrate the reversal procedure using the concept of the wavefront, defined as a hypothetical surface (or family of surfaces) of constant phase, $\phi(R) = \text{const}$. Normals to this surface coincide with the rays characterizing the local direction of the waves. Both direct and phase-conjugated waves have accurately coinciding wavefront surfaces, $\phi_{\text{conj}}(R) = -\text{const}$, and are counterpropagating (Fig. 1.3).

It is for this reason that the generation of a conjugated wave in Russian scientific literature is referred to as wavefront reversal (obrashenie volnovogo fronta).

In the theory of oscillations and waves it is convenient to use complex amplitudes $E(R) = |E(R)|\exp[i\phi(R)]$, so that

$$E_{\text{real}}(R, t) = \tfrac{1}{2}[E(R)e^{-i\omega t} + E^*(R)e^{i\omega t}] , \tag{1.1.3}$$

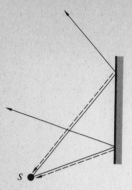

Fig. 1.4. A wave radiated by a point source and reflected by a conventional mirror (———) and by a phase-conjugate mirror (– – –)

where the asterisk denotes the complex conjugate. In terms of complex amplitudes, the phase sign reversal corresponds to a transition from $E(R)$ to the complex conjugate $E^*(R)$:

$$E_{\text{conj}}(R, t) = \tfrac{1}{2}[E^*(R)e^{-i\omega t} + E(R)e^{i\omega t}] , \tag{1.1.4}$$

with the relationship between the complex amplitudes of the direct and phase-conjugated waves being

$$E_{\text{conj}}(R) = E^*(R) , \tag{1.1.5}$$

or $\phi_{\text{conj}}(R) = -\phi(R)$. It is due to this that the term "optical phase conjugation" (OPC or PC for short) appears; the term "time reversal" is also used.

Wave propagation in a linear medium obeys the superposition principle. In addition to the wave $E_{\text{conj}}(R) = E^*(R)$, the field

$$E'_{\text{conj}}(R) = ce^{i\psi}E^*(R) \tag{1.1.6}$$

is also an exact solution of the wave equation, where $c \exp(i\psi)$ is an arbitrary complex constant. Therefore, any wave satisfying (1.1.6) is called a phase-conjugated wave, while (1.1.5) corresponds to the particular case $c = 1$, $\psi = 0$.

It is not difficult to reverse a plane wave. If we know in advance the direction n of its propagation, reversal can be obtained by placing a plane mirror strictly perpendicular to n. However, we cannot do this with a spherical wave. In this case, conjugation of a diverging spherical wave propagating from the source would yield a converging spherical wave propagating to the same source. When it is reflected from a plane mirror the spherical wave remains divergent (Fig. 1.4). To succeed in phase conjugation, we should use a mirror whose center of curvature is at the source. In general, the reversal of a wave of arbitrary structure requires a mirror whose profile strictly coincides with the wavefront profile, i.e., each wave requires a special mirror.

It is wonderful that methods of nonlinear optics and holography do enable us to realize a "magic mirror" automatically adjusting itself to the shape of any incident wave in order to reflect the signal in the form of a phase-conjugated wave.

The interest in the problem of optical phase conjugation is connected mainly with the extremely alluring prospects of its applications; the reader will learn of these in Sect. 1.3. To understand and evaluate these applications better, the reader not familiar with the problem of OPC should first get to know the major methods of generating a phase-conjugated wave (Sect. 1.2).

1.2 Ways of Generating a Phase-Conjugated Wave

1.2.1 OPC in Static and Dynamic Holography

To begin with, we shall consider the generation of a phase-conjugated wave in static holography, as this scheme is rather simple to understand. Suppose we have a photoplate (Fig. 1.5a) onto which a signal wave $E_3(r)$ and a reference wave $E_1(r)$, which is coherent with the signal wave, are incident (r stands for the position vector in the xy plane; by R we denote the position vector characterizing all three coordinates x, y, and z). The interference pattern of these fields is recorded in the photosensitive layer and then the hologram is developed. As a result, interference inhomogeneities are displayed as the modulation of optical properties of the layer (its absorptivity and refractive index). If the layer is thin enough, then its effect on the reconstructing wave may be described by introducing the complex transmission coefficient $t(r)$ for the field amplitude, $E_{\text{trans}}(r) = t(r)E_{\text{inc}}(r)$.

In a simple approximation, variations $\delta t(r)$ in the transmission are related to intensity variations of the recording field so that

$$\delta t(r) = \text{const}\,[E_1^*(r)E_3(r) + E_1(r)E_3^*(r)] \,. \tag{1.2.1}$$

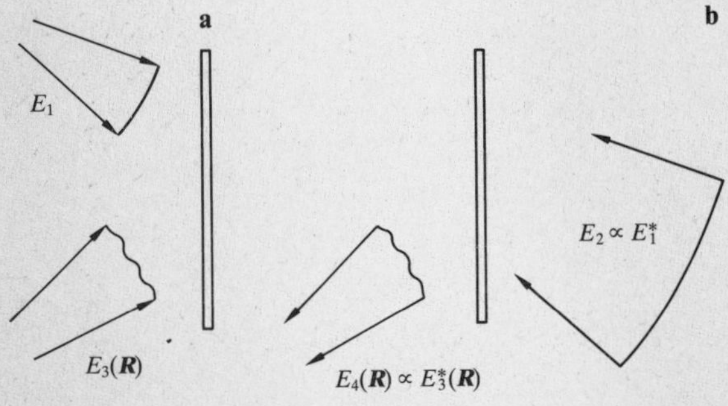

Fig. 1.5. (a) Recording the hologram of signal $E_3(R)$ with reference wave E_1. (b) Read-out of the phase-conjugated signal $E_4(R) \propto E_3^*(R)$ by counterpropagating reference wave E_2

Let us try to reconstruct the hologram, not in the ordinary way [when the reading wave $E_2(r)$ coincides with the recording reference wave, $E_2(r) \propto E_1(r)$], but so that the reading wave $E_2(r)$ and the recording reference wave $E_1(r)$ counterpropagate, $E_2(r) \propto E_1^*(r)$. This condition is usually not difficult to realize if the reference wave $E_1(r)$ is either plane or spherical (Fig. 1.5 b). Then passing the wave $E_2(r)$ through the hologram causes, owing to the second term in (1.2.1), the reconstruction of the field

$$E_4(r) \propto |E_1|^2 E_3^*(r), \qquad (1.2.2)$$

which propagates to meet the initial signal $E_3(r)$, and at $|E_1|^2 = $ const corresponds exactly to a phase-conjugated signal wave.

One of the disadvantages of this scheme is that a new hologram has to be recorded to conjugate each new wave $E_3(r)$, which is rather difficult. This disadvantage can be eliminated by using special media that do not require additional treatment, that is to say, those in which disturbances of optical properties [permittivity changes $\delta\varepsilon(r)$] occur immediately, in the presence of the interfering fields, and disappear after they are removed. Holography in such media is referred to as dynamic holography; wave interaction in such media is the subject of nonlinear optics, and the media themselves are called nonlinear.

Now suppose that a nonlinear medium is used for recording a volume hologram (Fig. 1.6a), and that three waves are simultaneously incident on the medium – the reference wave $E_1(r)$, the signal $E_3(r)$, and a second reference wave $E_2(r)$ obtained by reflection of the first one from a suitable mirror so that the condition $E_2(r) \propto E_1^*(r)$ is fulfilled. In such a medium, recording the holographic grating $\delta\varepsilon \propto E_1(r)E_3^*(r)$ and reading it $\delta\varepsilon(r)E_2(r)$ are time-coincident (Fig. 1.6b). Such a scheme is advantageous in many respects: it is not necessary to install the developed hologram in exactly the initial place, the hologram reverses every incident signal wave $E_3(r)$ by automatically adjusting itself to it, and the phase-conjugated wave $E_4(r) \propto E_3^*(r)$ is excited practically

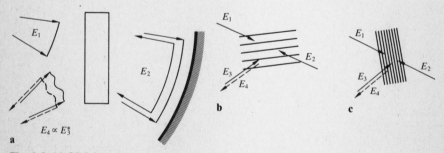

Fig. 1.6. (a) OPC by volume dynamic holography (OPC-FWM). Both holographic gratings – (b) for copropagating waves E_1 and E_3 and (c) for counterpropagating waves E_2 and E_3 – produce a phase-conjugated wave E_4

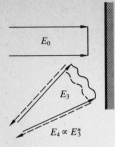

Fig. 1.7. Phase conjugation of signal E_3 by a nonlinear surface (OPC-S)

instantaneously, i.e., in real time. It should be noted that in sufficiently thick nonlinear media the OPC efficiency may be increased due to one more process: the reference wave $E_2(r)$ interfering with the signal $E_3(r)$ records the reflective holographic grating $\delta\varepsilon \propto E_2(r)E_3^*(r)$; its reading by the reference wave $E_1(r)$ also excites the phase-conjugated wave $E_4(r) \propto E_1(r)\delta\varepsilon(r) \propto E_3^*(r)$ (Fig. 1.6c). This method of phase conjugation in real time is referred to as OPC by four-wave mixing (OPC-FWM); the three incident waves E_1, E_2, E_3 and the fourth, i.e., the phase-conjugated wave E_4, are meant here. Sometimes this method is called OPC by dynamic holography.

Now imagine a reflecting surface whose amplitude reflectivity ϱ depends on the intensity of the incident radiation. If two waves — the reference wave $E_0(r)$ and the signal $E_3(R)$ (Fig. 1.7) — interfere on this surface, then the hologram $\delta\varrho(r)$ is recorded in the form of reflectivity modulation:

$$\delta\varrho(r) \propto E_0^*(r)E_3(r) + E_0(r)E_3^*(r) . \tag{1.2.3}$$

The reference wave $E_0(r)$ is reflected by the disturbed surface and becomes $E_{\mathrm{ref}}(r) = \varrho(r)E_0(r)$; owing to the second interference term it acquires the component

$$E_4(r) \propto E_0^2(r)E_3^*(r) . \tag{1.2.4}$$

If a plane reference wave $E_0(r)$ is incident normally on a plane mirror, then $E_0(r) = \mathrm{const}$ on its surface, and the wave $E_4(r)$ is a phase-conjugated wave with respect to the signal. This method, which also provides real-time phase conjugation, is referred to as OPC by a surface (OPC-S).

1.2.2 Parametric Methods of OPC

Suppose that a parameter of a medium (in optics, the permittivity ε) is time modulated to satisfy the harmonic equation

$$\delta\varepsilon(t) = A\cos(2\omega t + \phi) , \tag{1.2.5}$$

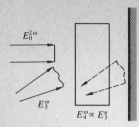

Fig. 1.8. OPC by parametric three-wave mixing

$E_4^\omega \propto E_3^*$

which has a frequency of 2ω − twice as large as the frequency of the signal $E_3(R, t) = \frac{1}{2}[E_3(R)\exp(-\mathrm{i}\omega t) + E_3^*(R)\exp(\mathrm{i}\omega t)]$ propagating in the medium. Then the signal induces in the medium an additional disturbance [in optical media, a dielectric displacement disturbance $\delta D(R, t)$] of the form $\delta D(R, t) = \delta\varepsilon(t)E_3(R, t)$. This disturbance contains, among others, the term

$$\delta D_4(R, t) = \frac{1}{4}[A\,\mathrm{e}^{-\mathrm{i}\phi}E_3^*(R)\,\mathrm{e}^{-\mathrm{i}\omega t} + \text{c.c.}]\,, \tag{1.2.6}$$

where c.c. stands for the complex conjugate of the explicitly written term.

If the values of A and ϕ are constant, i.e., the modulation $\delta\varepsilon(t)$ from (1.2.5) is synchronous in the entire volume, then disturbance (1.2.6) effectively radiates the wave $E_4(R) \propto E_3^*(R)$, which is phase conjugated with respect to the signal. For acoustic waves in piezoelectrics, the volume-synchronous modulation of the sound velocity can be obtained by applying an external homogeneous electric field.

In optics, the situation is somewhat more difficult. There exist media whose permittivity varies under the action of strong optical fields $E_0(R, t)$ by an amount proportional to the local field E_0, i.e., by $\delta\varepsilon \propto \chi^{(2)}E_0(R, t)$. Here $\chi^{(2)}$ is a coefficient characterizing the so-called second-order susceptibility of the medium. The idea consists in using a strong reference wave $E_0(R, t)$ at a frequency twice as large as that of the signal, Fig. 1.8. Unfortunately, the phase of the field $E_0(R, t)$ at a frequency 2ω varies strongly even over a distance of the order of the wavelength $\lambda \lesssim 10^{-4}$ cm, so that modulation synchronous throughout the volume is impossible. For instance, in the case of a traveling plane reference wave,

$$\delta\varepsilon(R, t) \propto \chi^{(2)}[E_0\exp(-2\mathrm{i}\omega t + 2\mathrm{i}kz) + \text{c.c.}] \tag{1.2.7}$$

and the phase-conjugated disturbance is

$$\delta D_4(R, t) \propto \chi^{(2)}[E_0 E_3^*(R)\exp(-\mathrm{i}\omega t + 2\mathrm{i}kz) + \text{c.c.}]\,. \tag{1.2.8}$$

If the signal E_3 propagates approximately in the direction of the reference wave, $E_3(R) \approx E_3\exp(\mathrm{i}kz)$, then the disturbance (1.2.8) represents a forward traveling wave with the same phase velocity, $\delta D_4 \propto [E_0 E_3^*\exp(-\mathrm{i}\omega t + \mathrm{i}kz) + \text{c.c.}]$, and, consequently, it excites effectively (synchronously in the entire volume) the wave E_4' with a conjugate phase. This wave E_4' propagates forwards; in order to obtain a backward propagating phase-conjugated wave, it is necessary to reflect it from a mirror installed perpendicular to the direction of travel of the reference wave.

An interesting fact is that parametric processes of the type considered here may contribute to OPC by four-wave mixing. This may occur if the nonlinearity of the medium is sufficiently fast, and permittivity variations $\delta\varepsilon \propto \chi^{(3)}E^2$ can follow the instantaneous (within a fraction of the light period) value of the field magnitude squared, $E^2 \propto \cos^2(\omega t + \phi)$. Here $\chi^{(3)}$ is a coefficient characterizing a third-order nonlinear susceptibility of the medium. In this case the term in E^2 proportional to the product of the field intensities of the two reference waves contains a component at the doubled light frequency:

$$\delta\varepsilon(R, t) \propto [E_1(R)E_2(R) \exp(-2i\omega t) + \text{c.c.}] . \tag{1.2.9}$$

It is interesting that in the case of exactly counterpropagating (i.e., mutually phase-conjugated) reference waves $E_1(R)$ and $E_2(R)$, where $E_2(R) \propto E_1^*(R)$, the phase of the modulation coefficient $E_1(R)E_2(R)$ in (1.2.9) is independent of coordinates, and, accordingly, the medium parameter modulation turns out to be synchronous throughout the whole volume.

If the surface reflectivity is modulated at the doubled frequency, $\delta\varrho(r, t) = \varrho_1 \cos(2\omega t + \phi)$, then the signal field $E_3(r, t) = \frac{1}{2}[E_3(r) \exp(-i\omega t) + \text{c.c.}]$ reflected by the surface contains the component $E_4(r, t) = \frac{1}{4}\varrho_1[\exp(-i\phi) \times E_3^*(r) \exp(-i\omega t) + \text{c.c.}]$. If the phase $\phi(R)$ of the modulation coefficient is constant along the surface, then this component will be phase conjugated relative to the signal.

1.2.3 What Stimulated Scattering Is

Before proceeding to OPC by stimulated light scattering (OPC – SS) we would like to remind the reader of what stimulated scattering (SS) is. Consider a situation with two monochromatic waves of different frequencies $E_L(R, t) = \frac{1}{2}[E_L(R) \exp(-i\omega_L t) + \text{c.c.}]$ and $E_S(R, t) = \frac{1}{2}[E_S(R) \exp(-i\omega_S t) + \text{c.c.}]$ propagating in a medium. Here, according to the established terminology, E_L is the laser pump wave and E_S is the Stokes wave. The SS process is possible if the frequency difference of the two waves $\Omega = \omega_L - \omega_S > 0$ is close to the frequency Ω_0 of certain intrinsic oscillations of the medium. In addition, a certain mechanism should exist which allows the traveling interference pattern of the fields $[E_L^*(R)E_S(R) \exp(i\Omega t) + \text{c.c.}]$ to excite these intrinsic oscillations and, moreover, which allows the latter to modulate the medium permittivity.

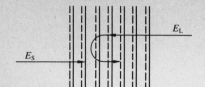

Fig. 1.9. Disturbances of density and permittivity $\delta\varepsilon$ are excited by the interference of the counterpropagating fields $E_L^* E_S$ at resonance, with a phase shift of $\pi/2$. (———) – fronts of $\delta\varepsilon$, (– – –) – fronts of "force" $E_L^* E_S$

In stimulated Brillouin scattering, for example, the intrinsic oscillations of the medium correspond to sound waves of extremely high frequencies, $\Omega_0/2\pi \sim 10^9$ Hz (the so-called hypersound), and the exciting mechanism is produced by electrostriction forces, which draw in substance to places with a larger local value of the field intensity $|E(R, t)|^2$.

At resonance, i.e., at $\Omega = \Omega_0$, in the steady-state condition, the density disturbances built up in the medium can be written as

$$\delta\varrho(R, t) = -iAE_L^*(R)E_S(R)e^{i\Omega t} + \text{c.c.} \tag{1.2.10}$$

It is important, in further considerations, that A is real, because at resonance, the built-up oscillations are $\pi/2$ out of phase [the factor $-i$ in (1.2.10)] with respect to the exciting force caused by interference of the fields E_L and E_S (Fig. 1.9). Density disturbances $\delta\varrho(R, t)$ in the medium vary the permittivity in such a way that it becomes modulated in space and time according to (1.2.10).

The traveling waves E_L and E_S are scattered by the inhomogeneities formed, and the term "stimulated scattering" emphasizes the fact that the scattering oscillations of the medium are stimulated by the interfering fields themselves. In particular, scattering of the laser wave $E_L(R)\exp(-i\omega_L t)$ caused by the first term in (1.2.10) gives rise to an additional term in the dielectric displacement at the Stokes wave frequency:

$$\delta D_S = E_S(R)\exp(-i\omega_S t)\delta\varepsilon_S, \quad \delta\varepsilon_S = -iA(\partial\varepsilon/\partial\varrho)|E_L|^2. \tag{1.2.11}$$

It should be noted, first, that the spatial structure of the oscillations built up in the medium is strictly coordinated with the interference pattern of the exciting light fields. Due to this, the disturbance δD_S is proportional to the Stokes wave field $E_S\exp(-i\omega_S t)$. So, the local result of the scattering $E_L \rightarrow E_S$ consists in the addition of $\delta\varepsilon_S$ to the permittivity for the Stokes field. Secondly, the additive $\delta\varepsilon_S$ is proportional to the laser pump intensity $|E_L|^2$. Thirdly, $\delta\varepsilon_S$ is pure imaginary. It can be proved that at $\Omega > 0$ the product $A(\partial\varepsilon/\partial\varrho)$ is positive, and, consequently, the imaginary part of $\delta\varepsilon_S$ is negative.

An interesting fact is that the interaction of the waves E_L and E_S by means of the medium leads to an exponential amplification of the wave E_S. Let us prove this, using simple considerations. The dependence of the complex amplitude of the traveling wave $E_S(z) \exp(-i\omega_S t)$ on the coordinate z along the propagation direction is of the form $\exp(ik_S z)$, where $k_S = (\omega_S/c)\sqrt{\varepsilon_S}$. The presence of the small additive $\delta\varepsilon_S$ leads to the correction $\delta k_S = \omega_S \delta\varepsilon_S / 2c\sqrt{\varepsilon_S}$ for the wave vector k_S. As $\delta\varepsilon_S$ is pure imaginary, δk_S gives rise not to a change in phase but to a change in the intensity of the wave E_S:

$$|E_S(z)|^2 \propto \exp(gz), \quad g = \frac{\omega_S}{c\sqrt{\varepsilon_S}} A\left(\frac{\partial\varepsilon}{\partial\varrho}\right)|E_L|^2. \tag{1.2.12}$$

The gain coefficient $g\,[\mathrm{cm}^{-1}]$ is proportional to the pump intensity: $g = G|E_L|^2$, where G is a positive constant characterizing the properties of the medium.

It is worth noting that the amplification of the Stokes signal is not due to the energy of the medium itself but due to the scattering $E_L \rightarrow E_S$ from the wave of higher frequency; the energy defect $\hbar(\omega_L - \omega_S) = \hbar\Omega_0$ is used to excite the phonon $\hbar\Omega_0$ in the medium.

For stimulated light scattering to be observed in a typical experimental situation, only a powerful laser pump E_L is sent into the medium. The Stokes wave E_S is initially excited owing to spontaneous scattering of the pump wave by thermal fluctuations of the same degrees of freedom of the medium which are involved in the interaction with the interfering light beams (for sound waves – by density fluctuations). Then, this priming Stokes wave is supported by the exponential amplification and can rise to such an extent that its intensity may become comparable with the intensity of the exciting pump. The priming signal E_S is usually weak, and, therefore, an effective wave re-scatter requires a total amplification by a factor $\exp(G|E_L|^2 z) \approx \exp(30)$.

Reaching this "threshold" gain $G|E_L|^2 z = 30$ requires large values of the pump intensity $|E_L|^2$ and amplification length z. The intensity is usually increased by using a pump beam strongly contracted in the transverse coordinate. In this case SS develops only in directions parallel or antiparallel to the pump propagation direction because for these directions the Stokes wave amplification length z is the largest.

1.2.4 Discrimination Mechanism of OPC-SS

OPC by stimulated Brillouin scattering (OPC-SBS) is usually realized in the geometry shown in Fig. 1.10. An intense laser pump beam E_L, following its passage through a distorter, is directed into an SBS-active medium. The purpose of the distorter (a phase plate, lens, amplitude mask) is to make the intensity distribution in the active medium rather non-uniform both in the transverse and longitudinal coordinates. The Stokes wave develops from

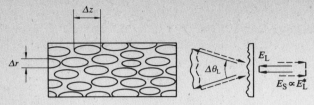

Fig. 1.10. A scheme for OPC stimulated scattering (OPC-SS). The dimensions of small-structure inhomogeneities in the pump intensity $|E_L(R, z)|^2$ are determined by the angular divergence $\Delta\theta_L$: $\Delta r \sim \lambda/\Delta\theta_L$, $\Delta z \sim \lambda/(\Delta\theta_L)^2$; a large number of inhomogeneities are present within the interaction volume. The scattered wave E_S is excited at the far (left) end of the cell by spontaneous noise and acquires, in the process of an exponential gain, the structure which is phase conjugate to the pump $E_L(R)$

spontaneous noise in a direction opposite to that of the pump wave and is exponentially amplified due to the SBS process as it propagates towards the window of the cell containing the medium (Fig. 1.10)[1].

It is a very surprising fact that with no specially prepared reference waves the amplified Stokes wave E_S assumes the transverse structure which is, to a high accuracy, phase conjugated with respect to the structure of the pump wave, i.e., $E_S(r) \propto E_L^*(r)$. It is in this connection that OPC-SS is sometimes referred to as optical phase self-conjugation.

Two properties of SS underlie the physical mechanism of OPC-SS: the very high *total* gain of the Stokes wave, $\exp(gz) \approx \exp(30)$, and the high spatial non-uniformity of the *local* gain $g(R) = G|E_L(R)|^2$ caused by inhomogeneities of the pump intensity $|E_L(R)|^2$ in the scattering volume. Owing to the very high total gain, even a moderate relative increase $\Delta g/g \sim 1$ may lead to radical changes. Spontaneous noise gives rise to scattered waves $E_S(R)$ with various transverse field structures. The effective gain for a given configuration $E_S(r, z)$ is determined in every section $z = $ const in the pump field $E_L(r, z)$ by the intensity overlap integral

$$g_{\mathrm{eff}} = \frac{G \int |E_L(r, z)|^2 |E_S(r, z)|^2 d^2r}{\int |E_S(r, z)|^2 d^2r}. \tag{1.2.13}$$

Therefore, a maximum of gain is observed for a wave $E_S(r)$ whose local maxima coincide with the maxima of the wave $E_L(r)$ everywhere in space. While propagating, each of the fields $E_L(r, z)$ and $E_S(r, z)$ varies its transverse structure due to diffraction and interference. If such variations are large enough, then the only possibility of keeping the intensity inhomogeneities coordinated throughout the whole volume in case of counterpropagation

---

1 The SBS process fails to develop in the propagation direction of the pump due to the extremely long settling time of the process

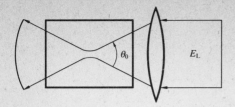

Fig. 1.11. The geometry of stimulated scattering at the focus of an ideal pump beam

consists in requiring the scattered field $E_S(r, z) \exp(ikz)$ to be phase-conjugated with respect to the exciting field $E_L(r, z) \exp(-ikz)$, i.e., $E_S(r, z) \propto E_L^*(r, z)$. As a result, it is the Stokes wave conjugated with respect to the pump that, under conditions of very high total gain, is primarily amplified and represented in the scattered radiation with an overwhelming weight. The rest of the waves are discriminated against owing to a lower gain. Thus, the physical mechanism of OPC-SS is based on amplification discrimination against the non-conjugated configurations of the backward-scattered wave in the field of an inhomogeneous pump wave.

As an example, consider the pump $E_L(r, z)$ with an irregular angular divergence $\Delta \theta_L$. Such a field possesses a speckle structure, i.e., fine-grained intensity inhomogeneities, which are well known to everyone able to see a rough surface illuminated by laser light. For such fields (obeying Gaussian statistics) the overlap integral (1.2.13) of a configuration $E_S(r)$ of the type $E_S(r) = \text{const} \cdot E_L^*(r)$ is, on average, twice as large as for any other configuration uncorrelated with the pump. Consequently, the total amplification of the intensity of this configuration is $\exp(15) \approx 3 \times 10^6$ larger than that of other configurations.

A second example corresponds to the focusing of the ideal pump beam into an SS-active medium (Fig. 1.11). The dimensions of the focal waist, where the intensity is maximum and where the SS is mainly developed, are $\Delta z_0 \times \Delta r_0 \sim (\lambda/\theta_0^2) \times (\lambda/\theta_0)$, where θ_0 is the angular convergence at the focus. Thus, the maximum gain $G|E_L|^2 z$ is observed within an area $(\Delta r_S)^2 \lesssim (\Delta r_0)^2$ for the Stokes waves which occupy a solid angle of opening angle $\theta_S \lesssim \Delta r_0/\Delta z_0 \approx \theta_0$. These conditions are satisfied only by the wave E_S with the diffraction divergence $\theta_S \approx \lambda/\Delta r_S \approx \theta_0$, which emerges from the waist. Thus, in this idealized situation, the scattered wave E_S may, as follows from simple geometric considerations, become conjugated with respect to the pump due to the amplification discrimination against non-conjugated waves.

1.2.5 Comparison of Various Methods

It should be noted that each of the above methods of OPC has both advantages and disadvantages of its own. An important advantage of OPC-SS, for instance, is the fact that optical phase self-conjugation is realized in this method. Due to this, arbitrary distortions that arise when the exciting wave passes through optically inhomogeneous elements and media have very little effect on OPC, where quality is consequently rather high. To the disadvantage of OPC-SS we should attribute the threshold nature of the SS process, as a result of which a sufficiently high power is required of the wave to be conjugated.

In holographic methods (static holography, OPC-FWM, OPC by a reflecting surface) this power requirement concerns the reference waves, and the wave to be conjugated may be much weaker. A great advantage of these methods is the possibility of phase conjugating a signal with a reflectivity more than unity, i.e., the possibility of amplification. In addition, these methods provide wider possibilities for signal selecting, control of the conjugated wave and choice of the speed of response. The main drawback of these methods consists in the stringent requirements on the optical quality of the reference waves and the nonlinear medium.

Although rather attractive from the general point of view, the parametric methods of PC have gained little ground in optics because of the difficulty in creating synchronous modulation at the doubled frequency in a large volume. Nowadays the majority of investigations are devoted to two basic methods: OPC-SS and OPC-FWM. It is to these methods that particular attention is given in the subsequent chapters of this book.

1.3 What Is OPC For?

1.3.1 Self-Compensation for Distortions

One of the most interesting applications of a phase-conjugated wave is connected with its ability to automatically restore its structure during the backward pass through an optically inhomogeneous medium (Figs. 1.1, 2).

Double-Pass Amplifier. One of the most important tasks of laser technology is the generation of intense highly directional beams. An optical amplifier is often used to increase the beam power. Unfortunately, a high power-takeoff is almost always accompanied by considerable degrading of the directivity of the radiation through the optical inhomogeneities in the amplifier's working medium. OPC allows self-compensation to be made for distortions, both static and dynamic (varying from pulse to pulse or even within a pulse), introduced by phase inhomogeneities of the amplifier. A double-pass self-compensation scheme is presented in Fig. 1.12. Suppose that a low-power but highly directional beam is fed to the input of an amplifier with optical inhomo-

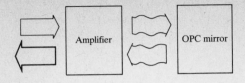

Fig. 1.12. Double-pass self-compensation scheme

geneities. Beam amplification during the first pass increases the angular divergence of the beam. Suppose the beam emerging from the amplifier undergoes the operation of OPC. Then the phase-conjugated wave will pass back into the amplifier and be additionally amplified therein. It is of importance that those same inhomogeneities which distorted the beam structure in the direct pass correct the amplified conjugated wave in the backward pass to give a high directivity of the initial low-power beam.

If two amplifiers with inhomogeneities of the same type are available, the scheme may be modified so that the radiation performs the second pass (after PC) through the second amplifier.

Resonators with Phase-Conjugate Mirrors (PCM). If the scheme described above fails to provide sufficient amplification in two passes, we may use a scheme with a larger number of double passes. Let a portion of the highly directional radiation corrected during each pair of passes return into the amplifier through reflection from an ordinary mirror. The latter and the PCM form an optical resonator. This resonator is operable both as a regenerative amplifier of the input signal and as an oscillator excited by the internal spontaneous noise. In order to obtain highly directional radiation in the latter case, one requires, as in a laser with ordinary mirrors, an aperture suppressing higher-order transverse modes. OPC facilitates the generation of an output beam of diffraction quality even if optical inhomogeneities are present in the resonator elements, due to self-compensation for distortions.

Furthermore, both longitudinal and transverse modes of resonators with PCMs possess a number of unique properties which is why OPC resonators are extremely interesting from the scientific point of view.

Compensation for Image Distortions in a Light Guide. Suppose that an image is applied to the input of a fiber light guide by a coherent monochromatic field $E_{in}(r)$ which excites a number of modes with different transverse indices. In the course of its propagation, the image is quickly distorted due to the difference in the phase velocities of the different transverse modes. If we conjugate the radiation after it has covered a certain length L, then the initial image will be obtained following the backward pass through the light guide, due to the self-compensation effect. Strictly speaking, we obtain a reconstructed field $E_{rec}(r) \propto E_{in}^*(r)$ which gives the same image, i.e., the pattern of intensity

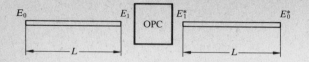

$I_{rec}(r) \propto |E_{in}(r)|^2$, as the initial field does. Indeed, a loss-free light guide may be considered as the distorting element (amplifier) shown in Fig. 1.12. Moreover, if the light guide is ideally uniform over its entire length, then the backward pass through this same light guide may be substituted by an equivalent one through another light guide of the same length L, Fig. 1.13. Thus, image transmission through a multimode fiber light guide can be made distortionless if two successive light guide sections of equal length are used with a reversal or phase-conjugation operation between them. If we want to reconstruct the field as well, then a second phase-conjugate device should be installed following the second light guide.

Compensation for Temporal Spread of Pulses. The propagation of pulses through a medium with dispersion corresponds, in a rough approximation, to the displacement of the envelope with group velocity $v_{gr} = \partial\omega/\partial k$. If a pulse $E_{in}(t)$ is of a sufficiently short duration τ, then the width of its spectrum $\Delta\omega \gtrsim \tau^{-1}$ is large, and, oven a large enough length, various spectral components arrive at different times because the group velocity $\partial\omega/\partial k$ is not constant within the limits of the pulse spectrum. As a result, a temporal spread of the pulse takes place. This phenomenon is particularly important for a long propagation distance through a single-mode fiber light guide. The idea of compensation for the dispersion spread consists in delaying faster components in the second passage and in accelerating the slower components so that the total time for their travel should be approximately equal. This can be achieved by inverting the position of the spectral components with respect to the central frequency ω_0 of the pulse. Such an inversion is possible, for example, through phase conjugation by parametric and four-wave methods; if $E_{in}(t) = a_p\exp[-i(\omega_0+p)t]$, then $E_{con}(t) = \exp(-2i\omega_0 t)\ E_{in}^*(t) = a_p^*$ $\times\exp[-i(\omega_0-p)t]$. As a result, the pulse at the output of the second light guide will be reconstructed with its initial duration so that $E_{rec}(2L, t) \approx E_{in}^*(t-2L/v_0)$, where $v_0 = \partial\omega/\partial k$ at the point $\omega = \omega_0$.

Compensation for temporal spread is also possible in two successive single-mode light guides of lengths L_1 and L_2, providing the condition $L_1 d^2k_1/d\omega^2 = L_2 d^2k_2/d\omega^2$ is satisfied[2].

---

2 If $L_1 d^2k_1/d\omega^2 = -L_2 d^2k_2/d\omega^2$, compensation for temporal spread is possible without a phase-conjugating device

1.3.2 Self-Targeting of Radiation

Focusing an intense light pulse on a small target is required in laser thermonuclear fusion (LTF). In LTF installations it is necessary to solve two tasks
simultaneously: first, to produce a powerful light pulse of low angular divergence and, second, to focus it accurately on a target. The application of OPC
is possible, first of all, in the solution of the first task (Sect. 1.3.1). It is worth
noting that the properties of a phase-conjugated wave allow one, in principle,
to solve both tasks simultaneously. A suitable scheme is shown in Fig. 1.14.
The pulse from an additional laser AL of moderate power illuminates a target.
A portion of the radiation reflected by the target gets into the aperture of a
high-power laser, passes an amplifier, and is then incident on an OPC device.
The phase-conjugated wave is amplified in the reverse pass through the laser
and takes off the stored energy. Distortions connected both with the amplifier
inhomogeneities and with imperfections of manufacture and the adjusting of
the focusing system are automatically compensated in the backward pass.
Consequently, the radiation is delivered exactly onto the target as if no errors
were present in the amplifier and focusing system. Moreover, if the addi-
tional-laser beam is sufficiently broad, there is no need to know the target
position in advance. The only condition is that the target should be within the
angle of vision of the OPC system. This scheme is called an OPC self-
targeting system. It is also possible to have a situation where the system target-
amplifier-OPC mirror forms an oscillator with self-excitation, without additional illumination from AL.

The self-targeting scheme can operate even if considerable phase inhomogeneities, say atmospheric ones, are in the path between the laser and target.
In particular, turbulent inhomogeneities of the refractive index largely affect
laser-beam communication through the atmosphere by limiting the communication distance due to irregular ray deviation. An OPC self-targeting
system may be useful in two respects: i) for automatic tracking of the correct
direction of communication including the case when the receiver and transmitter are traveling slowly; ii) for compensation of undesirable inhomogeneity
effects.

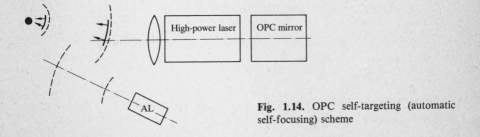

Fig. 1.14. OPC self-targeting (automatic
self-focusing) scheme

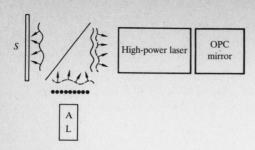

Fig. 1.15. OPC application in laser photo-lithography

Both in self-targeting and communication, a fast relative movement of the transmitter and receiver gives rise to a number of additional effects: a longitudinal motion with velocity $v_\parallel$ causes a Doppler shift of the reflected signal, $\Delta\omega/\omega = 2v_\parallel/c$, while a transverse motion with velocity $v_\perp$ leads to an angular error in self-targeting, $\alpha = 2v_\perp/c$. OPC makes it possible to compensate for and even use these effects.

The OPC self-targeting scheme may also be useful in photo-lithography (Fig. 1.15) where the rays of a high-power laser form the image of a complex transverse intensity distribution with a high spatial resolution on a given surface (for instance, on the surface of a blank for a microcircuit). In the conventional scheme, profiling the transverse intensity distribution in a laser beam by means of an amplitude mask is accompanied by a considerable loss of laser energy. Using OPC, the amplitude mask is placed in the path of low-power radiation from the additional laser, at a position coinciding with the reflection in a semi-transparent mirror of the target itself (Fig. 1.15). Such a scheme provides a more complete usage of the energy stored in the main amplifier and eliminates severe requirements on the quality of the optics and the focusing system of the high-power laser.

A high-temperature laser plasma is a good source of hard ultraviolet radiation and x-radiation. To obtain such a plasma, it is necessary to focus a laser pulse within a minimum area on the target. In this case OPC self-targeting is very effective and allows one to obtain practically ideal focusing even with low-quality optical elements. The scheme used is identical to that shown in Fig. 1.15 with the mask forming a finely localized beam.

1.3.3 Control of Spatial-Temporal Structure

In a number of tasks, the operation of precise OPC in the form

$$e^{-i\omega t}E_{\text{ref}}(r, t) = \text{const} \cdot E_{\text{in}}^*(r, t)e^{-i\omega t}$$

appears non-optimal, and, accordingly, it becomes necessary to modulate the phase-conjugated wave.

Spatial Modulation. The simplest example of a task which necessitates modification of precise OPC is the transport of light energy to a moving object. The lead angle required (to take into account the movement of the object during the time the light is traveling towards it) can be readily achieved by conjugating the input signal and simultaneously tilting it through an angle $\alpha = 2v_\perp/c$. To this end, the relationship between the incident and reflected waves should be $E_{ref}(r) = \text{const} \cdot \exp(ik\alpha \cdot r)E_{in}^*(r)$. If α is not too large, the tilted phase-conjugated wave is capable of compensating, to the same degree, for the effect of phase inhomogeneities of the medium. OPC with tilting can be realized in holographic methods by tilting one of the reference waves.

Transporting laser energy through the atmosphere is hindered by thermal self-defocusing of the beam as a whole. To compensate for this effect, additional spherical convergence and, maybe, tilting should be introduced into the phase-conjugated wave in the OPC self-targeting scheme. The corresponding transformation can be easily obtained by replacing one of the plane reference waves in the holographic OPC method by a spherical wave of the required curvature.

To transmit the image through inhomogeneous media, we can proceed as follows. Let a wave be directed from a point source, located in the region where the image should be delivered, towards a nonlinear medium, via the inhomogeneous medium. Information about the image may be present in the reference waves, for example, as the distribution of $E_2(r)$. Then the signal reflected in FWM acquires an additional spatial modulation:

$$E_{ref}(r) = \text{const} \cdot E_1 E_2(r) E_{in}^*(r) . \tag{1.3.1}$$

If the modulation of $E_2(r)$ is smooth enough, then the complex-conjugated wave $E_{in}^*(r)$ becomes regular after its reverse passage through inhomogeneities in the medium and it retains the modulation of $E_2(r)$. OPC where the reflected beam is tilted and acquires additional spherical convergence can also be described by (1.3.1), with a suitable choice of the spatial modulation.

Temporal Modulation. The simplest kind of temporal modulation of a phase-conjugated wave is a frequency shift. This may be of use, for instance, in the double-pass scheme of compensation for optical distortions caused by an amplifier with an inhomogeneously broadened gain line. During its forward and backward passes, the wave takes the energy stored by various groups of excited particles. It is therefore possible to achieve a better amplification of the input signal and more complete saturation and energy takeoff in the backward pass. A frequency shift by an amount $\Delta\omega = 2\omega_0 v_S/c'$ is automatically achieved in OPC-SBS; here v_S and c' are the velocities of sound and light, respectively, in an SBS medium. In FWM, the frequency shift of the reflected wave requires a shift of the frequencies ω_1 and ω_2 of the reference waves: $\omega_{\text{ref}} = \omega_1 + \omega_2 - \omega_{\text{in}}$. In parametric methods the sum $\omega_1 + \omega_2$ plays the role of the moldulation frequency of the medium parameter.

As in FWM, the time dependence of the phase-conjugated wave at the instantaneous response is given by

$$E_{\text{ref}}(t) \propto E_1(t) E_2(t) E_3^*(t) . \tag{1.3.2}$$

There are vast possibilities for coding the phase-conjugated wave, i.e., controlling its time variation through modulation of the reference waves $E_1(t)$ and $E_2(t)$. In communication tasks this allows one to transfer information from the reference waves to the reflected wave and in laser techniques, to obtain the required pulse time profile.

In installations designed for the study of laser thermonuclear fusion the problem arises, in addition to rigid requirements on the time profile, of how to increase the so-called "time contrast" of the pulse which irradiates the target. By this we mean the necessity of a considerable (more than $10^6 - 10^8$ times) suppression of the intensity of the leading edge of the pulse in comparison with its main portion. Owing to the threshold nature of stimulated light scattering, OPC by SS automatically provides a sharp increase in the contrast in the phase-conjugated wave.

If, in the case of OPC-FWM, the intensity of the phase-conjugated wave becomes comparable with that of the reference waves, then, as a result of depletion of the latter, the reflection may have two strongly differing stable regimes (optical bistability) even under given input conditions for incident waves. Transition from one regime to the other can be performed by means of relatively weak external actions. In this case the OPC cell operates as a typical bistable optical device. Such devices are considered promising in optical communication, information processing, and in designing optical computers.

Mathematical Operations on Fields. In a given section of a nonlinear medium the reflected signal generated is determined by the product

$$E_4(r) \propto E_1(r) E_2(r) E_3^*(r) . \tag{1.3.3}$$

Thus, FWM makes it possible to perform multiplication of three complex functions of the transverse coordinates $r = (x, y)$. For example, it is possible to "filter" the value of the third field in the region of intersection by assuming the first two factors constant, each within the limits of its part of the plane x, y. Furthermore, if the waves E_1, E_2, and E_3 in the plane of interaction are obtained from the initial waves $\tilde{E}_1$, $\tilde{E}_2$, $\tilde{E}_3$ using a spatial Fourier transform (for instance, in the focal plane of the corresponding lenses), then the transformed wave $\tilde{E}_4$ corresponds to convolution (operation without complex conjugation) and correlation (operation with complex conjugation) of the fields $\tilde{E}_1, \tilde{E}_2, \tilde{E}_3$:

$$\tilde{E}_4(r) \propto \int \tilde{E}_1(r_1) \tilde{E}_2(r_2) \tilde{E}_3^*(r_1 + r_2 - r) d^2r_1 d^2r_2 . \tag{1.3.4}$$

In fact, FWM makes it possible to realize all procedures and methods in the coherent optical processing of information which have been developed for static Fourier optics and holography. In the static case, however, one has to record a new hologram for each new pair of fields $E_1(r)$ and $E_3^*(r)$, whereas in FWM all operations are performed in real time. Moreover, in the dynamic case they can be brought into coincidence by applying temporal modulation to the beam.

Tunable Narrow-Band Filters. For FWM in a nonlinear medium of thickness L which is not too small or in a medium with lagging nonlinearity, the reflectivity of the signal $|E_4/E_3|^2$ has a resonant dependence on the frequency of the signal to be phase conjugated, Fig. 1.16.

Consider first a medium with zero-lag nonlinearity. In this case the coherent interference of all elementary waves emitted by various sections of the layer of thickness L in the direction of the phase-conjugated wave determines the width of this resonance: $\Delta \omega_3 \tau \sim 1$, where $\tau \sim nL/c$ is the time the light takes to travel through the layer. The position of the maximum $\omega_{3\,opt}$ is determined by the frequencies ω_1 and ω_2 of the reference waves and by the crossing angle, should the waves nor exactly counterpropagate. For exactly counterpropagating waves of the same frequency, $\omega_1 = \omega_2 = \omega_0$, the reflection will be optimum at $\omega_{3\,opt} = \omega_0$.

For media which respond slowly, the efficiency of dynamic recording of the holographic grating depends on the difference frequency of the interfering waves. Accordingly, the grating $\propto E_1 E_3^*$ is recorded effectively if the difference frequency $|\omega_1 - \omega_3|$ is less than the reciprocal of the relaxation time τ_{13}^{-1} of the grating. Similarly, the grating $\propto E_2 E_3^*$ is recorded only if $|\omega_2 - \omega_3| \lesssim \tau_{23}^{-1}$, where τ_{23} is the relaxation time of this grating. As a result, the dependence of the reflectivity on ω_3 may contain two maxima, at the frequencies $\omega_{3\,opt} = \omega_1$ and $\omega_{3\,opt} = \omega_2$, with widths τ_{13}^{-1} and τ_{23}^{-1}, respectively. There can

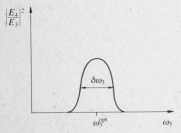

Fig. 1.16. Frequency-selective nature of the OPC reflectivity $|E_4/E_3|^2$ in FWM

exist more complex situations when the medium possesses natural oscillations of frequencies Ω_i small compared to the light frequency ω. This is the case with density hypersonic vibrations, the Rabi vibrations for the population of a two-level system in an intense field, vibrations and rotation of molecules, etc. In these cases resonances may also be observed as frequencies $\omega_{3\,\text{opt}} = \omega_1 \pm \Omega_i$, $\omega_{3\,\text{opt}} = \omega_2 \pm \Omega_i$.

Thus, OPC-FWM allows one to select incoming signals of the required frequency ω_3 from parasitic background noise at other frequencies. In other words, OPC-FWM serves as a narrow-band frequency filter, providing reflection with OPC. It is of great importance that the central frequency of this filter can be controlled by tuning the frequencies and directions of the reference waves. For a medium of length $L \sim 5\,\text{cm}$ the filter bandwidth $\Delta\omega/2\pi \sim 10^9$ Hz due to wave limitations; in resonant slowly responding media the bandwith can be even less, $\Delta\omega/2\pi \lesssim 10^8$ Hz.

In OPC-SBS frequency filtering of the signal to be conjugated is also possible. Suppose radiation of two components, one narrow-band (at a frequency ω) and one wide-band, enters an SBS-active medium. Then, due to the dependence of the SBS threshold on the linewidth, a situation may arise in which the narrow-band component excites SBS and is phase-conjugated, whereas the wide-band portion is under the threshold. In this case, selection of the narrow-band signal is possible even if it carries only a small portion of the signal energy. Here we may speak of nonlinear filtration (selection) of the phase-conjugated signal on the basis of its spectral brightness on a background of wide-band noise. Such nonlinear selection is required in a double-pass self-targeting scheme for weak phase-conjugated signals, which may be obscured by noise due to superluminescence of the amplifier.

1.3.4 Scientific Applications

Usage of OPC in Nonlinear Spectroscopy. To begin with, we should mention spectroscopic investigations of atomic and molecular transitions in inhomogeneously broadened lines. Nonlinear spectroscopy using OPC-FWM makes it possible to record the shape of the homogeneously broadened component of the Doppler contour of a gas absorption line. For example, the interference of waves E_1 and E_3 of the same frequency, $\omega_1 = \omega_3$, propagating at a small angle to one another excites, from the whole Maxwellian velocity distribution of the atoms, only a small layer of velocity space with spread $|\boldsymbol{k}_1 \cdot \boldsymbol{v} - \omega_0 + \omega_1| \lesssim \Gamma$. Here ω_0 is the central frequency of the atomic transition, $\boldsymbol{k}_1$ is the wave vector of the wave E_1, and Γ is the homogeneous width of the transition. Reading this interference pattern using a counterpropagating reference wave E_2, with $\boldsymbol{k}_2 = -\boldsymbol{k}_1$ and $\omega_2 = \omega_1$, will be resonant, i.e., effective, only if $|-\boldsymbol{k}_1 \cdot \boldsymbol{v} - \omega_0 + \omega_1| \lesssim \Gamma$.

This means that, in tuning the common frequency of the interacting waves, the intensity $|E_4|^2$ of the phase-conjugated wave will have a sharp resonance at $\omega_1 = \omega_0$ with a width $\sim \Gamma$. In this case, only atoms with $|\mathbf{k}_1 \cdot \mathbf{v}| \lesssim \Gamma$ will be excited, i.e., those which have a small velocity projection on the propagation direction of the reference waves. A large number of variations on FWM is possible in resonant media, with independent variation of the frequencies ω_1, ω_2, ω_3, propagation directions, and polarization states of the interacting waves. This makes it possible to obtain valuable additional information on the properties of atoms and their interactions.

FWM in the scheme of counterpropagating reference waves enables one to investigate low-frequency elementary excitations of non-resonant media due to the interference of pairs of waves, E_1 and E_3, or E_2 and E_3, at resonant frequencies $\omega_1 - \omega_3$ and $\omega_2 - \omega_3$ and wave vectors $\mathbf{k}_1 - \mathbf{k}_3$ and $\mathbf{k}_2 - \mathbf{k}_3$. When the reference waves propagate in the same direction, this method is called coherent anti-Stokes Raman spectroscopy (CARS). The specific feature of OPC-FWM consists in the fact that radiation is generated in a direction opposite to that of the signal only by the particular process in question, which provides a considerable decrease in the background level.

OPC Interferometry. The ability of the phase-conjugated wave to reconstruct the initial field structure requires two conditions: 1) inhomogeneities in the wave path should be of a pure phase nature, 2) these inhomogeneities should be the same in the forward and backward passes.

Thus, by observing distortions in the structure of the reconstructed wave we may register 1) weak absorption or amplification in the beam path and 2) fast-running processes accompanied by changes in optical properties.

Consider first the scheme presented in Fig. 1.17 in which the initial beam is split by a beam splitter M into two beams directed to an OPC device. With equal losses in each arm of the PC interferometer thus obtained, the phase-conjugated beams approaching the beam splitter have amplitudes and phases which exactly suppress the wave in the direction shown by the dotted line. The insertion of additional absorption or amplification into one of the arms will inevitably lead to a signal appearance in this direction. If OPC occurs with a

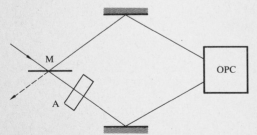

Fig. 1.17. Interferometric scheme using a phase-conjugate mirror (PCM)

frequency shift (such is the case, for example, in SBS), the exact suppression of the beam in the complementary direction requires that the difference ΔL of the optical paths of the two arms should satisfy the condition $\Delta k \Delta L = 2\pi m$, where m is an integer and Δk is the difference in the wave numbers of the forward and backward passes.

Registration of fast-varying phase disturbances of a medium is also possible with the scheme shown in Fig. 1.17 if, for the time of their travel from the element A (phase distorting in this case) to the PCM and back, the disturbances vary the optical path considerably (of the order of fractions of a wavelength). In this case, the spatial structure of the wave propagating in the additional direction carries information on the spatial distribution of the time-dependent part of the disturbances. A simpler scheme (Figs. 1.1, 2) is also possible wherein the initial beam, without being split, passes through the test object to the PCM; the nature of the fast-running processes in the distorting medium is judged from deviations from the ideal reconstruction of the initial beam.

The accuracy of OPC interferometry is limited, in the main, by the quality (i.e., accuracy) of phase conjugation in the PCM itself.

In this chapter we have discussed only the major methods of generation and application of phase-conjugated waves known at the time of writing. However, the list of methods and applications is expanding all the time.

1.4 Literature

A historical perspective of the phase-conjugated wave is set forth in a number of reviews [1.1 – 6] to which we refer the reader who is interested in the historical development of the subject. Here we mention some of the first reports on the major methods of OPC: in static [1.7 – 12] and dynamic [1.13 – 23] holography, OPC by stimulated light scattering [1.24 – 33], parametric methods [1.34 – 42]. Applications of OPC are considered in some detail in [1.2, 5, 43 – 47]. We also recommend some original works presenting the main applications of OPC: self-compensation for distortions in a double-pass scheme [1.25, 46, 48], in PC resonators [1.49 – 51], in light guides [1.52], compensation for pulse temporal spread [1.53], self-targeting of radiation [1.13, 28, 48, 54 – 56], spatial and time modulation [1.57 – 59], mathematical operations on fields [1.60], bistable optical devices [1.61, 62], tunable filters [1.63 – 67] and nonlinear selection [1.68, 69], nonlinear PC spectroscopy [1.70], PC interferometry [1.71 – 73].

A detailed account of those questions which have not been treated fully in this book is given in the three excellent reviews [1.5, 6a, 45].

2. Physics of Stimulated Scattering

In Sect. 1.2.3 a brief account was given of the fundamentals of the process of stimulated light scattering. However, a quantitative approach to the problem of optical phase conjugation by stimulated scattering requires a detailed theory of stimulated scattering (SS). It is to this theory that Chap. 2 is devoted.

2.1 Steady-State Picture of Stimulated Brillouin Scattering

Stimulated Brillouin scattering (SBS) is commonly used in optical phase conjugation because it has a relatively high gain constant and, consequently, a low SS excitation threshold, and also because of the low frequency shift $\Delta\omega/\omega \sim 2v/c \sim 10^{-5}$ to 10^{-6}, where v is the sound velocity in the medium. Both these properties are extremely valuable in applications.[1]

Suppose two plane monochromatic waves, an intense pump $E_L \exp(i k_L \cdot R - i \omega_L t)$ and a signal $E_S \exp(i k_S \cdot R - i \omega_S t)$ (which will be amplified), are propagating in a medium. The intensity $I(R, t)$ of the resultant field has the form of the traveling wave

$$I(R, t) = |E(R, t)|^2 = |E_L|^2 + |E_S|^2 + 2|E_L E_S|\cos(\Omega t - q \cdot R + \phi),$$
$$q = k_L - k_S, \quad \Omega = \omega_L - \omega_S, \quad \phi = \arg E_S - \arg E_L. \tag{2.1.1}$$

The resonant interaction of this traveling intensity wave with sound vibrations of the medium gives rise to SBS. To calculate the SBS gain, one should take into account the fact that the electrostriction force "draws" the substance into the region of larger intensity $|E|^2$; the addition to the pressure is given by [2.1]

$$\delta p_{str} = -\varrho_0 \left(\frac{\partial \varepsilon}{\partial \varrho}\right)_S \cdot \frac{\overline{E_{real}^2}}{8\pi} = -\varrho_0 \left(\frac{\partial \varepsilon}{\partial \varrho}\right)_S \cdot \frac{|E(R, t)|^2}{16\pi}. \tag{2.1.2}$$

The linearized hydrodynamic equations for the density $\varrho(R, t) - \varrho_0$ and velocity $v(R, t)$ of the medium are of the form

---

1 In Russian scientific literature this type of scattering is called Mandelstam-Brillouin scattering

$$\varrho_0\left(\frac{\partial v}{\partial t}\right)_i = -\frac{\partial}{\partial x_k}\left\{\left[p_0 + \left(\frac{\partial p}{\partial \varrho}\right)_s(\varrho - \varrho_0) + \delta p_{str}\right]\delta_{ik} - \sigma_{ik}^{vis}\right\}, \quad (2.1.3\,a)$$

$$\frac{\partial \varrho}{\partial t} + \varrho_0 \operatorname{div} v = 0, \quad (2.1.3\,b)$$

$$\sigma_{ik}^{vis} = \eta_1\left(\frac{\partial v_i}{\partial x_k} + \frac{\partial v_k}{\partial x_i} - \frac{2}{3}\delta_{ik}\frac{\partial v_l}{\partial x_l}\right). \quad (2.1.3\,c)$$

Here $(\partial p/\partial \varrho)_S = v^2$ is the square of the sound velocity, η_1 is the viscosity, and δ_{ik} is the Kronecker symbol. Equation (2.1.3a) represents a local form of Newton's second law, (2.1.3b) is a continuity equation, and (2.1.3c) is a viscous stress tensor for a liquid or gas [2.2] in the Navier-Stokes form. It should be emphasized that it is essential to take into account the viscous damping of sound because the rate of damping determines the amplitude of the resonantly built-up sound. From (2.1.3) we may obtain

$$\frac{\partial^2 \varrho}{\partial t^2} - v^2\nabla^2 - 2A\nabla^2\frac{\partial \varrho}{\partial t} = \nabla^2(\delta p_{str}) \equiv -\varrho_0\frac{\partial \varepsilon}{\partial \varrho}\frac{1}{16\pi}\nabla^2|E(R, t)|^2, \quad (2.1.4)$$

where $\nabla^2 = \partial^2/\partial x^2 + \partial^2/\partial y^2 + \partial^2/\partial z^2$ and $A = 2\eta_1/3\varrho_0$.

The steady-state solution of (2.1.4) taking $|E(R, t)|^2$ from (2.1.1) is given by

$$\varrho(R, t) - \varrho_0 = E_L^* E_S \exp(-i q \cdot R + i\Omega t)\varrho_0\frac{\partial \varepsilon}{\partial \varrho}\frac{1}{16\pi}\frac{q^2}{(\Omega_B^2 - \Omega^2 + 2i\Omega\Gamma)}$$

$$+ c.c., \quad (2.1.5)$$

$$\Omega_B(q) = |q|v, \quad \Gamma(q) = Aq^2. \quad (2.1.6)$$

Here Ω_B characterizes the frequency shift in SBS, and the constant $(2\Gamma)^{-1}$ characterizes the damping time of a free hypersonic wave with respect to its intensity. Γ also characterizes (at $\Gamma \ll \Omega_B$) the spontaneous scattering linewidth, $\Gamma = $ HWHM, i.e., the half-width at half-maximum intensity.[2]

---

2 In this consideration of damping only the shear viscosity has been taken into account. It is a good approximation for liquids. For gases, two more relaxation processes should be taken into account – the bulk (second) viscosity η_2 and the thermal conductivity Λ. Fortunately, these effects do not change the functional dependence $\Gamma = Aq^2$, [2.2, 3], they only renormalize the constant,

$$A = \left[\frac{2\eta_1}{3\varrho_0} + \frac{\eta_2}{2\varrho_0} + \frac{1}{2}\frac{\Lambda}{c_p}\left(\frac{c_p}{c_v} - 1\right)\right].$$

Here c_p, c_v are the specific heats at constant pressure and volume, respectively

Density modulation in the form of (2.1.5) leads, since $\delta\varepsilon = (\partial\varepsilon/\partial\varrho)$ $\times(\varrho - \varrho_0)$, to the appearance of a traveling volume diffraction grating in the permittivity. The scattering of the laser wave $E_L \exp(-i\omega_L t)$ due to the first term of (2.1.5) occurs with frequency shift $\omega_L \to \omega_S$ and gives rise to polarization at the signal frequency ω_S. It is remarkable that, irrespective of the values of ω_L, ω_S and the directions of the waves k_L and k_S this scattering produces a wave with the spatial dependence $\exp(ik_S \cdot R)$ corresponding to the signal direction, i.e., it automatically satisfies the phase matching condition or the Bragg condition for traveling gratings. This fact is a general property of the third-order susceptibility, which appears in the polarization of the medium in the form

$$P^{(3)} = \chi^{(3)} E_L E_L^* E_S. \tag{2.1.7}$$

However, the appearance of the component proportional to $\exp(ik_S \cdot R - i\omega_S t)$ in the polarization (or displacement) does not necessarily mean the transfer of energy in the direction of the signal E_S from the wave E_L caused by the grating $\delta\varepsilon \propto \exp(-iq \cdot R + i\Omega t)$. Indeed, the wave $E_S \exp(ik_S \cdot R - i\omega_S t)$ is already present in the medium, and, therefore, an additional field excited by the nonlinear polarization interferes with the initial field E_S. The interference result depends on the relation of these two contributions and may lead either to intensification or attenuation of the initial field or to an additional phase shift. The additional term in the electric displacement at ω_S can be written as

$$\delta D_S = \delta\varepsilon(R, t) E_L \exp(-i\omega_L t + ik_L \cdot R)$$

$$\equiv \delta\varepsilon_{\text{eff}}(\omega_S) E_S \exp(-i\omega_S t + ik_S \cdot R), \tag{2.1.8}$$

$$\delta\varepsilon_{\text{eff}}(\omega_S) = \varrho_0 \left(\frac{\partial\varepsilon}{\partial\varrho}\right)^2 \frac{|E_L|^2}{16\pi} \frac{q^2}{[\Omega_B^2 - (\omega_L - \omega_S)^2 + 2i\Gamma]}, \tag{2.1.9}$$

i.e., as far as the signal wave is concerned, its interaction with the pump leads only to a change in the permittivity by an amount $\delta\varepsilon_{\text{eff}}(\omega_S)$.

We should emphasize that the result of the interference of the laser wave E_L with the signal E_S depends on the phase of the field E_L, i.e., $\delta\varepsilon(R, t) \propto E_L^* E_S \exp(iq \cdot R - i\Omega t)$. However, the result of the scattering of the laser wave by the interference grating created,

$$\delta D \propto E_L \delta\varepsilon(R, t) \propto |E_L|^2 E_S \exp(-i\omega_S t + ik_S \cdot R),$$

does not depend on the phase of E_L and is determined only by the intensity $|E_L|^2$.

The amplitude of the wave E_S varies in space as

$$E_S \propto \exp\left(i \frac{\omega_S}{c} z \sqrt{\varepsilon_S}\right) \propto \exp\left(i \frac{\omega_S}{c} \sqrt{\varepsilon_0} z + i \frac{\omega_S}{2c} \frac{\delta\varepsilon_{eff}}{\sqrt{\varepsilon_0}} z\right),$$

and the intensity, for real ε_0, as

$$|E_S|^2 \propto \exp\left(-\frac{\omega_S}{c} \frac{Im\{\delta\varepsilon\}}{\sqrt{\varepsilon_0}} z\right) = \exp(-\alpha z), \tag{2.1.10}$$

where α is the coefficient of additional attenuation [cm^{-1}] of the intensity.

Equation (2.1.10) describes quantitatively the effect of phase relations on the energy exchange of the waves. Thus, for example, at $\omega_L - \omega_S > 0$, i.e., when the signal wave is Stokes shifted with respect to the pump, $Im\{\delta\varepsilon_{eff}(\omega_S)\} < 0$ and the signal is amplified.

The amplification is a maximum for $\omega_L - \omega_S = +\Omega_B$ when, at precise resonance, the phase shift between the electrostriction pressure δp and the response $\delta\varrho$ is equal to $\pi/2$ and when the amplitude of oscillations is a maximum (Fig. 1.9). On the contrary, at $\omega_L - \omega_S < 0$, i.e., for an anti-Stokes signal, $Im\{\delta\varepsilon_{eff}(\omega_S)\} > 0$ and the anti-Stokes signal is attenuated. This result is almost obvious, as in this case, the signal E_S plays the role of the pump with respect to the wave E_L.

Far from the Brillouin resonance, both in the Stokes and anti-Stokes region, $\delta\varepsilon_{eff}$ is practically real-valued. In this region, the interaction of the waves leads only to a change in the signal phase velocity.

The intensity gain coefficient g, where $|E(z)|^2 = E|(0)|^2 \exp(gz)$, can be written near the Stokes resonance as

$$g(\omega_S) = \frac{1}{1 + (\omega_L - \Omega_B - \omega_S)^2/\Gamma^2} \cdot \left(\varrho_0 \frac{\partial\varepsilon}{\partial\varrho}\right)^2 \frac{\omega q^2 |E_L|^2}{32\pi\Omega_B\Gamma c \varrho_0 \sqrt{\varepsilon_0}}. \tag{2.1.11}$$

It is convenient to present the gain at the optimum frequency shift in a form where the intensity is measured in MW/cm^2:

$$g[\text{cm}^{-1}] = G\left[\frac{\text{cm}}{\text{MW}}\right] I_L\left[\frac{\text{MW}}{\text{cm}^2}\right]. \tag{2.1.12}$$

We now discuss the dependence of G on the experimental geometry and the wavelength of the exciting radiation. To this end, we should take into account that if the direction of the scattered photon k_S forms an angle θ with the direction of the incident photon, then

$$|q| = (2\omega/c)\sqrt{\varepsilon_0} \sin(\theta/2). \tag{2.1.13}$$

Therefore $\Omega_B = qv \propto \omega_L \sin(\theta/2)$, $\Gamma \propto \omega_L^2 \sin^2(\theta/2)$ and

$$G(\omega_L, \theta) \propto (\omega_L)^0 [\sin(\theta/2)]^{-1}. \qquad (2.1.14)$$

Thus, the steady-state gain G for SBS is independent of light frequency.

In principle there is no threshold for most types of SS, and for SBS in particular. Indeed, in the absence of internal losses in the medium, the gain of the Stokes signal,

$$|E_S(z)|^2 - |E_S(0)|^2 = |E_S(0)|^2 [\exp(G|E_L|^2 z) - 1], \qquad (2.1.15)$$

is positive at an arbitrary small value of the pump intensity $|E_L|^2$. In fact, however, there exists an experimental threshold for the process. This is connected with the fact that no special Stokes signal is usually sent into the medium; in practice it originates from the pump wave, which is spontaneously scattered by fluctuational hypersonic waves. These are always present in a medium at thermodynamic equilibrium. Spontaneous scattering is usually very low (the extinction coefficient $dR/d\Theta \sim 10^{-7}\,\mathrm{cm}^{-1}\,\mathrm{sterad}^{-1}$). As a rule, due to the specific geometry of the amplifying region (Fig. 1.11) effective amplification occurs only near the axis in a narrow solid angle $\Delta\Theta \sim 10^{-4} - 10^{-6}$ sterad within which the signal path over the interaction region and, consequently, the gain (2.1.15) are both at a maximum. The length within which the priming signal is excited can be estimated by $l_{sp} \sim g^{-1} \sim 0.1$ cm. Then the intensity of the spontaneous priming is $I_{sp} \sim I_L (dR/d\Theta)\Delta\Theta l_{sp} \sim I_L \cdot (10^{-11}$ to $10^{-13})$. Thus, obtaining values of $I_S \sim I_L$ which can be registered experimentally requires a total gain as high as

$$\exp(G|E_L|^2 z) \approx 10^{13} \approx \exp(30), \qquad (2.1.16)$$

where z is the effective path of the signal in the interaction region. It is this enormous figure (10^{13}) that causes the experimental threshold of SS, $G|E_L|^2 z = 30$. In fact, for a fixed experimental geometry, a 20% decrease in the pump intensity below the threshold makes the signal intensity decrease by a factor of 400, i.e., it leads to a complete suppression of SS. In spite of the crude estimate and a certain arbitrariness in choosing numerical values, the "magic logarithm" $G|E_L|^2 z = 30$ proves to be practically universal for almost all types of SS.

From this, we can understand why SBS is almost always observed in the backward direction. Indeed, to reach the SBS threshold, either the beam has to be focused in a nonlinear medium or a sufficient interaction length has to be provided by directing the beam into a light guide. In both cases an appreciable amplification length is provided only in the longitudinal direction. Thus, the geometry permits either forward or backward scattering. Such is the case for stimulated Raman scattering (SRS). However, forward SBS is not observed. Indeed, the build-up time of SS is determined by the value of $\tau \sim \Gamma^{-1}$ which, for small scattering angles θ, turns out to be far greater than the typical pulse duration of intense lasers, since $\tau \propto \theta^{-2}$, (2.1.6, 13).

Typical numerical values of the parameters involved are as follows. For the radiation of a neodymium laser ($\lambda = 1.06$ μm) in a medium of refractive index $n = 1.5$, the value of $|k_L| \approx |k_S| = 0.89 \times 10^5$ cm^{-1}. For backward scattering, $q = 2k = 1.78 \times 10^5$ cm^{-1}, and at a sound velocity of $\sim 10^5$ cm/s we have $\Omega_B = 1.8 \times 10^{10}$ rad/s or $f_B = \Omega/2\pi = 2.9 \times 10^9$ Hz = 2.9 GHz, $f_B/c = 0.1$ cm^{-1}, in spectroscopic units. Such a high frequency, of the order of several GHz, corresponds to the so-called "hypersound", characterized by a very short phonon lifetime. For example, for benzene at the same q (i.e, backward SBS at $\lambda = 1$ μm) we obtain $\Gamma \sim 10^9$ s^{-1}, $\Gamma/2\pi c \approx 5 \times 10^{-3}$ cm^{-1}. A typical value of the "quality" factor Ω_B/Γ of a hypersonic wave is in the range $3 - 30$. The hypersound intensity damping α_s is very high: $\alpha_s = 2\Gamma/v_s \approx 10^4 - 10^3$ cm^{-1}. This means that the hypersound excited by the waves E_L and E_S has time to cover only a very small distance $l \lesssim 10^{-3}$ cm from the excitation place. This value is much smaller compared with both the characteristic values of the gain length g^{-1} in SBS and the characteristic longitudinal dimensions of inhomogeneity of the beams E_L and E_S due to diffraction. So, the mathematical problem of the interaction of the three waves, E_L, E_S, and hypersound $\varrho(R, t) - \varrho_0$, is considerably simplified. Indeed, the hypersound amplitude follows the local amplitude of the electrostriction force:

$$[\varrho(R, t) - \varrho_0] \propto -iE_L^*(R)E_S(R)e^{i\Omega t},$$

i.e., it is sufficient to consider the interaction of only two waves, E_L and E_S. As a matter of fact, we already employed this when, in calculating $\delta\varrho(R, t)$, we assumed k_L and k_S to be pure real-valued, taking into consideration the fact that $g \ll \alpha_s$.

Due to such a high hypersound damping, even for the spatially inhomogeneous beams E_L and E_S we may limit ourselves to the local interaction of these waves, i.e., we regard the relation

$$g_s(R) = G|E_L(R)|^2 \tag{2.1.17}$$

as being also valid for inhomogeneous fields $E_L(R)$ and $E_S(R)$. This fact is of great importance for the OPC-SS mechanism. Let us check the validity of (2.1.17) by considering the angular dependence of the gain at the fixed Stokes wave frequency $\omega_S = \omega_L - 2kv_s$, which corresponds to the gain maximum in the exact backward direction:

$$G(\beta = \pi - \theta) = G(0)(1 - \beta^2/8). \tag{2.1.18}$$

Table 2.1

Substance	$\Omega_B/2\pi$ [GHz]	$\Gamma/2\pi$ [GHz]	G [cm/MW]
CS_2	5.8	0.03	0.15
Acetone	4.6	0.095	0.022
Benzene	6.34	0.16	0.028
CCl_4	4.28	0.29	0.0058
CH_4, 140 atm	0.15	0.01	0.1

Thus, for beams with an angular divergence $2\beta \lesssim 0.1$ rad, the local relation (2.1.17) holds true to a 1% accuracy.

The largest known value of the constant G is for pure CS_2, where $G = 0.15$ cm/MW. It is worth noting that even slight contamination of CS_2 strongly increases the hypersound damping Γ and, consequently, decreases the gain to the same degree. Table 2.1 presents information on the constants which characterize SBS in widely used pure media, for an excitation $\lambda = 0.6943$ μm [2.4]; for parameters of compressed CH_4, including the pressure dependence, see [2.5].

2.2 Spontaneous and Stimulated Scattering; Quantum Representation of SS

In 1916, Einstein predicted theoretically that stimulated light emission should exist in addition to spontaneous emission and established the relationship between these two processes. Before ascertaining the relationship between spontaneous and stimulated light scattering, we first derive a similar relationship for light emission.

Suppose that two-level radiators (atoms, molecules, etc.) have population densities N_1 and N_2 [cm^{-3}] in the lower and upper levels, respectively, Fig. 2.1. We denote the total probability of spontaneous emission as A_{21} [s^{-1}]. This quantity is given by the integral of the differential probability over all solid angles $\Delta\Theta$ and frequencies $\Delta\omega$, and the sum of the polarizations e_1, e_2:

$$A_{21} = \int d\Theta\, d\omega \sum_{e_1,e_2} \frac{dA_{21}(e)}{d\Theta\, d\omega}. \tag{2.2.1}$$

Isotropically oriented systems are usually of interest: for this case,

$$\frac{dA(e)}{d\Theta\, d\omega} = A_{21} \frac{1}{8\pi} f(\omega), \tag{2.2.2}$$

where $f(\omega)$ is the normalized line special profile, $\int f(\omega)\, d\omega = 1$.

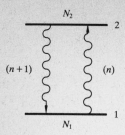

N_2

2

$(n+1)$ (n)

1

N_1

Fig. 2.1. Radiation transitions in a two-level system

Suppose that a light beam corresponding to a frequency interval $\Delta\omega$ and solid angle interval $\Delta\Theta$ propagates in a medium. Then, according to the postulates of quantum mechanics, the number of quanta emitted in the direction of the beam within the limits of its spectral bandwidth is equal to

$$W_+\left[\frac{\text{quanta}}{\text{cm}^3\,\text{s}}\right] = \Delta A_{21}N_2(n+1)\,. \tag{2.2.3}$$

Here $\Delta A_{21} = (dA_{21}(e)/d\Theta\,d\omega)\Delta\Theta\,\Delta\omega$ and n is the occupation parameter of the photon field, i.e., the number of quanta per oscillator (per cell of phase space). The first term in (2.2.3), $\propto n$, corresponds to the contribution of stimulated emission, and the second one to spontaneous emission. Absorption occurs as well:

$$W_- = N_1\Delta A_{21}n\,, \tag{2.2.4}$$

where we take into account the fact that $A_{21} = A_{12}$ for nondegenerate levels 1 and 2, i.e., the downward and upward transition probabilities are equal. As a consequence, the variation in the quantum flux intensity I [quantum/cm^{2}s] is

$$\frac{dI}{dz}\left[\frac{\text{quanta}}{\text{cm}^3\,\text{s}}\right] = (N_2-N_1)\Delta A_{21}\left(n + \frac{N_2}{N_2-N_1}\right)\,. \tag{2.2.5}$$

The term $\propto N_2/(N_2-N_1)$ describes spontaneous emission, and the term $\propto n$ gives the difference between stimulated radiation and absorption. To determine the rate of the resultant amplification, we express n in terms of I:

$$I\left[\frac{\text{quanta}}{\text{cm}^2\,\text{s}}\right] = v_{\text{gr}}\left[\frac{\text{cm}}{\text{s}}\right]n\left[\frac{\text{quanta}}{\text{f. osc.}}\right]\Delta N\left[\frac{\text{f. osc.}}{\text{cm}^3}\right] = n\frac{k^2}{(2\pi)^3}\Delta\theta\,\Delta\omega \tag{2.2.6}$$

where $v_{\text{gr}} = \partial\omega/\partial k$ is the group velocity, $N\equiv\Delta^3 k/(2\pi)^3 = k^2\Delta k\,\Delta\Theta/(2\pi)^3$ is the number of field oscillators in 1 cm^3 corresponding to the volume element $\Delta^3 k$ in wave-vector space, and $\Delta^3 k = k^2\Delta k\,\Delta\Theta = k^2\Delta\omega\cdot\Delta\Theta/v_{\text{gr}}$. As a result, we obtain

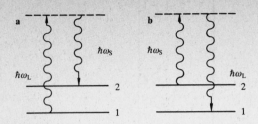

Fig. 2.2. (a) Transformation of incident quantum $\hbar\omega_L$ into scattered one $\hbar\omega_S$ and **(b)** the reverse process $\hbar\omega_S \rightarrow \hbar\omega_L$

$$\frac{dI}{dz} = \Delta A_{21} N_2 + gI, \tag{2.2.7}$$

$$g = \frac{(2\pi)^3}{k^2}(N_2 - N_1)\frac{dA_{21}(e)}{d\Theta\,d\omega} \equiv \left(\frac{\pi}{k}\right)^2 (N_2 - N_1) A_{21} f(\omega), \tag{2.2.8}$$

where g has the dimensions of inverse length. The second expression in (2.2.8) is for isotropic systems. If levels 1 and 2 have degeneracy factors g_1 and g_2, respectively, then the term $(N_2 - N_1)$ in (2.2.8) is substituted by $(N_2 - g_2 N_1/g_1)$.

We now proceed to deriving a similar relationship between the processes of spontaneous and stimulated scattering. In quantum-mechanical terms the scattering of photons from a beam (ω_L, k_L) into a beam (ω_S, k_S) is described by the scheme in Fig. 2.2a. For this process the value of W_+ (quanta/cm^3 s) is

$$W_+ = |M|^2 n_L (n_S + 1) N_1, \tag{2.2.9}$$

where $|M|^2$ is the product of the square of a matrix element and the density of states, and n_L and n_S are the occupation parameters of the photon fields. The inverse process, Fig. 2.2b, is characterized by the rate

$$W_- = |M|^2 (n_L + 1) n_S N_2. \tag{2.2.10}$$

As a result, the variation in the quantum flux intensity I_S is

$$\frac{dI_S}{dz} = -N_2 |M|^2 n_S + n_L |M|^2 (N_1 - N_2)\left(n_S + \frac{N_1}{N_1 - N_2}\right). \tag{2.2.11}$$

The first term in (2.2.11) characterizes attenuation (extinction) of the signal beam I_S due to spontaneous scattering, which is usually insignificant. The term $\propto n_L - n_S$ represents the gain we require, of the signal I_S due to the pump I_L, and the term $\propto n_L N_1/(N_1 - N_2)$ describes spontaneous scattering of the pump beam into the signal beam.

Let us write a similar relation proceeding from the macroscopic point of view. Spontaneous scattering is characterized by the extinction coefficient (differential with respect to frequency, solid angle, and polarization) $dR(e)/d\Theta\,d\omega$:

$$\left(\frac{dI_S}{dz}\right)_{sp} = \frac{dR(e_S)}{d\Theta_S d\omega_S} \cdot \Delta\Theta_S \Delta\omega_S I_L , \qquad (2.2.12)$$

where I_L and I_S are expressed in quanta/cm^{2}s. On the other hand, owing to stimulated scattering we may expect that

$$\left(\frac{dI_S}{dz}\right)_{st} = g I_S = g n_S \frac{k_S^2}{(2\pi)^3} \Delta\Theta_S \Delta\omega_S . \qquad (2.2.13)$$

By comparing the ratio of (2.2.12) to (2.2.13) with (2.2.11), we can obtain

$$g = \frac{(2\pi)^3}{k_S^2} \cdot \frac{dR(e_S)}{d\Theta_S d\omega_S} I_L \left(\frac{N_1 - N_2}{N_1}\right) . \qquad (2.2.14)$$

Equation (2.2.14) is also valid for a medium not in thermodynamic equilibrium. However if the medium is in equilibrium and its temperature is $k_B T$ in energy units, then

$$\frac{N_1 - N_2}{N_1} = 1 - \exp(-\Delta E/k_B T) = 1 - \exp(-\hbar\Omega/k_B T) , \qquad \Omega = \omega_L - \omega_S .$$

As a consequence, taking into account that $P_L[\mathrm{erg/cm^2 s}] = I_L \hbar \omega_L$, we obtain the desired relationship:

$$g[\mathrm{cm^{-1}}] = \frac{(2\pi)^3}{k_S^2} \cdot \frac{dR(e_S)}{d\omega_S d\Theta_S} \cdot P_L \frac{1}{\hbar\omega_L} [1 - \exp(-\hbar\Omega/k_B T)] . \quad (2.2.15)$$

An important conclusion can be drawn from the above and from (2.2.15): each type of spontaneous light scattering has a corresponding stimulated analogue. Furthermore, for a medium in equilibrium the signal is amplified ($g_S > 0$) only when it is Stokes-shifted with respect to the pump, $\Omega = \omega_L - \omega_S > 0$. Also, at a given value of $dR/d\Theta$ the gain is inversely proportional to the scattering linewidth. The gain turns out to be proportional to the pump intensity. If the difference between the energies of quanta is small compared with the temperature, $\hbar\Omega \ll k_B T$, then Planck's constant $\hbar$ cancels in (2.2.15). We have already been convinced of the validity of all these properties in Sect. 2.1 while considering SBS.

The above quantum-mechanical consideration has considerable generality. However, substitution of explicit expressions shows that, in most problems, the final result for g does not contain Planck's constant, i.e., *stimulated light scattering is of a classical nature*. This fact allows one to find an adequate and clearly evident classical description for almost all types of SS in terms of waves interacting via excitations of the medium. Moreover, a complete account of non-stationarity, non-monochromaticity, and diffraction of light beams is possible only in terms of classical interacting waves. This is connected with the fact that, according to the relationship already known, the state of a field with a strictly definite number of quanta possesses a completely indefinite phase, whereas many effects of nonlinear optics depend largely on the phase relationships of fields. That is why we began this chapter with a classical approach to SBS (Sect. 2.1), and we shall keep to that approach from now on. Nevertheless, in a number of cases, a quantum-mechanical approach facilitates understanding of the structure of solutions of classical wave problems.

2.3 Other Types of Stimulated Scattering

In this section we shall discuss various types of stimulated scattering, giving particular attention to those which are of interest in the OPC problem.

2.3.1 Stimulated Raman Scattering (SRS) by Molecular Vibrations

Among all types of stimulated scattering, stimulated Raman scattering (SRS) was the first to be discovered. Its spontaneous analogue is scattering by fluctuations of that component of the molecular polarizability which varies in step with the molecular vibration coordinate Q, i.e., $\delta\alpha(t) = (\partial\alpha/\partial Q)Q(t)$. In quantum-mechanical terms (Fig. 2.2a), the initial and final states 1 and 2 are the ground and the first excited vibrational levels, respectively, of the lower electron term of a molecule. The effective energy of a molecule with a polarizability α in the light field is given, in the dipole approximation, by $U_{\text{eff}} = -d \cdot E_{\text{real}}/2 = -\alpha|E|^2/4$, where $d = \alpha E$ is the induced dipole moment. In the stimulated process the interference of waves

$$E_L \exp(-i\omega_L t + i k_L \cdot R) \quad \text{and} \quad E_S \exp(-i\omega_S t + i k_S \cdot R)$$

builds up molecular vibrations through the effective force

$$f = -\partial U_{\text{eff}}/\partial Q = 0.25 \, (\partial\alpha/\partial Q)|E|^2.$$

The amplitude of steady-state vibrations is

$$Q(R, t) = \frac{1}{4M} \cdot \frac{\partial\alpha}{\partial Q} \cdot \frac{E_L^* E_S \exp(i\Omega t - i q \cdot R)}{\Omega_0^2 - \Omega^2 + 2i\Omega\Gamma}, \tag{2.3.1}$$

where Ω_0 is the natural frequency, M is the generalized vibrating mass, and Γ is the damping factor. For typical values of E_L and E_S, this amplitude turns out to be far less than the mean-square amplitude $(\langle \delta Q^2 \rangle)^{1/2} = (\hbar/2M\Omega_0)^{1/2}$ of the zero-point fluctuations causing spontaneous scattering. However, stimulated vibrations of all molecules are phased by external fields, (2.3.1), and consequently the amplitudes of the contributions of individual molecules are added, not the intensities as in the spontaneous case. Because of this, the energy exchange of the waves in SS turns out to be far greater than in spontaneous scattering.

In other respects, consideration of SRS is anlogous to that of SBS. The traveling spatial-temporal permittivity grating has the form $\delta\varepsilon(R, t) = 4\pi N(\partial\alpha/\partial Q) \cdot Q(R, t)$, where N is the number of molecules per unit volume. The gain coefficient at the line center (i.e., at $\Omega = +\Omega_0$) is

$$g = \frac{(\omega_L - \Omega_0)}{c\sqrt{\varepsilon_0}} \cdot \frac{1}{2M\Omega_0\Gamma} \cdot \left(\frac{\partial\alpha}{\partial Q}\right)^2 |E_L|^2. \tag{2.3.2}$$

It is of importance that Ω_0 and Γ are independent of both the pump frequency ω_L and the scattering angle θ. We consider typical values for media with the largest values of G at the exciting wavelength $\lambda = 0.69$ μm: CS_2, $G = 10^{-2}$ cm/MW, $\Omega_0/2\pi c = 656$ cm^{-1}, $\Gamma/2\pi \approx 5$ GHz; liquid nitrogen, $G \approx 1.6 \times 10^{-2}$ cm/MW, $\Omega_0/2\pi c = 2326$ cm^{-1}, $\Gamma/2\pi \approx 6$ GHz. We can derive the dependence of G on the frequency ω_L of the exciting radiation from (2.3.2) taking into account that Γ, Ω_0, $(\partial\alpha/\partial Q)$, and M are almost independent of frequency.

2.3.2 Other Types of SRS

In molecular gases a second type of SRS can be observed, namely, modulation of the anisotropic part $\alpha_{ik} - \frac{1}{3}\delta_{ik} \text{Tr} \, \hat{\alpha}$ of the polarizability by molecular rotation. In quantum-mechanical terms, this type of SRS takes place with a change in the discrete rotational quantum number of a molecule. The energy of the rotation quantum is large for light molecules; for example, the transition $J = 0 \rightarrow J = 2$ for a molecule of H_2 corresponds to $\Omega_{rot}/2\pi c \approx 350$ cm^{-1}. The stimulated analogue of spontaneous rotational RS is the process in which molecular rotation is synchronously built up by the interference of the fields E_L and E_S of differing frequencies.

In principle, SRS is possible with a change of electronic state. This process is known as stimulated electronic Raman scattering (SERS). As far as gases are concerned, the process occurring with the participation of two quanta, $\hbar\omega_L$ and $\hbar\omega_S$, is allowed in the dipole approximation only between states 1 and 2 of the same parity. Moreover, if population inversion occurs in the medium, then it is a signal with an anti-Stokes frequency shift, $\omega_S = \omega_L + |\omega_{12}|$, that has a positive gain coefficient in the presence of the pump field E_L.

2.3.3 Stimulated Rayleigh-Wing Scattering (SRWS)

With increasing density, the rotational Raman scattering in gases and, moreover, in liquids loses its structure of discrete lines due to frequent molecular collisions and exhibits a continuous spectral distribution corresponding to the so-called Rayleigh-line anisotropic wing. In this case, light is also scattered by orientational fluctuations, but now their time behavior corresponds not to free rotation but to viscous relaxation of orientation.

In the stimulated process, light fields lead to a certain ordering of the molecular orientation. The corresponding effective energy is $U = -0.25$ $\alpha_{ik} E_i^* E_k$, where α_{ik} is the polarizability tensor at the light frequency. As a result, in a strictly monochromatic field the molecules are so aligned that the major axis of the tensor $\hat{\alpha}$ coincides with the direction of the field E, and the effective refractive index for the field itself is increased. This is the so-called orientational optical nonlinearity or the optical Kerr effect. The effect of the nonstationary interference pattern of the fields E_L and E_S on the orientation can be described within the framework of the simplest model with one relaxation time Γ^{-1}. We write the equation for the traceless part of the permittivity tensor $\tilde{\varepsilon}_{ik} = \varepsilon_{ik} - \frac{1}{3}(\mathrm{Tr}\,\hat{\varepsilon})\delta_{ik}$ characterizing anisotropy as

$$\frac{d\tilde{\varepsilon}_{ik}}{dt} + \Gamma\tilde{\varepsilon}_{ik} = A(E_i^* E_k + E_i E_k^* - \tfrac{2}{3}(E\cdot E^*)\delta_{ik}),\qquad(2.3.3)$$

where A is a constant, and the tensor structure of the right-hand side is uniquely defined by the conditions $\tilde{\varepsilon}_{ik} = \tilde{\varepsilon}_{ki}$, $\sum_i \tilde{\varepsilon}_{ii} \equiv \mathrm{Tr}\,\hat{\tilde{\varepsilon}} = 0$. For the fields E_L and E_S with frequencies ω_L and ω_S, the steady-state amplitude of the orientation grating is given by

$$\tilde{\varepsilon}_{im}(R,t) = \exp(\mathrm{i}\Omega t - \mathrm{i}q\cdot R)\frac{A}{\mathrm{i}\Omega + \Gamma}(E_{Li}^* E_{Sm} + E_{Si} E_{Lm}^* - \tfrac{2}{3}\delta_{im} E_L^*\cdot E_S)$$
$$+ \text{c.c.}\ .\qquad(2.3.4)$$

The scattering of the pump field E_L by this grating leads to a change of the effective tensor $\hat{\varepsilon}$ for the signal by an amount

$$\delta\varepsilon_{ik}(\omega_S) = \frac{A(E_L\cdot E_L^*)}{\mathrm{i}\Omega + \Gamma}(a_i^* a_k + \delta_{ik} - \tfrac{2}{3}a_i a_k^*),\qquad(2.3.5)$$

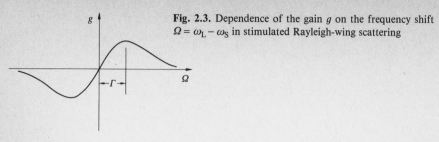

Fig. 2.3. Dependence of the gain g on the frequency shift $\Omega = \omega_L - \omega_S$ in stimulated Rayleigh-wing scattering

where $E_L = a E_L$ and $(a^* \cdot a) = 1$. We shall discuss the general polarization properties of SRWS in Sect. 2.8.

Consider here a linearly polarized pump, $a = a^*$, and suppose that the polarization direction for the signal coincides with that for the pump, $E_S = a E_S$. Then the gain coefficient $g(\omega_S)$ is given by

$$g(\omega_S) = \frac{\omega}{c} \, \frac{\mathrm{Im}\{\tilde{\varepsilon}_{ik}(\omega_S)\} a_i a_k}{\sqrt{\varepsilon_0}} = \frac{4 A \, \omega |E_L|^2}{6 c \Gamma \sqrt{\varepsilon_0}} \cdot \frac{2 \Gamma \Omega}{\Gamma^2 + \Omega^2} \,, \tag{2.3.6}$$

which reaches its maximum value g_m at $\Omega = +\Gamma$, i.e., at the half-width of the Lorentz line for spontaneous scattering, Fig. 2.3. The value of this maximum depends on the parameter A and can be expressed in terms of the self-focusing constant ε_2.

When considering the self-focusing of linearly polarized light, one usually writes

$$\varepsilon = \varepsilon_0 + \tfrac{1}{2}\varepsilon_2 |E|^2 \,. \tag{2.3.7}$$

Such a steady-state dependence can be obtained for a monochromatic field, in this case $\varepsilon_2 = 8 A/3 \Gamma$. Then g_m is given by

$$g_m = \frac{\omega}{c} \, \frac{\varepsilon_2 |E_L|^2}{4 \sqrt{\varepsilon_0}} \equiv \left[\frac{d\phi(E_L)}{dz} \right]_{\mathrm{NL}} \,, \tag{2.3.8}$$

i.e., g_m coincides numerically with the rate of nonlinear phase change for the pump due to the effect of steady-state self-action in the medium with orientational nonlinearity.

For CS_2, $\varepsilon_2 \approx 4 \cdot 10^{-11}$ cm^3/erg, and at $\lambda = 1$ μm, $G \approx 2.3 \times 10^{-3}$ cm/MW; in this case the width of the main (narrow) part of the orientational scattering is $\Gamma/2\pi \approx 70$ GHz.

These two mechanisms of nonlinearity, SRS and SRWS, are characterized by their locality. Due to this, the linewidth Γ and the gain g are independent of the wave vector q of the interference grating and, accordingly, of the scattering angle, and $g(\omega) \propto \omega^1$. In contrast, in SBS and stimulated temperature scattering (Sect. 2.3.4), $\Gamma \propto q^2$ and the gain turns out to be dependent on the scattering angle.

2.3.4 Stimulated Temperature Scattering Due to Absorption (STS-II)

Suppose two waves, $E_L \exp(i k_L \cdot R - i \omega_L t)$ and $E_S \exp(i k_S \cdot R - i \omega_S t)$, are propagating in a weakly absorbing medium. The heat release $\alpha n c |E|^2 / 8\pi$ [erg/cm^{3}s] due to absorption turns out to be spatially and time modulated as $|E_L + E_S|^2$. The induced temperature grating $T(R, t)$ obeys the heat-transfer equation

$$\varrho c_p \left(\frac{\partial T}{\partial t} - \chi \Delta T \right) = \alpha \frac{nc}{8\pi} E_L^* E_S \exp(-i q \cdot R + i \Omega t) + \text{c.c.} . \qquad (2.3.9)$$

Here ϱ, c_p, and χ are the density, specific heat, and thermal diffusivity of the medium, respectively, α is the intensity absorptivity [cm^{-1}], and n is the refractive index. The steady-state solution of (2.3.9) is of the form

$$T(R, t) = E_L^* E_S \exp(-i q \cdot R + i \Omega t) \cdot \frac{\alpha n c}{8\pi \varrho c_p (\Gamma + i \Omega)} + \text{c.c.} , \qquad (2.3.10)$$

$$\Gamma = \chi q^2 .$$

The dielectric susceptibility is temperature dependent, $\delta \varepsilon = (\partial \varepsilon / \partial T) \delta T$, and therefore a traveling susceptibility grating is recorded inside the medium. Further, as for other types of SS, the scattering of the pump field by this grating leads to a change in the effective susceptibility at the signal frequency ω_S. As a result, the gain is

$$g = -\alpha + \frac{\alpha \omega |E_L|^2}{16\pi \Gamma \varrho c_p} \cdot \left(\frac{\partial \varepsilon}{\partial T} \right) \cdot \frac{2\Gamma\Omega}{\Gamma^2 + \Omega^2} \equiv \alpha \left(\frac{P_L}{P_1} - 1 \right) . \qquad (2.3.11)$$

Here we take account of the original absorption of the wave E_S. For most substances $\partial \varepsilon / \partial T$ is negative, as the main mechanism for $\partial \varepsilon / \partial T$ is thermal expansion. Therefore, a positive gain is reached when the anti-Stokes shift occurs, $\Omega < 0$. Also, as the force of the nonlinear interaction is proportional to the absorptivity α, the resultant gain is positive only if $P_L = nc |E_L|^2 / 8\pi > P_1$. If the point of maximum gain is chosen, i.e., $\Omega = -\Gamma$, then P_1 is a constant for a given material:

$$P_1 = \frac{2\Gamma \varrho c_p n c}{\omega |\partial \varepsilon / \partial T|} = \frac{8\chi n^3 \varrho c_p \omega \sin^2(\theta/2)}{c |\partial \varepsilon / \partial T|} , \qquad (2.3.12)$$

where the scattering angle can be expressed in terms of q using (2.1.13). It is important that the value of P_1 is independent of α and, hence, can be termed as the physical threshold of STS gain.

All these properties of STS which are due to absorption, i.e., the presence of the physical threshold, the proportionality between the gain and the absorptivity, and the amplification in the anti-Stokes region, contradict (2.2.15). The situation is like this.

There exists a very weak temperature scattering not related to absorption. Its mechanism is caused by an electrocaloric effect, and the gain can be expressed by (2.2.15) in terms of the cross section for scattering by temperature (entropy) fluctuations. Such a stimulated scattering, known as STS-I, is practically never observed on the background of a stronger nonlinear process which is caused by absorption. As the macroscopic description of the latter is identical to that of SRS, SRWS, SBS, etc., the process caused by absorption is referred to as stimulated temperature scattering of the second kind (STS-II). We may say that STS-II has no spontaneous analogue in the sense of (2.2.15). For detailed information on specific features of SS caused by absorption refer to [2.6].

Like SBS, and for the same reasons, STS-II can be observed mainly in the backward direction where $\theta \approx 180°$. We present here typical figures. At $\chi \sim 10^{-3}$ cm²/s and $\lambda = 0.69$ μm, $n = 1.5$, we have $\Gamma/2\pi \approx 12$ MHz, this being much less than the Brillouin scattering linewidth even for ideally pure CS$_2$. If we take $\partial\varepsilon/\partial T \approx -2\times10^{-3}$ K^{-1}, $\varrho c_p \sim 1.5$ J cm^{-3} K^{-1}, then we have $P_1 \sim 2$ MW/cm².

2.3.5 General Properties of Stimulated Scattering

In spite of the variety of physical mechanisms of SS, certain general regularities are inherent in all of them.

First, for the signal to be amplified (Im$\{\varepsilon_{\text{eff}}\} < 0$), a phase shift of the grating $\delta\varepsilon(R, t)$ recorded in the medium is required with respect to the interference pattern ($E_{\text{L}}^* E_{\text{S}}$+c.c.) which excites the grating. This phase shift is provided only by relaxation of the medium's excitations. If the signal and pump are of the same frequency, no relaxation occurs in the steady-state case. *Therefore, steady-state SS without a frequency shift is not observed.*

Secondly, the amplitude of the built-up disturbance is directly proportional to the accumulation time τ, i.e., inversely proportional to the relaxation constant Γ. As a consequence, all types of SS possess a general property: $G \propto \Gamma^{-1}$, i.e., *the longer the lag in SS, the stronger the SS is.*

Thirdly, except for processes caused by absorption *the gain is positive in the Stokes region.*

Finally, the characteristic spatial scale of attenuation of a medium's excitations (the largest being $l \leq 10^{-3}$ cm in SBS) is always less than the inhomogeneity dimensions of the pump field $|E_{\text{L}}(R)|^2$. Therefore, we may consider that practically always, for all types of SS, there exists a local relationship between the gain $g(R)$ of the Stokes wave and the pump intensity: $g(R) = G |E_L(R)|^2$.

As far as the applications of OPC to the compensation of distortion in powerful laser beams and the self-targeting of radiation are concerned, those types of SS with a small frequency shift are most preferable. First of all, we should mention STS, SBS, and, to a certain extent, SRWS. On the contrary, the Stokes shift in SRS is usually so large that scattered light cannot enter the amplification line of a laser medium and is not amplified on the return pass.

SRWS has a relatively high threshold and almost always occurs in the presence of self-focusing of light. That is why there are only a few publications in which SRWS, excited by picosecond pulses, was reported to reproduce the incident light structure.

SBS is the most suitable for the purpose of OPC, whereas STS is used rather seldom. The reason is apparently that the characteristic duration ($\tau \sim 5 - 50$ ns) of pulses to be conjugated, generated by Q-switched lasers, is shorter than the time of STS buildup. As a result, during non-steady-state operation the actual threshold of STS is much larger than that of SBS. On the contrary, SBS operates in the nearly steady-state condition and its threshold is low enough. Also, SBS develops in transparent media (as distinct from STS-II), and, accordingly, the problem of optical breakdown in SBS media is not so acute.

2.4 Effect of Pump Depletion in Stimulated Scattering

When we make the approximation that the pump field E_L is undepleted, the intensity $|E_S|^2 = I_S$ of the scattered field is given by

$$I_S(z) = I_S(0) \exp(G |E_L|^2 z) . \tag{2.4.1}$$

Here $I_S(0)$ is the intensity of the input priming field, the coordinate z is chosen in the direction of the Stokes wave, and the waves E_L and E_S are considered plane and monochromatic. The region of applicability of (2.4.1) is rather narrow. Indeed, if $G |E_L|^2 z \gtrsim 30$, i.e., the pump intensity exceeds the threshold, an appreciable portion of the pump intensity is transformed into the Stokes wave, and the undepleted pump approximation fails to work. On the other hand, at $G |E_L|^2 z \lesssim 23$ the intensity of the scattered field is very small, and the problem is of no interest.

In almost all real experiments the pump depletion effect, i.e., retroaction of the Stokes wave on the pump, is observed. Naturally, even if the time-integrated transformation coefficient is small so that, say, the reflected energy is 10% of the incident energy, then due to a smooth temporal change of the pump intensity $|E_L(t)|^2$, the instantaneous transformation coefficient is close to zero both at the beginning and at the end of the pulse and approaches 100% at its maximum. Thus, the pump depletion effect should be taken into account.

As the pump possesses the exact resonant anti-Stokes shift with respect to the scattered wave, we may use the results given in the previous sections. Stimulated scattering is caused by the appearance of the traveling spatial-temporal permittivity grating in the medium:

$$\delta\varepsilon(R, t) = A E_L^* E_S \exp(i\Omega t - iq \cdot R) + A^* E_L E_S^* \exp(-i\Omega t + iq \cdot R).$$

$$(2.4.2)$$

The scattering of the laser field due to the first term in (2.4.2) gives rise to the Stokes wave gain:

$$g_S = -\frac{\omega_S}{c} \frac{\text{Im}\{\delta\varepsilon_{\text{eff}}(\omega_S)\}}{\sqrt{\varepsilon_0}} = -\frac{\omega_S |E_L|^2}{c\sqrt{\varepsilon_0}} \text{Im}\{A\} = G |E_L|^2.$$

The retroaction on the pump is described by scattering of the Stokes wave due to the second term in (2.4.2) and leads to the pump attenuation

$$\alpha_L = \frac{\omega_L}{c} \frac{\text{Im}\{\delta\varepsilon_{\text{eff}}(\omega_L)\}}{\sqrt{\varepsilon_0}} = \frac{\omega_L |E_S|^2 \text{Im}\{A^*\}}{\sqrt{\varepsilon_0}} = \frac{\omega_L}{\omega_S} G |E_S|^2. \qquad (2.4.3)$$

As a result, the system of equations for the intensities of the interacting waves, $I_L(z) = |E_L(z)|^2$ and $I_S(z) = |E_S(z)|^2$, in backward SBS ($\omega_L \approx \omega_S$) is of the form

$$\frac{dI_S}{dz} = G I_L I_S, \qquad \frac{dI_L}{dz} = G I_L I_S = \frac{dI_S}{dz}. \qquad (2.4.4)$$

In this case we consider the pump to be propagating in the $(-z)$ direction. It is the second equation in (2.4.4) that describes the attenuation of the pump along its path. In SRS, when ω_S differs considerably from ω_L, taking into consideration (2.4.3) we can obtain a similar system of equations for the flux of quanta, $j_{L,S} = nc|E_{L,S}|^2/8\pi\hbar\omega_{L,S}$. Thus, in all cases, $dj_S/dz = dj_L/dz$. This relation, or the equality $\alpha_L = \omega_L g_S/\omega_S$ equivalent to it, expresses the so-called Manley-Row relation following from the real-valued nature of the grating $\delta\varepsilon(R, t)$ in (2.4.1). From a quantum-mechanical viewpoint, conservation of the number of quanta in the elementary scattering process $\hbar\omega_L \rightarrow \hbar\omega_S$, both spontaneous and stimulated, corresponds to the Manley-Row relation.

The boundary condition for the pump is specified at the input end of the medium (Fig. 2.4),

$$I_L(z = L) \equiv |E_L(z = L)|^2 = I_0, \qquad (2.4.5)$$

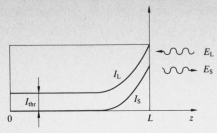

Fig. 2.4. Distribution of the plane pump intensity I_L and the scattered wave intensity I_S in backward SS at saturation

where I_0 is the input pump intensity. It is a bit more difficult to specify the boundary condition for the Stokes wave in the section $z = 0$. Usually the input Stokes signal is very low and formed as a result of the spontaneous scattering of the pump wave into the Stokes wave. We need to take into account only those Stokes waves which have maximum gain after passing through the whole medium; such waves occupy a small solid angle $\Delta\Theta$ and a small frequency band $\Delta\omega$. Here we face a formal contradiction with the supposition that the Stokes wave is plane and monochromatic. As, however, the final results are weakly (logarithmically) dependent on the priming intensity, the inaccuracy introduced is negligible. To describe such a situation, it would be better to modify the equation for the Stokes intensity to

$$\frac{dI_S}{dz} = GI_L I_S + \Delta R \cdot I_L, \tag{2.4.6}$$

where $\Delta R = (dR/d\Theta\, d\omega)\Delta\Theta \cdot \Delta\omega$ is the coefficient of pump-beam extinction for scattering into the Stokes beam. Near $z = 0$, where the Stokes signal is weak and the spontaneous scattering ΔR should be taken into account, the pump intensity varies only very slightly, $I_L(z) \approx I_L(z = 0)$. Then (2.4.6), with the exact boundary condition $I_S(z = 0) = 0$, can be explicitly integrated, and we have

$$I_S(z) = \frac{\Delta R}{G} \{\exp(GI_L(0)z] - 1\}. \tag{2.4.7}$$

At a distance $z \gtrsim [GI_L(0)]^{-1}$ the factor unity in (2.4.7) may be neglected, and expression (2.4.7) is equivalent to the unmodified (2.4.3) with the input Stokes signal specified as $I_S(z = 0) = \Delta R/G$. This result can also be obtained from the following considerations. Under conditions of exponential gain, the main contribution to noise is made by the input layer of effective width l, the reciprocal of the gain, $l = g^{-1} = [GI_L(z = 0)]^{-1}$.

The noise accumulated in this layer is the result of the scattering of the pump I_L with an extinction coefficient ΔR, i.e., $I_S = \Delta R \cdot l \cdot I_L = \Delta R/G$. It is interesting to note that the effective input Stokes signal turns out to be independent of the value (unknown beforehand) of the intensity $I_L(z = 0)$ of the pump transmitted through the cell. Finally, allowing for (2.2.15), we have the boundary condition for I_S:

$$I_S(z=0) = I_{S0} \approx \frac{\Delta R}{G} = \frac{k^2}{(2\pi)^3} k_B T \frac{\omega}{\Omega} \Delta\Theta\Delta\omega \left[\frac{erg}{cm^2 s}\right]. \qquad (2.4.8)$$

For typical experimental conditions the ratio $I_{S0}/I_0 = I_S(z=0)/I_L(z=L)$ is very small, of the order of $10^{-12} - 10^{-13}$.

The general solution of (2.4.4) can be readily obtained using the law of conservation of energy (strictly speaking, of the number of quanta) which follows from $I_L(z) - I_S(z) = N = I_L(z=0) - I_S(z=0)$. As the input Stokes signal is extremely low compared to the pump signal, the value of N equals, to a very high accuracy, the pump intensity transmitted through the cell, $N \approx I_L(z=0)$. The solution is

$$I_L(z) = N[1 - C\exp(NGz)]^{-1}, \qquad I_S(z) = I_L(z) - N. \qquad (2.4.9)$$

The two unknown constants, N and C, should be determined from the boundary conditions (2.4.5, 8) which give a system of transcendental equations for N and C. To solve this system, we should take into account the fact that the input Stokes noise is low, $\leq 10^{-12}$ of the pump intensity. Then we have $C \approx I_{S0}/N = \exp(-D)$. As will be seen further on, in the region interesting from a practical viewpoint — in the vicinity of and far above the SS threshold — the transmitted pump intensity N varies only slightly with the incident intensity $I_L(z=L) = I_{L0}$. Therefore, the value of $\exp(-D)$ may be considered constant, $\exp(-D) = \exp(-30)$, and independent both of the incident intensity I_{L0} and the transmitted intensity N. As a result, it becomes possible to relate the reflectivity

$$\eta = \frac{I_S(z=L)}{I_{L0}} = 1 - \frac{N}{I_{L0}} \qquad (2.4.10)$$

to the dimensionless pump intensity x:

$$I_{L0} = x \cdot \frac{D}{GL} = \frac{30}{GL} \cdot x. \qquad (2.4.11)$$

With these variables, the transcendental equation for η is of the form

$$x = (1 + D^{-1}\ln\eta)/(1 - \eta). \qquad (2.4.12)$$

The dependence $\eta(x)$ of the solution of this equation on x can be easily plotted using the graph $x(\eta)$, see Fig. 2.5.

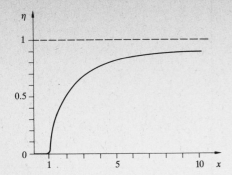

Fig. 2.5. Dependence of the reflectivity η on the pump intensity I reduced to the steady-state threshold value: $x = I/I_{\text{thr}}$

The value $x = 1$, $GI_{\text{L}0}L = 30$ is taken as the threshold. Then below the threshold, at $GI_{\text{L}0}L \lesssim 25$ for example, the reflectivity depends exponentially on the pump intensity:

$$\eta \approx I_{\text{S}0} \exp(GI_{\text{L}0}L) = e^{-D} \cdot e^{Dx}. \tag{2.4.13}$$

When the threshold is considerably exceeded, $x \gtrsim 2$, we have

$$\eta \approx 1 - \frac{1}{x}, \qquad I_{\text{tr}} = I_{\text{L}0}(1 - \eta) \approx I_{\text{thr}} = \frac{30}{GL}. \tag{2.4.14}$$

From (2.4.14) an important conclusion can be made: at saturation in the steady-state case, the pump intensity transmitted through the cell is equal to the threshold value.

2.5 Dynamics of Stimulated Scattering

In the previous sections we have dealt with the steady-state SS condition for the monochromatic pump and Stokes waves. In a number of experimental situations, however, non-stationarity and non-monochromaticity play an essential role. Even the local response of the medium $\delta\varepsilon \propto E_{\text{L}}^* E_{\text{S}}$ reaches a steady state for a time interval of the order of the reciprocal of the spontaneous scattering linewidth, $\tau_{\text{loc}} \sim \Gamma^{-1} \sim 10^{-8}$ to 10^{-9} s.

It is convenient to consider the dynamics of SS in stages, separating the more complicated physical aspects of the phenomenon into simpler ones. To begin with, we estimate the SS linewidth and the velocity of the SS build-up wave for a given monochromatic pump wave. This estimation shows that SS with an exponential gain $\sim \exp(30)$ lags considerably compared to the local process of hypersound excitation. Then we present the dynamics of SS, i.e., the time behavior of the Stokes wave intensity. On the basis of the formulas obtained we shall determine the threshold characteristics of the pump pulse for SS in the nonstationary condition. Finally, we shall discuss the over-threshold behavior of SS, which is determined, in the main, by saturation effects. Attention will also be given to experimental methods of investigation of nonstationary SS.

2.5.1 Spectral Width of Stimulated Scattering

In the case of a monochromatic pump wave, non-steady-state effects may be considered in a linear approximation in terms of the gain of the separate spectral components of the Stokes field. For SRS or SBS problems, we have, according to the results of Sects. 2.1, 3,

$$\exp(-i\omega_S t)E_S(\omega_S, z) = \exp(-i\omega_S t)E_S(\omega_S, z = 0)$$

$$\times \exp\left[ik(\omega_S)z + \frac{1}{2}g\frac{\Gamma z}{[\Gamma - i(\omega_S - \omega_1)]}\right]. \quad (2.5.1)$$

Here $\omega_1 = \omega_L - \Omega_B$ is the central frequency of the Stokes line, $g = G|E_L|^2$ is the steady-state gain in the line center, and $k(\omega_S)$ is the undisturbed wave number. The gain is maximum for spectral components in the proximity of the line center, and, therefore, the SS linewidth can be readily estimated on the basis of the expansion

$$g(\omega_S) \equiv g[1 + (\omega_S - \omega_1)^2/\Gamma^2]^{-1} = g[1 - (\omega_S - \omega_1)^2/\Gamma^2 + \cdots], \quad (2.5.2a)$$

$$I_S(\omega_S, z) = I_S(\omega_S, 0) \exp(gz) \exp\left[-\left(\frac{\omega_S - \omega_1}{\Gamma/\sqrt{gz}}\right)^2\right]. \quad (2.5.2b)$$

Thus, the SS linewidth (half-width at $1/e$ of the maximum intensity) is given by

$$\Delta\omega_S[HWe^{-1}M] = \Gamma/\sqrt{gz} \quad (2.5.2c)$$

and, in the steady-state condition, is a factor of $\sqrt{gz} \approx 5$ less than the spontaneous scattering linewidth. It can be seen that the propagation of the Stokes wave under conditions of high gain ($gz \sim 30$) considerably increases the build-up time of the process as a whole in comparison with that of the local response.

A more detailed consideration of the time behavior of SS may be given in the framework of the system of equations

$$\frac{\partial E_S}{\partial z} + \frac{1}{v_0}\frac{\partial E_S}{\partial t} = iAE_L Q(z, t), \quad (2.5.3)$$

$$\frac{\partial Q}{\partial t} + \Gamma Q(z, t) = -iBE_L^* E_S(z, t) .$$
(2.5.4)

Here E_L = const for the monochromatic pump, v_0 is the group velocity of the Stokes wave in the absence of the pump, $E_S(z, t)$ is the slowly varying complex amplitude of the Stokes field, and $Q(z, t)$ is the slowly varying coordinate of the medium's excitations, for instance, the amplitude of hypersonic waves. In view of the large spatial damping of the latter, the term of the form $v_S \partial Q/\partial z$ is omitted in (2.5.4). The solutions of system (2.5.3, 4) for spectral components $E_S(z) \exp[-i(\omega_S - \omega_1)t]$ have the form of (2.5.1), the gain g at the maximum being expressed in terms of constants A and B: $g = 2AB|E_L|^2/\Gamma = G|E_L|^2$.

2.5.2 Stimulated Scattering Build-Up Wave

Equations (2.5.3, 4) have an exact analytical solution, see below. First, however, we shall find an approximate solution and discuss its qualitative features. As is seen from (2.5.3), the main contribution to the scattered field is due to spectral components of E_S in a narrow (as compared to Γ) frequency band. This means that the complex amplitude $E_S(z, t)$ varies slowly over a time $\sim \Gamma^{-1}$, and (2.5.4) can be solved by iteration with respect to Γ^{-1}:

$$Q(z, t) = \frac{1}{\Gamma}\left(-iBE_L^* E_S - \frac{\partial Q}{\partial t}\right) \approx -i\frac{BE_L^*}{\Gamma}E_S + i\frac{BE_L^*}{\Gamma^2}\frac{\partial E_S}{\partial t} + \cdots .$$
(2.5.5)

Substituting (2.55) into (2.5.3), we obtain

$$\frac{\partial E_S}{\partial z} + \left(\frac{1}{v_0} + \frac{g}{2\Gamma}\right)\frac{\partial E_S}{\partial t} = \frac{1}{2}gE_S(z, t) .$$
(2.5.6)

Thus, the lag in the local response, i.e., the finite build-up time leads to a decrease of the group velocity of those spectral components which are amplified with maximum gain g.

If we define the delay of pulse arrival by the usual formula $t = z/v_{gr}$, then it follows from (2.5.6) that SS gives rise to a delay of the leading front of the Stokes wave by an amount

$$\Delta t = z\left(\frac{1}{v_0} + \frac{g}{2\Gamma}\right) .$$
(2.5.7)

We may assert that the expression $(v_0^{-1} + g/2\Gamma)^{-1}$ represents the velocity of the leading front of the SS build-up wave. This expression can also be derived from

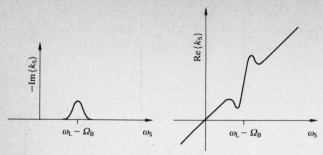

Fig. 2.6. Spectral dependence of the imaginary and real parts of the wave vector $k_S(\omega_S)$

$$k(\omega_S) = \frac{\omega_S}{c} n + \frac{1}{2i} g \frac{\Gamma}{\Gamma - i(\omega_S - \omega_1)} \qquad (2.5.8)$$

by using the well-known formula $v_{gr}^{-1} = d \operatorname{Re}\{k(\omega)\}/d\omega$. The group delay is connected with the anomalous dispersion of the signal in the region of maximum gain, Fig. 2.6. As a matter of fact, however, the SS build-up wave fails to travel as far as the output section of the SS cell because saturation effects then come into action.

2.5.3 Calculation of SS Dynamics

As it is the dynamics of SS that is of particular interest, system (2.5.3, 4) should be solved more accurately. The exact solution may be represented using a Green's function:

$$E_S(z, t') = \int_{-\infty}^{t'} E_S(z = 0, t'') G(z, t' - t'') dt'' ,$$

$$G_S(z, t) = \frac{1}{2\pi} \int_{-\infty}^{+\infty} \exp\left(-ipt + i\frac{p}{v_0} z + \frac{g\Gamma z}{2(\Gamma - ip)}\right) dp . \qquad (2.5.9)$$

The integral in (2.5.9) can be calculated in the explicit form

$$G_S(z, \tau) = \delta(\tau) + \theta(\tau)(2\tau)^{-1} e^{-\Gamma\tau} \sqrt{2gz\Gamma\tau} \, J_1(i\sqrt{2gz\Gamma\tau}) , \qquad (2.5.10)$$

where J_1 is a Bessel function, $\delta(\tau)$ is the Dirac delta function, and $\theta(\tau)$ is Heaviside's step function ($\theta = 0$ at $\tau < 0$, $\theta = 1$ at $\tau > 0$). In the situation of interest the argument $\sqrt{2gz\Gamma\tau}$ of the Bessel function is large, the value of the Bessel function is exponentially large, and, accordingly, the factor before the exponential may be omitted. To calculate the asymptotic expression for the Green's function, it is convenient to use (2.5.9) directly. In the asymptotic limit, the main contribution to the integral in (2.5.9) is due to those points in the complex plane p where the exponential factor varies slowly (the saddle-point or stationary-phase method):

$$\frac{d}{dp}\left(-ip\tau+\frac{gz\Gamma}{2(\Gamma-ip)}\right)=0\rightarrow p_{1,2}=-i\Gamma\pm i\sqrt{\frac{gz\Gamma}{2\tau}}. \qquad (2.5.11)$$

The contribution of point p_2 is exponentially small, and

$$G(z,\tau)\propto\exp\left(-ip_1\tau+\frac{1}{2}\frac{gz\Gamma}{\Gamma-ip_1}\right)=\exp(-\Gamma\tau+\sqrt{2gz\Gamma\tau}). \qquad (2.5.12)$$

Equation (2.5.12) describes (to an accuracy of the unimportant pre-exponential factor) the output amplitude of the Stokes wave when a δ-shaped pulse is applied to the input. The maximum of this response is reached at $t_{\mathrm{m}}=(gz/2\Gamma)+z/v_0$, this being in full agreement with the results of the simplified analysis; the response value $|G(z,\tau_{\mathrm{m}})|^2\propto\exp(gz)$ corresponds to the steady-state gain g. The duration Δt of the intensity response near its maximum for (2.5.12) is

$$\Delta t[\mathrm{HW\,e}^{-1}\mathrm{M}]=\Gamma^{-1}\sqrt{gz}, \qquad (2.5.13)$$

in strict accordance with the spectral width (2.5.2 c) of the amplified signal.

Of particular interest is the response of the system to spontaneous noise, which, together with the pump, is switched on at $t=0$. Spontaneous primings can be considered as a continuous sequence of incoherent δ-shaped pulses, and then

$$I_{\mathrm{S}}(z,t)\propto\int_0^t|G(z,t-t')|^2dt',$$

$$I_{\mathrm{S}}(z,t)\propto\begin{cases}\exp(-2\Gamma\tau+2\sqrt{2gz\Gamma\tau}), & \tau<gz/2\Gamma, \\ \exp(gz), & \tau>gz/2\Gamma.\end{cases} \qquad (2.5.14)$$

In a real situation the pump field is not switched on instantaneously, but smoothly, and it is not strictly monochromatic. To obtain the analytical solution to this problem, we resort to the fact that in most cases the delay due to the undisturbed group velocity is small: $z\sim10\,\mathrm{cm}$, $v_{\mathrm{L}}\sim v_{\mathrm{S}}\sim2\times10^{10}\,\mathrm{cm/s}$, $z/v\sim5\times10^{-10}\mathrm{s}\lesssim\Gamma^{-1}$. Then we may consider the pump field $E_{\mathrm{L}}(z,t)$ homogeneous over the interaction volume, $E_{\mathrm{L}}(z,t)=E_{\mathrm{L}}(t)$, and assume $v_0^{-1}=0$ in (2.5.3). In this approximation it is convenient to make the substitution

$$E_{\mathrm{S}}(z,t)=\mathrm{e}^{-\Gamma t}E_{\mathrm{L}}(t)\,Y(z,t),$$

$$Q(z,t)=\mathrm{e}^{-\Gamma t}H(z,t), \qquad (2.5.15)$$

$$T=\int_{-\infty}^t|E_{\mathrm{L}}(t')|^2dt',$$

so that the system (2.5.3, 4) is of the form

$$\frac{\partial Y}{\partial z} = iAH, \qquad \frac{\partial H}{\partial T} = -iBY. \tag{2.5.16}$$

This form is exactly equivalent to the initial form with $v_0 = 0$, $\Gamma = 0$, and, consequently, the Green's function for it can be derived from (2.5.10) by substitution. Returning to the initial variables, we obtain for the intensity of the response to be input δ-shaped signal, omitting the unimportant preexponential factor,

$$|E_S(z, t, t')|^2 \propto \exp\left\{ -2\Gamma(t - t') + 2\left[2\Gamma G z \int_{t'}^{t} |E_L(t'')|^2 dt'' \right]^{1/2} \right\}. \tag{2.5.17}$$

For a constant intensity pump, (2.5.17) is reduced to the square of (2.5.12).

2.5.4 Determination of Threshold in Nonstationary Condition

Proceeding from (2.5.17) we can answer two basic questions concerning SS in the nonstationary condition: i) at what moment of time does the intensity of the scattered wave become comparable with that of the pump, i.e., when is the SS threshold reached, and ii) which parameters of the pulse determine the threshold value in an essentially nonstationary condition?

Consider first a pump intensity pulse $|E_L(t)|^2$ of rectangular shape. Then two dimensionless parameters are at our disposal: the steady-state gain index regardless of saturation, $G|E_L|^2 z = gz$, and the ratio of the pulse duration to the local nonlinearity relaxation time, $t_p/\tau_0 = \Gamma t_p$. The threshold condition, as earlier, needs a noise intensity gain of $\exp(30)$:

$$-2\Gamma t + \sqrt{8\Gamma t G |E_L|^2 z} \approx 30. \tag{2.5.18}$$

Demanding that the threshold should be reached by the end of the pulse at least, we obtain

$$G|E_L|^2 z \gtrsim \begin{cases} (15 + \Gamma t_p)^2/2\Gamma t_p, & \Gamma t_p < 15, \\ 30, & \Gamma t_p > 15. \end{cases} \tag{2.5.19}$$

For sufficiently long pulses, $\Gamma t_p \gtrsim 7$, the threshold condition differs only slightly from the steady-state condition $G|E_L|^2 z = 30$. On the contrary, at $\Gamma t_p \sim 1$, the threshold value of $|E_L|^2$ increases as t_p^{-1}, and in this regime the threshold is determined by the pulse energy density

$$\int |E_L|^2 dt \gtrsim \frac{(30)^2}{8\Gamma G z}. \tag{2.5.20}$$

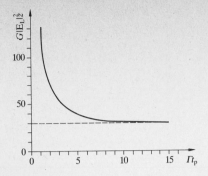

Fig. 2.7. Dependence of the threshold intensity $G|E_L|^2z$ on the pump pulse duration t_p multiplied by the linewidth Γ. Dashed lines denote the steady-state threshold

This formula is also valid, at $\Gamma t_p \sim 1$, for an arbitrary pulse shape. For intermediate cases the dependence of the threshold intensity $G|E_L|^2z$ upon Γt_p is plotted in Fig. 2.7.

For an arbitrary pulse shape $|E_L(t)|^2$ and intermediate $\Gamma t_p \sim 5$, the SS threshold is determined by the existence of two instants t_1 and t_2, for which the exponent of the intensity amplification factor in nonsteady-state conditions should satisfy

$$-2\Gamma(t_2-t_1)+\sqrt{8Gz\Gamma\int_{t_1}^{t_2}|E_L(t')|^2dt'} \gtrsim 30. \tag{2.5.21}$$

The analysis of this condition requires the specification of a particular pulse shape and is usually made by using numerical methods [2.6]. An analytical approach is simple with the time behavior of the pump intensity linearly rising from zero, starting from $t = 0$. It is convenient to characterize the rate of this rise by a time τ_r during which the intensity increases to reach the steady-state threshold:

$$I_L(t) = \text{const} \cdot t = \frac{30}{Gz} \cdot \frac{t}{\tau_r}. \tag{2.5.22}$$

For the pump wave (2.5.22), it is possible to find a value of t_2 depending on t_1 such that (2.5.21) becomes an equality. A minimum value of t_2 corresponds to the onset of SS:

$$t_2 = \tau_r[1+\sqrt{1+(30/\Gamma\tau_r)}]/2. \tag{2.5.23}$$

From (2.5.23) it follows that the onset of SS is delayed relative to the instant $t = \tau_r$ corresponding to the steady-state threshold. Accordingly, reaching the SS threshold in the nonsteady-state condition requires that the steady-state threshold level be exceeded by a factor of $t_2/\tau_r = \frac{1}{2}(1+\sqrt{1+30/\Gamma\tau_r})$.

2.5.5 Above-Threshold Behavior of SS and Loop Scheme

In stimulated scattering the Stokes pulse always arises after a certain delay with respect to the pump. For long pulses, $\Gamma t_p \gg 1$, this delay can be estimated either by (2.5.18) or by (2.5.23). Once the threshold is exceeded, a noticeable depletion of the pump intensity occurs. However, at the initial stage the saturation is not described by the formulas of the steady-state theory from Sect. 2.4. Unfortunately, there is no analytical theory of nonsteady-state saturation, and only numerical calculations for a particular case can be found in [2.6].

There exists a simple and effective experimental scheme to study these effects [2.5]. Suppose that a signal proportional to the time dependence of the pump intensity $I_L(t)$ incident on the cell is applied to the horizontal-deflecting plates of an oscilloscope. The signal for the transmitted intensity $I_{out}(t) = I_L(t) - I_S(t)$ is applied, without any time delay, to the vertical-deflecting plates; here $I_S(t)$ is the intensity of the reflected wave. In the absence of SS and other nonlinear processes, a straight line with a slope of 45° (dashed line 1 in Fig. 2.8a) should be traced on the screen. This line is traced on two passes, namely, during the rise and the fall of the pump pulse intensity. However, SS decreases the transmitted intensity, and in the steady-state case, dashed curve 2 should be traced and, according to the results of Sect. 2.2, be stabilized at a trans-mitted-signal level equal to the threshold value of the intensity. This same trajectory should also be traced on both passes.

In contrast, under nonsteady-state saturation, the trajectory of I_{out} (I_{inc}) is loop-shaped (solid curve 3 in Fig. 2.8a). The corresponding time dependence of the intensities $I_L(t)$ and $I_S(t)$ is shown in Fig. 2.8b. In the rising section, the loop becomes separated from line 1 at a point corresponding to the nonsteady-state threshold (2.5.23). Saturation develops above the threshold. In this case, the steady-state regime (curve 2) is approached, under essentially unsteady conditions, in the section of falling pump intensity. The lower part of the loop merges with line 1 near the intensity value corresponding to the steady-state threshold. Thus, the loop scheme enables one to measure the steady-state threshold of the pump even in nonsteady-state operation.

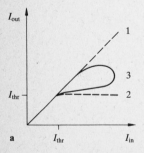

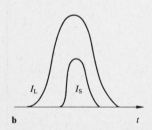

Fig. 2.8. (a) "Loop" scheme of registration of above-threshold nonsteady-state effects in SS. (b) Typical time dependence of a gigantic laser pulse, $I_L(t)$, and of reflected light, $I_S(t)$

2.6 Stimulated Scattering of Non-Monochromatic Radiation

In the previous section we discussed the case of a single pump pulse with a smooth temporal profile. However, very often one has to deal with lasers simultaneously generating several modes with different longitudinal mode indices. Non-monochromaticity of this radiation manifests itself in a number of spikes on the intensity time-dependence curve due to the interference of contributions from different modes. For non-locked modes the existing amplitude fluctuations are of a random nature. Such non-monochromatic pump fields will be given particular attention in this section.

In spontaneous scattering of non-monochromatic incident light $E_L(t)$ by a fluctuating molecular parameter $Q(t)$ of the medium, the scattered field $E_S(t)$ is of the form

$$E_S(t) = \text{const} \cdot Q(t) E_L(t) . \tag{2.6.1}$$

As a result, the spectrum of the scattered light is given by convolution of the incident light spectrum with the fluctuation spectrum $Q(t)$. Standard spectroscopy uses a sufficiently monochromatic field $E_L(t)$ so that the scattered light spectrum should coincide with that of the fluctuations Q. On the contrary, when $\Delta\omega_L \gg \Delta\omega_Q$, we have $\Delta\omega_S \approx \Delta\omega_L$. Then $Q(t)$ may be regarded as a slow function of time, and from (2.6.1) it follows that the scattered field $E_S(t)$ reproduces basically the time behavior of the incident field $E_L(t)$. It turns out that in stimulated scattering, when the simple relation (2.6.1) is not valid, various regimes may occur; in some cases the time behavior of the pump is basically reproduced, in others it is essentially distorted.

2.6.1 Quasi-Static Case

Consider first a quasi-static case in which the spectral width $\Delta\omega_L$ of the laser is much smaller than the width Γ of the spontaneous scattering line, $\Delta\omega_L \ll \Gamma$. Then scattering is independent for each spike of the pump intensity. Under non-saturation conditions,

$$|E_S(t)|^2 \propto \exp[Gz|E_L(t)|^2] . \tag{2.6.2}$$

Even moderate relative fluctuations of the pump intensity, $\langle\Delta I_L^2\rangle \sim \langle I_L\rangle^2$, typical of radiation with non-synchronized modes, are exponentially emphasized. Consequently, the fading of the Stokes intensity appears for all time intervals when the pump intensity is below the threshold. Saturation should be taken into account above the threshold, and the simple formula (2.6.2) fails to work. However, fading and underlying pump intensity spikes still remain when saturation is taken into account.

2.6.2 Broad-Band Pump in a Short Medium

Another limiting case of particular interest is when $\Delta \omega_L \gtrsim \Gamma$. Consider first this case in terms of time behavior supposing $v_0 \to \infty$, i.e., $\Delta \omega_L z / v_0 \ll 1$. Here we may use the results of Sect. 2.5.3 and write, for non-saturation conditions, omitting preexponential factors,

$$E_S(z, t) = E_L(t) y(z, t) ,$$

$$y(z, t) \propto \int dt' Q_{sp}(t') \exp \left\{ -\Gamma(t-t') + \left[2\Gamma G z \int_{t'}^{t} dt'' |E_L(t'')|^2 \right]^2 \right\} .$$

$$(2.6.3)$$

The complex quantity y is the instantaneous amplitude diffraction efficiency of the traveling volume grating from which the laser wave is reflected. The time behavior of $y(z, t)$ is determined by the behavior of two factors under the integral over dt': the spontaneous priming signal $Q_{sp}(t')$ and the exponential factor.

For a strictly constant pump intensity, $|E_L(t'')|^2 = \text{const}$, the width of the exponential function is of the order of $\Delta t \sim \Gamma^{-1}(G|E_L|^2 z)^{1/2} \sim 5\Gamma^{-1}$ (at the SS threshold), see (2.5.13). Therefore $y(z, t)$ represents noise filtered off in a frequency band $\Delta \omega_y = \Gamma(Gz|E_L|^2)^{-1/2} \approx \Gamma/5$ in accord with (2.5.2c), and the characteristic correlation time is $\Delta t_y \sim 5\Gamma^{-1}$. In this case the phase modulation of the pump field $E_L(t)$ could be arbitrarily strong. It is of importance that if $E_S(z, t) \propto E_L(t) y(z, t)$ with a slow $y(z, t)$, then the interference term in the electrostriction pressure $\delta p \propto E_L^* E_S \propto |E_L(t)|^2 y(z, t)$ also varies slowly with time, i.e., it is highly monochromatic. Thus, a striking fact occurs: in this approximation, the spectral width, $\Delta \omega_y \approx \Gamma/5$, of excited hypersonic waves may turn out to be far smaller than the pump spectrum width $\Delta \omega_L$. Under such conditions the field $E_S(t)$ reproduces the time behavior of the pump $E_L(t)$.

If the pump intensity $|E_L(t)|^2$ varies sufficiently quickly, all these conclusions remain valid even in the presence of amplitude modulation. This will hold true if fluctuations of the exponential factor in (2.6.3) in the vicinity of its maximum are small compared to unity. For a pump beam with fluctuations $\langle \Delta I_L^2 \rangle = \langle I_L \rangle^2$, this gives the condition $\Delta t_L = (\Delta \omega_L)^{-1} \lesssim (2\Gamma)^{-1}$, i.e., the hypersound present in the medium should not be appreciably damped during the pump "fading" time.

Hence, under the assumptions made, i.e., for $|E_L|^2 = \text{const}$ or for $(\Delta I_L / \Delta I_L) \cdot 2\Gamma \Delta t_L \lesssim 1$, we can draw an important conclusion: the time behavior of the scattered field follows that of the pump, and molecular vibrations at the medium outlet are rather monochromatic $- \Delta \omega_y \sim \Gamma/5$ independent of the pump spectrum bandwidth. Under these conditions the gain is given by $g = G\langle|E_L|^2\rangle$ as in the case of a monochromatic pump of the same intensity $\langle|E_L|^2\rangle$. This is direct consequence of the time behavior reproduction, $E_S(t) \propto E_L(t)$. On the contrary, the gain of the specially formed mono-chromatic signal is determined by the convolution of the pump spectrum with the gain line from (2.1.11). In most experiments, however, SS develops from spontaneous noise, and the reproduction takes place provided the above conditions are satisfied.

2.6.3 Effect of Group Velocity Detuning

In our previous consideration we have neglected the time lag due to the finiteness of the velocity v_0 of light in the medium. If each of the waves $E_L(t)$ and $E_S(t)$ propagated in the medium with its own group velocity, the interference force building up the vibrations of the medium would be proportional to $E_L^*(t-z/v_L)E_S(t-z/v_S)$. For counterpropagating waves in backward SBS we should consider $v_S = v_0$, $v_L = -v_0$. Thus, the waves $E_L(t-z/v_L)$ and $E_S(t-z/v_S)$ could have a matched temporal structure of the type $E_S(t) \propto E_L(t)$ only over the limited length $\Delta z_{mis} \approx (\Delta\omega_L)^{-1}v_Lv_S/|v_L-v_S|$, called the mismatch length. For counterpropagating waves this is about half as long as the coherence length of radiation in the medium.

If the length L of wave interaction exceeds the mismatch length Δz_{mis}, it may seem that the build-up force $E_L^*E_S$ cannot be monochromatic in the entire volume and, consequently, time behavior reproduction cannot occur. However, as will be shown below, the condition $L < \Delta z_{mis}$ is sufficient but not necessary.

It is convenient to consider this question in terms of a sufficiently visual "stick" model of pump radiation. In the framework of this model the pump spectrum is considered to be a set of discrete monochromatic components ω_n, the intervals $|\omega_n - \omega_m|$ between which are far larger than the scattering linewidth:

$$E_L(z, t) = E_L(t-z/v_L) \equiv \sum_n C(\Omega_n) \exp[-i\Omega_n(t-z/v_L)] \,. \tag{2.6.4}$$

Then the system (2.5.3, 4) should be solved taking (2.6.4) into account. The latter leads us to seek the solution for the Stokes field in the form

$$E_S(z, t) = \sum_n S(\Omega_n, z) \exp[-i\Omega_n(t-z/v_L)] \,. \tag{2.6.5}$$

It should be noted that the slowly varying amplitude $E_S(z, t)$ is read out after separation of the central frequency factor $\omega_S = \omega_L - \Omega_{SS}$. Let us discuss the specific feature of seeking the solution in this form. If we wanted to solve the free equation for the Stokes wave (i.e., at $Q \equiv 0$), it would be natural to write exponents in a form, dependent on $t - z/v_S$, corresponding to the Stokes wave group velocity. However, it is the search for the solution in the form (2.6.5), dependent on $t - z/v_L$, that ensures timematch to the pump in the whole interaction volume.

The solution of (2.5.4) for the medium's disturbance amplitude is of the form

$$Q(z, t) = -iB \sum_n \sum_m \frac{S(\Omega_n, z) C^*(\Omega_m)}{\Gamma - i(\Omega_n - \Omega_m)} \exp[-i(\Omega_n - \Omega_m)(t - z/v_L)] . \quad (2.6.6)$$

The condition $|\Omega_m - \Omega_n| \gg \Gamma$ enables one to neglect all terms with $n \neq m$, and hence

$$Q(z, t) \approx -\frac{iB}{\Gamma} \sum_n S(\Omega_n, z) C^*(\Omega_n) . \quad (2.6.7)$$

In other words, under this condition, medium disturbances are recorded only by the interference of corresponding spectral components mutually shifted by the same resonant amount, $\omega_L - \omega_S = \Omega_{SS}$. In accord with Sect. 2.6.2, the disturbance wave turns, in this approximation, to be monochromatic, i.e., the slowly varying amplitude $Q(z, t)$ from (2.6.7) is independent of time. Substituting (2.6.5, 7) into (2.5.3), we obtain a system (at each value of Ω_n),

$$\frac{\partial S}{\partial z} - i\Omega_n \beta S(\Omega_n, z) = \frac{1}{2} GC(\Omega_n) \sum_m C^*(\Omega_m) S(\Omega_m, z) , \quad (2.6.8)$$

where $\beta = v_S^{-1} - v_L^{-1}$ characterizes the group velocity detuning.

At $\beta = 0$ one of the solutions of (2.6.7) is of the form

$$S(\Omega_n, z) = e^{gz/2} C(\Omega_n) , \quad g = G \sum_n |C(\Omega_n)|^2 = G \langle |E_L|^2 \rangle ,$$

$$\exp(-i\omega_S t) E_S(z, t) = \exp[-i(\omega_L - \Omega_{SS}) t] E_L(z, t) \cdot \text{const} \exp(gz/2) . \quad (2.6.9)$$

This solution describes the exact reproduction of the pump time behavior and, in accord with Sect. 2.6.1, has an amplitude gain $\mu = g/2$, as in the case with a monochromatic pump of the same intensity. At small β we can find an approximate solution of (2.6.7) [which may be converted into (2.6.8) as $\beta \to 0$] from perturbation theory [2.7]:

$$S(\Omega_n, z) = \text{const} \cdot e^{\mu z} \frac{C(\Omega_n)}{\mu - i\beta\Omega_n} ,$$

$$\mu = \tfrac{1}{2} g - 2\beta^2 \overline{\Omega^2}/g + \cdots , \quad \overline{\Omega^2} = \sum_n |C(\Omega_n)|^2 \Omega_n^2 / \sum_m |C(\Omega_n)|^2 , \quad (2.6.10)$$

where it is supposed that $\bar{\Omega} = 0$. Owing to the smallness of β, the denominators in (2.6.10) can be represented as

$$(\mu - i\beta\Omega_n)^{-1} \approx 2g^{-1}(1 + 2i\beta\Omega_n/g),$$

and then we have, in terms of the time dependence,

$$E_S(z, t) \approx \text{const} \exp(\mu z) E_L\left(t - \frac{z}{v_L} - \frac{2\beta}{g}\right). \tag{2.6.11}$$

Thus, due to group velocity detuning, (2.6.11) turns out to be, to first order with respect to β, slightly shifted relative to the pump. Owing to this nonsynchronism, the efficiency with which medium oscillations are built up becomes worse and the amplitude gain μ decreases.

At $\beta^2\overline{\Omega^2} \sim g^2$ the amplitude gain μ considerably decreases [2.8]. The condition $|\beta|\Delta\omega_L \sim g$, at which the reproduction of the time structure fails and, accordingly, the SS threshold sharply increases, can be rewritten as

$$g\Delta z_{mis} \sim 1. \tag{2.6.12}$$

Thus, obtaining effective SS of broad-band radiation with a low threshold requires that the mismatch length $\Delta z_{mis} \sim (\Delta\omega_L)^{-1}v_Lv_S|v_L - v_S|^{-1}$ should exceed the gain length g^{-1}. It is clear that at the threshold $gL \sim 30$ this condition limits the value of $\Delta\omega_L$ by a factor of 30 less than the primitive condition $\Delta z_{mis} \gtrsim L$ does, where L is the total length of the interaction region.

A striking feature of (2.6.9, 11) should be noted. The spatial dependence of the complex amplitude of the Stokes wave is given by $\exp(gz/2)$. A slower spatial-temporal modulation $\propto E_L(t - z/v_L)$ is superimposed on this very sharp dependence. In the case of counterpropagating interacting waves the Stokes wave envelope profile travels synchronously with the pump wave, i.e., *in opposition* to the energy flux in the Stokes field. We may assert that the group velocity of the Stokes wave "has changed its sign" in the presence of the pump.

2.7 Excitation of Anti-Stokes Components — Four-Wave Processes

2.7.1 Presentation of Nonlinear Polarization

Both Stokes ($\omega_L - \omega_S > 0$) and anti-Stokes ($\omega_L - \omega_S < 0$) components are present in the spontaneous scattering spectrum of light. In SS, as has been shown in quantum and classical terms, an anti-Stokes component is not

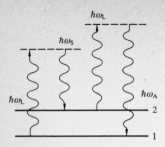

amplified but, on the contrary, is attenuated by transfer of its energy of the pump wave. Thus, in the framework of the two-photon processes of SS (Fig. 2.2), $\hbar\omega_L \rightarrow \hbar\omega_S$, described above, the appearance of intense components with $\omega_S > \omega_L$ is forbidden. However, a four-photon competitive process is possible which arises in the same order with respect to nonlinearity. If fields with frequencies ω_L, ω_S, where $\omega_L > \omega_S$, are present in the medium, then in the four-wave process

$$\hbar\omega_L + \hbar\omega_L = \hbar\omega_S + \hbar\omega_A, \qquad (2.7.1)$$

an anti-Stokes component with a frequency $\omega_A = 2\omega_L - \omega_S$ (Fig. 2.9) can be intensely excited.

To describe this four-wave process from a macroscopic viewpoint, consider a permittivity disturbance $\delta\varepsilon(R, t)$ built up in the medium by the interference of waves E_L and E_S:

$$\delta\varepsilon(R, t) = A(\Omega)E_L^* E_S \exp(-iq \cdot R + i\Omega t)$$
$$+ A^*(\Omega)E_L E_S^* \exp(iq \cdot R - i\Omega t), \qquad (2.7.2)$$

where, for most types of SS, $A(\Omega)$ possesses a resonant dependence on Ω. Usually we confine ourselves to the contribution of the scattering of the pump field E_L due to the first term in (2.7.2), which leads to a gain in the signal E_S. The scattering of E_S due to the second term is responsible for pump attenuation under saturation. Finally, the scattering of the wave $E_L \exp(ik_L \cdot R - i\omega_L t)$ due to the second term in (2.7.2) results in the non-linear polarization

$$P^{NL} = \exp[-i(2\omega_L - \omega_S)t + i(2k_L - k_S) \cdot R]A^*(\Omega)E_L E_L E_S^*, \qquad (2.7.3)$$

which possesses the anti-Stokes component frequency corresponding to (2.7.1). However, as distinct from two-photon processes, the spatial dependence $P^{NL}(R, t)$ does not, in general, satisfy the wave equation. The effective excitation of the anti-Stokes component requires the so-called phase-matching condition

$$\frac{\omega_A}{c} n(\omega_A) = |2k_L - k_S| \tag{2.7.4a}$$

or the equivalent law on conservation of momentum of the interacting photons:

$$\hbar k_L + \hbar k_L = \hbar k_S + \hbar k_A . \tag{2.7.4b}$$

We should emphasize that, in the two-photon process $\hbar\omega_L \to \hbar\omega_S$, $\hbar k_L \to \hbar k_S$, neither energy nor momentum *of the light field* are conserved since the generation of elementary excitations of the medium with energy $\hbar(\omega_L - \omega_S)$ and momentum $\hbar(k_L - k_S)$ occurs in this process. In four-wave processes the higher energy levels of the medium are excited only virtually.

2.7.2 Excitation of Anti-Stokes Components Under Forward Scattering

In a non-dispersive medium $n(\omega_L) = n(\omega_S) = n(\omega_A)$, and the condition (2.7.4) is strictly satisfied only in the case of colinear copropagating waves E_L and E_S. An anti-Stokes component should be radiated in the same direction only.

A frequency shift $\Delta\omega \sim 0.01 - 0.2 \ \omega_L$ is typical of SRS and dispersion is important. If $d^2k/d\omega^2 > 0$, which is observed in most media, the geometry shown in Fig. 2.10 can satisfy the phase-matching condition (2.7.4). For small frequency shifts Ω the angle $\theta_S \approx \theta_A$, determined by (2.7.4), is given by

$$\theta \approx |\Omega| \sqrt{\frac{1}{k} \frac{d^2k}{d\omega^2}} \equiv \frac{|\Omega|}{\omega} \sqrt{\frac{\lambda^2}{n} \frac{d^2n(\lambda)}{d\lambda^2}} , \tag{2.7.5}$$

where λ is the wavelength in vacuum.

A unidirectional pump beam is usually used in SRS experiments. In this case, the anti-Stokes radiation in the far field zone represents a ring around the central pump direction.

For a high-intensity pump wave the spectrum of the scattered light also contains higher-order components, $\omega_L \pm n\Omega$, in addition to the first Stokes and anti-Stokes components $\omega_L \pm \Omega$. Higher anti-Stokes components $\omega_L + n\Omega$ can be excited by a four-photon mechanism,

$$\hbar[\omega_L + (n-1)\Omega] + \hbar[\omega_L + (n-1)\Omega] \to \hbar(\omega_L + n\Omega) + \hbar[\omega_L + (n-2)\Omega] , \tag{2.7.6}$$

Fig. 2.10. Phase matching condition for four-wave generation of an anti-Stokes component

provided that the phase matching condition is satisfied. The generation of higher Stokes components by process (2.7.6) is complemented by successive normal SS processes, $\hbar(\omega_L - \Omega) \to \hbar(\omega_L - 2\Omega) \to \cdots \hbar(\omega_L - n\Omega)$, provided the intensity of the $(n-1)$th component exceeds the SS threshold.

2.7.3 SBS of Counterpropagating Waves

As far as OPC is concerned, the generation of anti-Stokes components in SBS is of particular interest. As SBS occurs in the backward direction, the phase matching condition (2.7.4) for a unidirectional pump is not satisfied. Hence, the four-wave generation of anti-Stokes components in SBS is possible only in the field of counterpropagating pump waves,

$$E_L(R) \exp(-i\omega_L t) = [E_+ \exp(ik_L z) + E_- \exp(-ik_L z)] \exp(-i\omega_L t) .$$
$$(2.7.7)$$

If one pump quantum of each of the two waves takes part in the elementary four-wave process, $2\omega_L \to \omega_S + \omega_A$, then the phase matching condition (2.7.4b) takes on the form

$$0 = k_L^+ + k_L^- = k_S + k_A . \qquad (2.7.8)$$

This condition can be satisfactorily met if the Stokes and anti-Stokes components are also counterpropagating and only slightly differ in frequency. Indeed, in the case of SBS,

$$p = |k_S + k_A| = |(\omega_L - \Omega)n/c - (\omega_L + \Omega)n/c| = 2\Omega n/c \sim 0.1 - 1 \text{ cm}^{-1} .$$

In backward SS with a small frequency shift, the counterpropagating pump wave E_- is effectively scattered into an anti-Stokes wave $a_+ \exp(-i\omega_A t + ik_A z)$ by a traveling grating recorded in the medium in the course of the SS interaction of the Stokes wave $s_- \exp(-i\omega_S t - ik_S z)$ and the pump wave E_+. The system of equations for coupled waves a_+ and s_- is, in accord with (2.7.2, 3), of the form

$$\frac{ds_-}{dz} = \frac{1}{2}(g_+ s_- + g_\pm a_+^* e^{-ipz})[1 + i(\Omega - \Omega_B)/\Gamma]^{-1} ,$$
$$(2.7.9)$$
$$\frac{da_+^*}{dz} = -\frac{1}{2}(g_- a_+^* + g_\pm^* s_- e^{ipz})[1 + i(\Omega - \Omega_B)/\Gamma]^{-1} .$$

The system for mutually coupled waves s_+ and a_- is of a similar form. Here Γ is the line half-width, $g_+ = G|E_+|^2$, $g_- = G|E_-|^2$, $g_\pm = GE_+ E_-$. The term $\propto g_\pm$ in (2.7.9) describes the parametric four-wave interaction of the pair

ω_S, ω_A with the pump, the term $\propto g_+$ describes the amplification of the Stokes wave s_- in the pump field $|E_+|^2$, and the term $\propto g_-$ describes the attenuation of the anti-Stokes wave a_+ in the pump field $|E_-|^2$.

If $p \gg g_+, g_-$ (such is the case, for instance, in SRS), the term containing an exponent strongly oscillates over the scale of changes of the functions s_-, a_+, and, therefore, practically has no effect on the solution. Omitting this term, we obtain the independent gain of the Stokes field, $s_- \propto \exp(-g_+ z/2)$, and the attenuation of the anti-Stokes field, $a_+ \propto \exp(-g_- z/2)$. The system can be solved precisely by substituting the variables $s_-(z) \equiv S_-(z) \exp(-ipz/2)$, $a_+^*(z) = A_+^* \exp(ipz/2)$. Usually in SBS $p \lesssim g$, and we give the solution for a limiting case, $p \ll g$. Then, supposing $p = 0$, we have

$$s_-(0) = \frac{(g_+ + g_-) s_-(L) + g_\pm [1 - \exp(\mu L)] a_+^*(0)}{g_- + g_+ \exp(-\mu L)},$$

$$(2.7.10)$$

$$\mu = (g_+ + g_-)/2\alpha, \qquad \alpha = 1 + i(\Omega - \Omega_B)/\Gamma.$$

From (2.7.10) it follows, first, that at $\alpha = 1$ (in the center of the line) and at $g_+/g_- \gg 1$, i.e., even when the counterpropagating pump intensity $|E_-|^2$ is low, its presence strongly influences SS of the type $|E_+|^2 \rightarrow |s_-|^2$:

$$s_-(0) \approx \frac{(g_+/g_-) \exp(g_+ L/2)}{(g_+/g_-) + \exp(g_+ L/2)} \cdot s_-(L).$$

$$(2.7.11)$$

If the intensity of one of the pump waves is small enough, say $|E_-|^2 < |E_+|^2 \times \exp(-g_+ L/2) \sim |E_+|^2 \exp(-15)$, then the parametric coupling may be neglected, and we observe the usual exponential stimulated scattering $E_+ \rightarrow s_-$. However, already when $|E_-|^2/|E_+|^2 \gtrsim \exp(-15) \approx 3 \times 10^{-7}$, an exponential gain is suppressed due to the parametric interaction with the anti-Stokes component: $|s_-(0)|^2 = (|E_+|^2/|E_-|^2) |s_-(L)|^2$.

From a physical viewpoint, the reason for this is negative feedback due to the four-wave process. Consider first the Stokes wave amplification at $|E_+|^2 \gg |E_-|^2$, $\Omega = \Omega_B$, neglecting the coupling to the anti-Stokes wave. Then we have $s_-(z = 0) \approx \exp(0.5 L g_+) s_-(L)$. The amplified Stokes signal excites the wave $a_+(0) \propto (g_\pm/g_+) s_-(0)$ in the region $z \approx 0$. This anti-Stokes component propagates towards the end of the cell at $z = L$ practically unchanged and is scattered backwards into a Stokes component by a four-wave process. This secondary scattering gives the signal $\delta s_-(L)$, in antiphase with respect to the initial signal $s_-(L)$, producing a large negative feedback.

The situation in which the counterpropagating pump waves have a complicated spatial structure within the interaction volume requires special consideration. As the efficiency of the four-wave interaction is determined by the degree of spatial correlation of the reference wave structure, we may

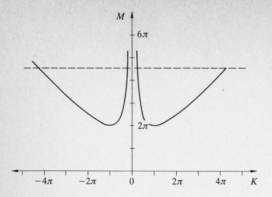

Fig. 2.11. Dependence of the parametric SS-generation threshold in the field of counterpropagating pump waves on the logartihm of their intensity ratio, K. $(---)$ denote the threshold level for a unidirectional pump

expect the condition for SBS suppression to be $|E_+|^2/|E_-|^2 \leq H \cdot 3 \times 10^6$, where H is the square of the normalized overlapping amplitude.

The solution (2.7.10) is also indicative of the possibility of scattering the pair of counterpropagating waves $\omega_L + \omega_L$ into the pair $\omega_S + \omega_A$ by parametric generation, i.e., the possibilty of having nonzero output amplitudes $s_-(z=0)$ and $a_+(z=L)$ under zero input conditions $s_-(z=L) = 0$ and $a_+(z=0) = 0$. The threshold for this process is determined by making the complex denominator in the right side of (2.7.10) vanish, see Fig. 2.11:

$$M = \pm \pi \frac{\Gamma}{\Omega - \Omega_B} [1 + (\Omega - \Omega_B)^2/\Gamma^2] ,$$

$$(\Omega - \Omega_B)/\Gamma = \pm \sqrt{(M/|K|) - 1} .$$

(2.7.12)

Here $M = 0.5\, G(|E_+|^2 + |E_-|^2)\, L$ is the amplitude gain at the lenght L, as calculated for the total intensity of the counterpropagating waves, and G is taken for the center of the Brillouin line: $K = \ln(|E_+|^2/|E_-|^2)$.

It follows that for self-sustained oscillation to occur at a precise frequency shift $\Omega = \Omega_B$ would require an infinitely high intensity. The availability of detuning, $\Omega - \Omega_B \neq 0$, gives a reactive nonlinear component $\mathrm{Im}\{\mu\} \neq 0$, which makes self-sustained oscillation possible. At equal intensities $(k \to 0)$ the threshold also rises to infinity.

It can be readily seen that even in the presence of a weak counterpropagating pump wave with an intensity $|E_-|^2/|E_+|^2 = \exp(-\pi) \approx 0.043$ (reflection from one glass face), the threshold for parametric generation corresponds to $M \approx 2\pi \approx 6.3$, this being about half as great as that of the usual SS excited by a unidirectional pump wave from spontaneous noise, where $M_{SS} \approx 12 - 15$.

Thus, an account of the four-wave parametric SBS interaction is of importance both for estimating the intensity of Stokes waves and for determining the threshold of SS. When four-wave parametric generation occurs, both the Stokes and anti-Stokes components are phase conjugated with respect to each other, compare with (2.7.9). Therefore, SS in the field of two counter-propagating pump waves is of great interest for obtaining OPC with a high (far larger than unity) reflectivity, see also Sect. 7.4.

2.8 Polarization Properties of Stimulated Scattering

The polarizations of the interacting waves have been regarded so far as linear and coinciding with one another. In this section we shall give a detailed account of the polarization properties of different types of SS.

In liquids and gases the main scattering mechanism, SBS and STS, are associated with the buildup of a scalar parameter of the medium, viz., its density. The part of the density disturbance $\delta\varrho$ in which we are interested is built up due to the interference term $\propto E_L^* E_S$. In an isotropic medium the scalar parameter $\delta\varrho$ may be proportional only to the scalar product of the fields,

$$\delta\varrho(R, t) \propto E_L^* \cdot E_S \exp(-i q \cdot R + i \Omega t) . \tag{2.8.1}$$

The disturbance of the dielectric susceptibility tensor is of scalar form, $\delta\varepsilon_{ik} = \delta_{ik}(\partial\varepsilon/\partial\varrho)\delta\varrho$, where δ_{ik} is the Kronecker delta. The scattering of the pump field E_L by the disturbance $\delta\hat{\varepsilon}$ gives rise to the dielectric displacement

$$\delta D_S = A E_L (E_L^* \cdot E_S) . \tag{2.8.2}$$

Stimulated scattering mechanisms with the nonlinear dielectric displacement (2.8.2) are called scalar mechanisms. Now consider the consequences of (2.8.2).

In forward and backward SS, which are of particular interest, δD_S from (2.8.2) automatically satisfies the condition of being orthogonal to the wave vector, and the shortened equation for E_S is of the form (in the center of the SBS line)

$$\frac{dE_S}{dz} = \frac{1}{2} G E_L (E_L^* \cdot E_S) . \tag{2.8.3}$$

Let the field E_L be completely polarized, $E_L = e_1 E_L$, where e_1 is the complex unit vector of polarization. It is convenient to project the field E_S onto e_1 and onto another unit vector e_2 defined by the condition $e_1^* \cdot e_2 = 0$:

$$E_S(z) = s_1(z) e_1 + s_2(z) e_2 . \tag{2.8.4}$$

Then from (2.8.3) we can obtain

$$E_S(z) = s_2(0) e_2 + s_1(0) e^{gz/2} e_1 , \quad g = G |E_L|^2 . \tag{2.8.5}$$

This means that the orthogonal component $s_2 e_2$ of the Stokes field does not interfere with the pump from (2.8.1), i.e., it does not build up a density disturbance, and, according to (2.8.2), the pump is not scattered into this orthogonal component. In contrast, the component $s_1(z) e_1$, with coinciding polarization, is amplified in full. Therefore, in scalar-type SS, the signal tends for reproduce the polarization unit vector of the laser wave. This statement is valid for arbitrary (generally speaking, for elliptic) polarization of the pump, both in foward and backward SS. It should be emphasized that in scalar-type backward SS, the circularly polarized pump $e_1 = (e_x + i e_y)/\sqrt{2}$ is reflected into the Stokes wave as if from a conventional mirror, $E_S \propto (e_x + i e_y)/\sqrt{2}$, i.e., a right-hand circularly polarized photon of the pump is converted into a left-hand circularly polarized photon of the counterpropagating Stokes wave. As light waves are transverse, we should, for non-colinear k_L and k_S, project the dielectric displacement (2.8.2) onto a plane perpendicular to the direction of k_S, and then the solution becomes rather more complicated than (2.8.5). It is important that in this case the Stokes wave polarized orthogonal to the pump is not amplified either.

Polarization properties of Rayleigh-wing stimulated scattering (RWSS) may be considered in an analogous way starting from (3.3.3, 5). It is interesting to note that in the case of RWSS for a given pump polarization, both "eigenpolarizations" of the Stokes field are amplified, but by different amounts. The best amplification is produced by a circularly polarized pump, e.g., $e_L = (e_x + i e_y)/\sqrt{2}$; for such a pump the larger amplitude gain is obtained for the Stokes wave of the opposite circularity, $e_S = (e_x - i e_y)/\sqrt{2} = e_L^*$. For a linearly polarized pump wave, the larger amplitude gain is $\frac{2}{3}$ of the larger gain in the circularly polarized pump case, and is achieved for $e_S = e_L$.

We now proceed to the case of SRS. In not-too-dense gases the vibrational-rotational RS transitions obey the selection rule $\Delta J = \pm 2$ or $\Delta J = 0$ for the rotational quantum number J. Transitions with $\Delta J = \pm 2$ correspond to symmetrical traceless scattering, because for transitions between different J the matrix element of the scalar $\text{Tr}\,\hat{\alpha}$ becomes zero, where $\hat{\alpha}$ is the polarizability tensor. Transition with $\Delta J = 0$ contain both the scalar and the symmetrical traceless parts.

2.9 Literature

The list of work on stimulated light scattering and its applications is long enough. Here we confine ourselves to references which illustrate the material of this chapter.

The possibility of the existence of stimulated light scattering was considered, in a general way, as early as the 1930 s [2.9]. Stimulated light scattering is considered in a number of reviews: [2.4, 10 – 13]. We mention here some earlier work devoted to the investigation of SRS [2.14, 15], SBS [2.15, 17], SRWS [2.18, 19], STS [2.20 – 23]. The relation between spontaneous and stimulated scattering was considered in [2.24]; see also the reviews [2.11, 13]. Stimulated scattering under steady-state saturation was investigated in [2.25]. The nonstationary theory of SS was first formulated in [2.26]; see also [2.27 – 30] and the review [2.13]. The quasi-static condition of nonstationary SRS is considered in [2.31]. Many questions of the theory of nonstationary SS in time-random pump fields are elucidated in [2.32]. The "stick" model of spectrum of exciting and scattered light was used in the consideration of non-stationary processes in SS in [2.7, 33 – 36]. Generation of anti-Stokes components and the corresponding phase-matching conditions in SRS are discussed in [2.37 – 39]. SBS in the presence of two counterpropagating pump waves is discussed in [2.40]. Polarization effects in SS considered in many works, of which [2.41] deals with an experiment of SBS.

3. Properties of Speckle-Inhomogeneous Fields

As has been mentioned in Chap. 1, the quality of OPC by SS is high when the pump field has a developed speckle structure in the interaction volume. Here we shall try to explain what a speckle structure is, under what conditions it is observed, and what properties the speckle fields possess. It is very easy to see a speckle structure: a wall surface, a sheet of paper and, in general, almost any object illuminated by a laser beam looks to be covered with a large number of irregular bright speckles. Their positions change as the position of the observer, the light source, or of the object itself changes. Figure 3.1 presents a photograph of a speckle-field section.

Fig. 3.1. Transverse intensity distribution in the speckle field (positive photograph)

The appearance of a speckle structure in the visual image of a uniformly illuminated surface is connected with the high coherence of laser radiation and also with the roughness of the diffusively reflecting surface. Due to this roughness, different points of the surface reflect radiation with randomly distributed phases. Owing to coherence, the contributions of different points add not by intensity but by amplitude, and they give a very broken, irregular interference pattern. In OPC experiments, speckle fields are usually obtained not by reflection from a diffusive surface, but by transmission of light through an inhomogeneous phase mask (phase plate). In most cases the particular realization of the speckle picture is not essential; it is only the statistical characteristics which are of importance. We begin the study of these characteristics by considering a probability distribution, i.e., the speckle-field statistics.

3.1 Central Limit Theorem. Gaussian Statistics of Speckle Fields

The complex amplitude of a speckle field at a given point is the result of the linear addition of the contributions from different points of the scatterer.

If the number of contributions is large enough and if they are independent, then the resultant signal will possess quite definite (Gaussian) statistics. This statement represents the central limit theorem of probability theory. In view of the extreme importance of speckle-field statistics for OPC problems, we shall outline the proof of the theorem and adduce, in passing, a number of statements of probability theory.

3.1.1 Characteristic Function

To begin with, we consider the case of a real random quantity x, which is fully characterized if the probability distribution, $dP = W(x)\,dx$, is specified, where the density $W(x)$ satisfies three conditions: (1) $W(x) = W^*(x)$, (2) $W(x) \geqslant 0$, (3) $\int_{-\infty}^{+\infty} W(x)\,dx = 1$. The average value $\langle f(x) \rangle$ of a function $f(x)$ is

$$\langle f(x) \rangle = \int W(x)\,f(x)\,dx . \tag{3.1.1}$$

In accord with Fourier's theorems, the function $W(x)$ can be completely defined by its Fourier transform for which a special term, the characteristic function $\chi(\mu)$, is introduced in the theory of probability:

$$\chi(\mu) = \int_{-\infty}^{+\infty} W(x)\,e^{i\mu x}\,dx \equiv \langle e^{i\mu x} \rangle , \tag{3.1.2}$$

where $W(x) = (2\pi)^{-1} \int \chi(\mu) \exp(-i\mu x)\,d\mu$.

We should like to mention one more important notion — the independence of two or more random quantities. Quantities x and y are considered independent if the joint probability density $dP = W(x, y)\,dx\,dy$ can be represented as the product of separate factors, each dependent on one quantity: $W(x, y) = W_1(x)\,W_2(y)$. The characteristic function is very convenient in handling independent random quantities. Indeed, for $x = x_1 + x_2 + \cdots + x_N$ with independent x_i, the characteristic function $\chi_x(\mu)$ of x,

$$\chi_x(\mu) = \langle \exp(i\mu x) \rangle = \langle \exp[i\mu(x_1 + x_2 + \cdots + x_N)] \rangle$$
$$= \chi_1(\mu) \cdot \chi_2(\mu) \ldots \chi_N(\mu) , \tag{3.1.3}$$

is the product of the characteristic functions of the separate summands.

The moments (i.e., the means $\langle x^n \rangle$) of a random quantity x determine an expansion of $\chi(\mu)$ in a Taylor series at a point $\mu = 0$; indeed, from (3.1.2) it follows that

$$\chi(\mu) = 1 + i\mu\langle x \rangle - \frac{\mu^2}{2}\langle x^2 \rangle + \ldots . \tag{3.1.4}$$

Below in almost all cases, we shall deal with random quantities having zero means. It should be noted what otherwise, the reference point can be easily shifted by introducing $\tilde{x} = x - \langle x \rangle$. Under such conditions, the quantity $\langle (x - \langle x \rangle)^2 \rangle = \langle \tilde{x}^2 \rangle$ which is also referred to as the square of the variance, σ^2, characterizes the width of the distribution function $W(x)$. According to Fourier analysis (or the uncertainty relation) the width $\Delta\mu$ of a region in which $\chi(\mu)$ is markedly different from zero is given by $\Delta\mu \sim \sigma^{-1}$.

3.1.2 Central Limit Theorem

Now consider a quantity x which is the sum of a large number of small independent summands with zero mean values, $x = x_1 + x_2 + \cdots + x_N$. It can be easily seen that

$$\langle x \rangle = 0, \qquad \sigma^2 = \langle x^2 \rangle = \sum_{i=1}^{N} \sigma_i^2, \qquad (3.1.5)$$

i.e., the squares of variances are added. For the characteristic function $\chi(\mu)$ of a quantity x, the values $|\mu| \lesssim \sigma^{-1} \sim (\sigma_i\sqrt{N})^{-1}$ are of most interest. At such values of μ, the expansion (3.1.4) converges fairly well. It is convenient to make further calculations on the assumption that all σ_i are equal (it is easy to check that the final result does not depend on this assumption). In such a case we have

$$\chi(\mu) \approx \left[1 - \frac{\mu\sigma_i^2}{2} \right]^N \approx \exp\left(-\frac{\mu^2 N\sigma_i^2}{2} \right) = \exp\left(-\frac{\mu^2\sigma^2}{2} \right) \qquad (3.1.6)$$

at $N \to \infty$ using definition of e. Thus, we can see that at large N, $\chi(\mu)$ has a Gaussian form with respect to μ. Applying the inverse Fourier transform, we obtain

$$W(x) = \frac{1}{\sqrt{2\pi\sigma^2}} \exp(-x^2/2\sigma^2). \qquad (3.1.7)$$

A real-valued random quantity with the probability distribution given by (3.1.7) is called a Gaussian random quantity. Thus, the sum x of a large number of small independent summands x_i at $N \to \infty$ obeys Gaussian statistics (3.1.7). It should be emphasized that the form of (3.1.7) does not depend on the form of the distributions $W_i(x)$ of the separate summands; only their independence and the finiteness of their moments are of importance[1].

---

1 It is easy to estimate the accuracy of (3.1.6). The term neglected in the expansion (3.1.4) is $-i\mu^3\langle x_i^3 \rangle/6$. If we introduce the parameter $\xi = \langle x_i^3 \rangle(\langle x_i^2 \rangle)^{-3/2}$, then $|\xi| \lesssim 1$ for not very

In optics, one has to deal mainly with complex random quantities $E = \operatorname{Re}\{E\} + i\operatorname{Im}\{E\}$ for which it is convenient to introduce the probability distribution $dP = W(\operatorname{Re}\{E\}, \operatorname{Im}\{E\})\,d\operatorname{Re}\{E\}\,d\operatorname{Im}\{E\}$. As a rule, $W(\operatorname{Re}\{E\}, \operatorname{Im}\{E\})$ is independent of the phase of the field E. It is therefore convenient to replace the variables $\operatorname{Re}\{E\}$, $\operatorname{Im}\{E\}$ by "independent" variables E, E^*. As in the case of real-valued quantities we can also introduce the characteristic function $\chi(\eta, \eta^*)$:

$$\chi(\eta, \eta^*) = \langle \exp(\eta^* E - \eta E^*) \rangle. \tag{3.1.8}$$

This facilitates the calculation of moments of the random quantity E:

$$\langle E^{*m} E^n \rangle = \left(-\frac{\partial}{\partial \eta} \right)^m \left(\frac{\partial}{\partial \eta^*} \right)^n \chi(\eta, \eta^*) \Bigg|_{\eta = \eta^* = 0} \tag{3.1.9}$$

so that

$$\chi(\eta, \eta^*) = 1 + \eta^* \langle E \rangle - \eta \langle E^* \rangle + \frac{\eta^2}{2} \langle E^{*2} \rangle + \frac{\eta^{*2}}{2} \langle E^2 \rangle$$

$$- \eta \eta^* \langle |E|^2 \rangle + \cdots. \tag{3.1.10}$$

The condition for phase randomness involves the equalities $\langle E \rangle = \langle E^* \rangle = 0$, $\langle E^2 \rangle = \langle E^{*2} \rangle = 0$, etc. Hence, $\chi(\eta, \eta^*) \approx 1 - \langle |E|^2 \rangle \eta \eta^*$ at small η, η^*. From now on, when dealing with complex random quantities, we shall consider the phase randomness condition to be satisfied. From this condition it follows, in particular, that $\chi(\eta, \eta^*)$ depends only on the bilinear combination $\eta \eta^* = |\eta|^2$ and also that $\operatorname{Re}\{E\}$ and $\operatorname{Im}\{E\}$ are uncorrelated and included in the probability density via $|E|^2$.

Repeating these considerations, which were made above for a real-valued random quantity x, we can show that the sum $E = E_1 + E_2 + \cdots + E_N$ of a large number of independent complex summands with $\langle E_i \rangle = 0$, $\langle E_i^2 \rangle = 0$ also obeys Gaussian statistics at large values of N:

$$\chi(\eta, \eta^*) \approx \exp(-I_0 \eta \eta^*), \quad I_0 = \langle |E|^2 \rangle,$$

$$dP = W(E)\,d\operatorname{Re}\{E\}\,d\operatorname{Im}\{E\}$$

$$= (\pi I_0)^{-1} \exp(-|E|^2/I_0)\,d\operatorname{Re}\{E\}\,d\operatorname{Im}\{E\}. \tag{3.1.11}$$

---

exotic cases. The contribution of these terms to (3.1.6) leads to a factor $\exp(-i\mu^3 \xi N \sigma_i^3/6)$. Expressing σ_i in terms of σ using $\sigma_i \sim \sigma N^{-1/2}$, we obtain the correction factor, $\exp(-i\mu^3 \sigma^3 \xi N^{-1/2}/6)$. In the essential region, $\mu \sim \sigma^{-1}$, and the correction factor is of the order $\exp(-i\xi N^{-1/2}/6)$, i.e., very close to unity at $N \to \infty$

The probability distribution for $\mathrm{Re}\{E\}$ and $\mathrm{Im}\{E\}$ given by (3.1.11) represents the product of the Gaussian functions $\exp[-(\mathrm{Re}\{E\})^2/I_0]$ $\cdot \exp[-(\mathrm{Im}\{E\})^2 I_0]$. In a number of cases it is convenient to describe this same physical situation in terms of phase ϕ and amplitude A: $E = A\exp(i\phi)$. Then A and ϕ are independent, ϕ is distributed uniformly from 0 to 2π, and the amplitude obeys the Rayleigh distribution

$$dP = \frac{2A}{I_0}\exp(-A^2/I_0)dA \equiv W_A(A)dA .\tag{3.1.12}$$

Finally, the probability distribution for the intensity $I = A^2 = |E|^2$ is of an exponential form:

$$dP \equiv W_I(I)dI = I_0^{-1}\exp(-I/I_0)dI .\tag{3.1.13}$$

It should be emphasized that all three distributions (3.1.11 – 13) describe the same physical situation but in terms of different variables.

Thus, we have come to the conclusion that the speckle field E, which is always the sum of a large number of small independent complex summands, obeys Gaussian statistics. The characteristic function given by (3.1.11) makes it possible to readily calculate the intensity moments $\langle I^n \rangle \equiv \langle |E|^{2n} \rangle$ of the speckle field with a complex amplitude E. Indeed, the substitution of (3.1.11) into (3.1.9) yields

$$\langle I^n \rangle = n!\, I_0^n .\tag{3.1.14}$$

A most important example of the application of this formula is the calculation of the overlap integral of the pump intensity with the intensity of the phase-conjugated Stokes wave. For the pump speckle field this procedure consists in calculating $\langle I_L I_S \rangle = \mathrm{const} \cdot \langle I_L^2 \rangle$. Using (3.1.14) one obtains

$$\langle I_L^2 \rangle = 2\langle I_L \rangle^2 .\tag{3.1.15}$$

3.1.3 Gaussian Statistics of Speckle Fields

We have dealt so far with the statistics of the speckle field at a given point. To characterize the field as a whole, it is necessary to specify the joint probability distributions of field values at several points: $E_1 = E(R_1)$, $E_2 = E(R_2),\ldots,$ $E_m = E(R_m)$, where m is a finite fixed number. The values of $E_1, E_2, \ldots, E_m$ are obtained by adding contributions of a large number $(N \to \infty)$ of summands, these contributions having different amplitudes and phases for different observation points. Here it is also convenient to consider separately real-valued and complex random quantities. In the first case, at $N \to \infty$, the characteristic function χ, i.e., the Fourier transform of an m-dimensional joint probability distribution $W_m(x_1, \ldots, x_m)$, is

$$\chi(\mu_1, \ldots, \mu_m) \equiv \langle \exp(i\mu_1 x_1 + \ldots i\mu_m x_m) \rangle = \exp(-R_{ij}\mu_i\mu_j/2) \,,$$

$$R_{ij} = \langle x_i x_j \rangle \,, \tag{3.1.16}$$

where the summation is taken over repeated indices i, j from 1 to m. The probability distribution is then also of a Gaussian form:

$$W(x_1, \ldots, x_m) \propto \exp(-T_{ij}x_i x_j/2) \,, \qquad \hat{T} = \hat{R}^{-1} \,, \tag{3.1.17}$$

i.e., the matrix $\hat{T}$ is the reciprocal of the matrix $\hat{R}$. It is not difficult to obtain the general algorithm for calculating the moments. Thus, for example, at $\langle x_i \rangle = 0$,

$$\langle x_1 x_2 x_3 x_4 \rangle = \langle x_1 x_2 \rangle \langle x_3 x_4 \rangle + \langle x_1 x_3 \rangle \langle x_2 x_4 \rangle + \langle x_1 x_4 \rangle \langle x_2 x_3 \rangle \,, \tag{3.1.18}$$

from which, in particular, it follows that for a Gaussian real-valued quantity $x = x_1 = x_2 = x_3 = x_4$ the relationship $\langle x^4 \rangle = 3 \langle x^2 \rangle^2$ holds true.

In the case of complex random quantities the central limit theorem leads to the following:

$$\chi(\eta_1, \ldots, \eta_m, \eta_1^*, \ldots, \eta_m^*) \equiv \langle \exp(\eta_i^* E_i - \eta_i E_i^*) \rangle$$

$$= \exp(-K_{ij}\eta_i \eta_j^*) \,, \qquad K_{ij} = \langle E_i^* E_j \rangle \,. \tag{3.1.19}$$

The probability distribution is similar to the form of (3.1.17). Random quantities obeying (3.1.16) or (3.1.19) are called *joint Gaussian quantities*. Thus, the set of values $E_1, \ldots, E_m$ of the speckle field at different points obeys joint Gaussian statistics.

The most important property of a set of joint Gaussian quantities is the possibility of completely characterizing it, i.e., finding all its moments, by means of a pairwise correlation matrix $\langle E_i^* E_j \rangle$ of order 2. Thus, for example,

$$\langle E_1^* E_2^* E_3 E_4 \rangle = \langle E_1^* E_3 \rangle \langle E_2^* E_4 \rangle + \langle E_1^* E_4 \rangle \langle E_2^* E_3 \rangle \,. \tag{3.1.20}$$

Equation (3.1.20) may be considered as a particular case of (3.1.18) if $\langle E_i E_j \rangle = \langle E_i^* E_j^* \rangle = 0$.

A second important property of joint Gaussian random quantities should be noted. Suppose we have m such quantities $E_j, j = 1, \ldots, m$, each with zero mean, $\langle E_j \rangle = \langle E_j^* \rangle = 0$, and that these quantities are uncorrelated with one another, i.e.,

$$\langle E_i^* E_j \rangle = 0 \qquad \text{at} \quad i \neq j \,. \tag{3.1.21}$$

Then from (3.1.19) it follows that their joint characteristic function is factorized and can be represented as the product of separate cofactors with respect to each pair of variables η_i, η_i^*:

$$\chi(\eta_1,\ldots,\eta_m,\eta_1^*,\ldots,\eta_m^*) = \prod_{j=1}^{m} \exp(-|\eta_j|^2 \langle |E_j|^2 \rangle) . \tag{3.1.22}$$

From this it follows that the multidimensional joint probability distribution is also factorized:

$$W_{2m}(\mathrm{Re}\{E_1\},\ldots,\mathrm{Im}\{E_m\})$$

$$= A \prod_{j=1}^{m} \exp[-(\mathrm{Re}\{E_j\})^2/\langle |E_j|^2 \rangle] \exp[-(\mathrm{Im}\{E_j\})^2/\langle |E_j|^2 \rangle] , \tag{3.1.23}$$

$$A^{-1} = \pi^m \prod_{j=1}^{m} \langle |E_j|^2 \rangle .$$

In other words, *uncorrelation* of joint Gaussian random quantities involves their complete statistical *independence*.

We have considered the set of values of the field $E(\boldsymbol{R})$ at arbitrary discrete points $\boldsymbol{R} = \boldsymbol{R}_j, j = 1,\ldots, m$. It is clear that by displacing these points continuously, we shall obtain a statistical description of the entire random field. In the general case, this field is a smooth function of coordinates and the joint probability distributions of the field values at different points are Gaussian distributions. Thus, the speckle field, as a whole, is a smooth random function with complex Gaussian statistics.

3.2 Parabolic Wave Equation

Below we deal with the propagation of highly directional monochromatic light beams. In such cases, the initial Helmholtz equation,

$$\nabla^2 E(\boldsymbol{R}) + k^2 E(\boldsymbol{R}) = 0 , \tag{3.2.1}$$

with $k = \omega\sqrt{\varepsilon}/c$ can be approximately reduced to a simpler parabolic equation. Let us represent the field $E(\boldsymbol{R})$ in the form of a rapidly varying factor $\exp(ikz)$ corresponding to the central direction of propagation of the beam along the z axis and a slow, on the scale $\lambda = 2\pi/k$, amplitude $E(r,z)$, where $r = (x,y)$:

$$E(\boldsymbol{R}) = e^{ikz}E(r,z) . \tag{3.2.2}$$

Then (3.2.1) can be identically transformed as

$$-\frac{i}{2k}\frac{\partial^2 E}{\partial z^2}+\frac{\partial E}{\partial z}-\frac{i}{2k}\nabla_\perp^2 E(r,z)=0\,, \qquad (3.2.3)$$

where $\nabla_\perp^2=(\partial^2/\partial x^2+\partial^2/\partial y^2)$. Let the transverse size of inhomogeneities of the field $E(r,z)$ be $\sim a\gg\lambda$; then $\nabla_\perp^2 E\sim E/a^2$. Suppose that the term $\propto\partial^2 E/\partial z^2$ in (3.2.3) may be omitted. Then from (3.2.3) it follows that $\partial E/\partial z\sim E/(ka^2)$, i.e., the field changes along z by about $\Delta z\sim(ka^2)$. Thus, the term with $\partial^2 E/\partial z^2$ is a factor of (ka^2) smaller than the remaining terms, and for fields with $a\gg\lambda$ we obtain a parabolic wave equation

$$\frac{\partial E(r,z)}{\partial z}-\frac{i}{2k}\nabla_\perp^2 E(r,z)=0\,. \qquad (3.2.4)$$

It is of interest to ascertain the form of solutions for plane and spherical waves in the parabolic approximation. For plane waves, from (3.2.4) we have

$$E(R)=e^{ikz}E(r,z)=\exp[iq\cdot r+i(k-q^2/2k)z]\,, \qquad (3.2.5)$$

i.e., (3.2.4) corresponds to choosing, from the two possible values of $k_z=\pm(k^2-q^2)^{1/2}$ following from the exact solution of (3.2.1), the single value $k_z=+k-q^2/2k$ in the approximation, allowing only for the first term in the Taylor series with respect to $\theta^2=q^2/k^2$.

For waves from a point source, we can obtain from (3.2.4)

$$E(R)=\frac{1}{z}\exp(ikz+ikr^2/2z) \qquad (3.2.6)$$

instead of the field $E(R)=|R|^{-1}\exp(ik|R|)$ satisfying the exact equation (3.2.1). Here the factor $|R|=\sqrt{z^2+r^2}$ in the preexponential is replaced by z, and in the exponential, allowing for the first non-vanishing correction,

$$|R|\approx z+r^2/2z\approx z(1+\theta^2/2)\,.$$

Thus, for not very large distances, the small parameter determining the applicability of the parabolic equation is $\theta^2=\lambda^2/a^2$.

3.3 Characteristic Dimensions of Speckle-Structure Inhomogeneities

In the parabolic approximation, the field $E(R)$ can be represented as

$$E(R)=e^{ikz}E(r,z)=e^{ikz}\sum_\theta\tilde E(\theta)\exp(ik\theta\cdot r-ik\theta^2z/2)\,, \qquad (3.3.1)$$

where $E(r, z)$ is the slowly varying amplitude. Let $E(r, z)$ be a speckle field, i.e., a random function of the coordinates (r, z); the mean $\langle E \rangle = 0$. Consider the correlator of field values at two points: $\langle E^*(r_2, z_2) \cdot E(r_1, z_1) \rangle$. This can be readily found under the assumption that the statistical properties of the field do not vary when the origin is moved. For such statistically uniform fields,

$$\langle E^*(r_2, z_2) E(r_1, z_1) \rangle = K(r_1 - r_2, z_1 - z_2) , \tag{3.3.2}$$

so that the correlation function depends only on the argument difference. The representation of the field in the form of expansion (3.3.1) with respect to angular components leads to

$$\langle E^*(r_2, z_2) E(r_1, z_1) \rangle = \sum_{\theta_1} \sum_{\theta_2} \langle \tilde{E}^*(\theta_2) \tilde{E}(\theta_1) \rangle$$
$$\times \exp[ik\theta_1 \cdot (r_1 - r_2) - ik\theta_1^2 (z_1 - z_2)/2]$$
$$\times \exp[ikr_2 \cdot (\theta_1 - \theta_2) - ikz_2(\theta_1^2 - \theta_2^2)/2] . \tag{3.3.3}$$

In order to satisfy the condition of statistical uniformity expressed by (3.3.2), we should consider the right side of (3.3.3) independent of r_2, z_2 at fixed values of $r_1 - r_2$ and $z_1 - z_2$. For this purpose, the amplitudes of different angular components should be considered uncorrelated:

$$\langle \tilde{E}^*(\theta_1) \tilde{E}(\theta_2) \rangle = \langle |E(\theta_1)|^2 \rangle \delta(\theta_1 - \theta_2) ,$$

where $\delta(\theta_1 - \theta_2)$ is the discrete Kronecker symbol. Taking this into account, the correlation function is

$$\langle E^*(r_2 z_2) E(r_1, z_1) \rangle = \sum_{\theta} \langle |\tilde{E}(\theta)|^2 \rangle \exp[ik\theta \cdot (r_1 - r_2) - ik\theta^2(z_1 - z_2)/2] . \tag{3.3.4}$$

The mean field intensity at a point (r, z), i.e., $\langle |E(r, z)|^2 \rangle$ is the sum of the mean intensities of separate angular components: $\langle I \rangle = \sum \langle |\tilde{E}(\theta)|^2 \rangle$.

It is convenient to introduce the normalized correlation function (the complex degree of coherence):

$$\gamma(r_2, z_2; r_1, z_1) = \frac{\langle E^*(r_2, z_2) E(r_1, z_1) \rangle}{\sqrt{\langle |E(r_2, z_2)|^2 \rangle \langle |E(r_1, z_1)|^2 \rangle}} . \tag{3.3.5}$$

For coinciding arguments, $\gamma(r, z; r, z) = 1$, and for statistically uniform fields, γ depends only on the difference $(z_1 - z_2, r_1 - r_2)$.

In a number of cases it is convenient to use an expansion with respect to continuous Fourier components in place of the discrete Fourier series (3.3.1). Let us introduce the normalized angular intensity spectrum $j(\theta) \propto \langle |\tilde{E}(\theta)|^2 \rangle$, which is such that

$$\int j(\theta)\, d\theta_x d\theta_y = 1 . \tag{3.3.6}$$

Then (3.3.4) can be written as

$$\gamma(r_1 - r_2, z_1 - z_2) = \int j(\theta)\, d\theta_x d\theta_y \exp[ik\theta \cdot (r_1 - r_2) - ik\theta^2(z_1 - z_2)/2] \tag{3.3.7}$$

Equation (3.3.7) at $z_1 = z_2$ is equivalent to the Van Zittert-Zernike theorem which reads: the correlation function of the complex amplitude of the field of a distant quasi-monochromatic ($k = \text{const}$) source is equal to a two-dimensional Fourier transform of the angular distribution $j(\theta)$ of the waves coming to the observer.

According to Fourier analysis it follows from (3.3.7) that the speckle-field correlation with divergence $\sim \Delta\theta$ has a transverse dimension $\Delta r \sim (k\Delta\theta)^{-1}$ and a longitudinal dimension $\Delta z \sim 1/k(\overset{\rightarrow}{\Delta}\theta)^2 \sim \lambda(\Delta\theta)^2$. More accurate estimates can be obtained provided the form of the angular spectrum is specified. Suppose that $j(\theta) \propto \exp(-\theta^2/\theta_0^2)$, where $\theta_0 = \text{HWe}^{-1}\text{M}$ is the half-width at the intensity level $\exp(-1)$ times smaller than at the center. Calculations then yield

$$\gamma(r, z) = \frac{1 - i\alpha}{1 + \alpha^2} \exp\left[-\frac{1}{4}\frac{k^2 r^2 \theta_0^2}{1 + \alpha^2}(1 - i\alpha) \right], \qquad \alpha = kz\theta_0^2/2 . \tag{3.3.8}$$

In a given cross section ($z_1 - z_2 = z = 0$) the correlation radius determined by a decrease of $|\gamma(r)|^2$ down to the level $\exp(-1)$ is given by

$$r_{\text{cor}}(\text{HWe}^{-1}\text{M}) = \frac{\sqrt{2}}{k\theta_0}\frac{\lambda}{\pi\sqrt{2}\,\theta_0} , \tag{3.3.9}$$

where the values of λ and θ_0 are the parameters of the field in the medium. The field correlator along the axis ($r_1 - r_2 = 0$) decreases, depending on $z_1 - z_2$, according to Lorentz's law, so that a decrease in $|\gamma|^2$ by a factor of 0.5 takes place within the dimension

$$z_{\text{cor}}(\text{HWHM}) = \frac{2}{k\theta_0^2} = \frac{\lambda}{\pi\theta_0^2} . \tag{3.3.10}$$

As the speckle-field obeys Gaussian statistics, the correlation function $\gamma(r, z)$ for complex field amplitudes makes it possible to judge the picture of intensity fluctuations. Indeed, assuming $E_1 = E_3 = E(R_1)$, $E_2 = E_4 = E(R_2)$ in (3.1.20), we obtain for the speckle-field intensity correlator:

$$\langle I(r_1, z_1) I(r_2, z_2) \rangle = \langle I \rangle^2 [1 + |\gamma(r_1 - r_2, z_1 - z_2)|^2] . \tag{3.3.11}$$

In other words, intensity fluctuations become practically independent at points which are distant from one another by more than r_{cor} with respect to the transverse coordinate or z_{cor} with respect to the longitudinal coordinate.

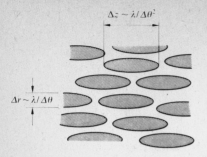

$\Delta z \sim \lambda / \Delta \theta^2$

$\Delta r \sim \lambda / \Delta \theta$

Fig. 3.2. Transverse and longitudinal scales of the speckle field

A schematic of the "cut" of the speckle-field intensity by a plane (x, z) is shown in Fig. 3.2. Regions of higher intensity have the shape of long "cucumbers", of length $\sim \lambda / \Delta \theta^2$ and transverse size $\sim \lambda / \Delta \theta$. In view of the conservation of energy flux, these regions represent, spatially, the set of "serpents" of varying thickness stretched along the axis which merge and become entangled with one another.

3.4 Focused Speckle Field

Locally, speckle fields can be considered, practically always, to the statistically uniform. However, such uniformity is only an approximation for real spatially limited beams. The mean intensity of the field and its angular spectrum usually vary continuously from point to point. If the dependence within a certain cross section $z_0 = $ const is given, the wave equation will enable us to completely determine the above parameters throughout the entire volume of propagation.

To formulate these statements quantitatively, let us write the equation for a bilinear combination:

$$B(r_1, r_2, z) = E^*(r_2, z) \cdot E(r_1, z) . \tag{3.4.1}$$

The parabolic equation for the slow complex amplitude is (3.2.4), from which it follows that if $r = (r_1 + r_2)/2$, $\varrho = r_1 - r_2$, the function B satisfies the equation

$$\frac{\partial B}{\partial z} - \frac{i}{k} \frac{\partial^2}{(\partial \varrho \cdot \partial r)} B(r, \varrho, z) = 0 . \tag{3.4.2}$$

Using the Fourier transform with respect to the variable ϱ,

$$B(r, \varrho, z) = \int \tilde{B}(r, \theta, z) \exp(ik\theta \cdot \varrho) d^2\theta , \tag{3.4.3}$$

we obtain the "continuity equation"

$$\frac{\partial \tilde{B}(r,\theta,z)}{\partial z} + \theta \frac{\partial \tilde{B}(r,\theta,z)}{\partial r} = 0 . \qquad (3.4.4)$$

The quantity $\tilde{B}(r,\theta,z)$ can be interpreted as the quantum flux density at a point (r,z) in a direction forming an angle $\theta = (\theta_x, \theta_y)$ with the z axis. The general solution of (3.4.4) is

$$\tilde{B}(r,\theta,z) = \tilde{B}(r - \theta(z - z_0), \theta, z_0) . \qquad (3.4.5)$$

It should be emphasized that although (3.4.4) and its solution (3.4.5) have the same form as photon transfer equations in geometrical optics, they involve a description of all the diffraction and interference effects (including a description of speckle-structure details) to the same extent that the initial parabolic equation does.

This section has dealt so far with a particular "realization" of an inhomogeneous field. If we deal with speckle fields, it is expedient to average (3.4.1) over an ensemble of small-structure inhomogeneities. It is obvious that the solution for the averaged function is also given by (3.4.5).

In experiments, a situation is often met in which a speckle beam of restricted transverse size is focused by a lens into the operating volume. The so-called "doubly Gaussian" speckle beam is a suitable model in such a case, see Fig. 3.3. As an example, let us consider that in the plane $z = 0$, where the cross section is smallest, the averaged function $\tilde{B}(r,\theta,z)$ is the product of two Gaussians:

$$\langle \tilde{B}(r,\theta,z = 0) \rangle = \frac{I_0}{\pi \theta_0^2} \exp\left(-\frac{r^2}{a_0^2}\right) \exp\left(-\frac{\theta^2}{\theta_0^2}\right) . \qquad (3.4.6)$$

Then in an arbitrary cross section $z = $ const,

$$\langle \tilde{B}(r,\theta,z) \rangle = \frac{I_0}{\pi \theta_0^2} \exp[-(r - \theta z)^2/a_0^2 - \theta^2/\theta_0^2] , \qquad (3.4.7)$$

and applying Fourier transform (3.4.3), we obtain

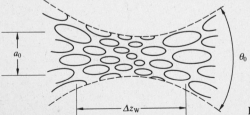

Fig. 3.3. Focused speckle beam

$$\langle E^*(r_2, z) E(r_1, z) \rangle$$

$$= \frac{I_0 a_0^2}{a^2(z)} \exp\left(- \frac{r^2 - ikz\theta_0^2 \varrho \cdot r + k^2 a_0^2 \theta_0^2 \varrho^2/4}{a^2(z)} \right), \tag{3.4.8}$$

$$a^2(z) = a_0^2 + \theta_0^2 z^2. \tag{3.4.9}$$

If we assume $r_1 - r_2 = \varrho = 0$, it follows from (3.4.8) that the intensity distribution remains Gaussian in all cross sections,

$$\langle I(r, z) \rangle = \frac{I_0 a_0^2}{a^2(z)} \exp[-r^2/a^2(z)], \tag{3.4.10}$$

and the beam radius $a(z)$ varies according to (3.4.9). The intensity along the axis decreases as $1/a^2(z)$ in accord with the Lorentz function, and the waist length at half maximum intensity is $\Delta z(\text{HWHM}) = a_0/\theta_0$.

The beam radius at the waist is $\Delta r(\text{HWe}^{-1}\text{M}) = a_0$ and the angular divergence $\Delta\theta(\text{HWe}^{-1}\text{M}) = \theta_0$. Later on, we shall need the parameter $\xi = k a_0 \theta_0$ characterizing the ratio of the beam divergence θ_0 to its diffraction limit $(k a_0)^{-1}$ at a given transverse size a_0.

Let us determine the characteristic transverse sizes of speckles. From (3.4.9) it follows that in any cross section $z = $ const the decrease of the correlator as a function of $\varrho = r_1 - r_2$ is approximately ξ times faster than the decrease of the envelope. This means that any cross section contains approximately $\xi \times \xi = \xi^2$ (in two transverse coordinates) speckles, each of size $\Delta r_{sp} \sim a(z)/\xi$. The longitudinal sizes of the speckles can be evaluated from (3.3.10) if $\Delta\theta(z) = \theta_0 a_0/a(z)$ is substituted for the local angular divergence. Then we have $\Delta z_{sp} \sim (2/k\theta_0^2) \cdot [a^2(z)/a_0^2]$. There are approximately $2\Delta z_w/\Delta z_{sp} \sim \xi$ speckle inhomogeneities along the waist length.

It is interesting to note that at $\xi = 1$, (3.4.6 − 9) involve the description of the ideal Gaussian beam, whose field is of the form

$$E(r, z) = \frac{\sqrt{I}}{1 - iz/k a_0^2} \exp\left(- \frac{r^2}{2 a_0^2(1 - iz/k a_0^2)} \right). \tag{3.4.11}$$

The value $\xi < 1$ have no physical sense as in this case the normalized correlation function would exceed unity.

The formulae of this section are used in considering OPC-SS of focused speckle beams. In addition, from a methodological point of view, the "doubly Gaussian" speckle beam is a good example of a complete analytical solution of the speckle field problem of diffraction where one need not assume statistical uniformity.

3.5 Dislocations of Wavefront

The concept of wavefront is often used in optics. This concept is useful for visualization rather than for actual calculations. Thus, for instance, in OPC the wavefront of the phase-conjugated wave exactly coincides with that of the incident signal. For a narrow monochromatic beam of the form $E(R) = \exp(ikz)E(r, z)$, the wavefront is defined as a surface of constant phase of the total complex amplitude $E(R)$:

$$kz + \arcsin \frac{\mathrm{Im}\{E(r, z)\}}{\{[\mathrm{Re}\{E(r, z)\}]^2 + [\mathrm{Im}\{E(r, z)\}]^2\}^{1/2}} = \mathrm{const}. \tag{3.5.1}$$

The arcsin term in (3.5.1) is the phase (argument) of the slowly varying complex amplitude $E(r, z)$. Because the phase of a plane wave is determined to an accuracy of an integer multiple of 2π, the wavefront is a family of parallel planes perpendicular to the propagation direction and a distance λ from one another. The wavefront of a spherical wave is represented by a set of sections of concentric spheres also a distance λ from one another. In the case of an arbitrary narrow monochromatic beam the wavefront is a family of smooth bent surfaces approximately λ apart from one another. Hence, the rays corresponding to the local direction of propagation are normal to the wavefront surface. In the course of propagation along the z axis, the shape of these surfaces gradually varies in accord with the diffraction law (3.2.4) and with (3.5.1).

This section is devoted to the following statement: for fields of a complicated structure, and for speckle fields in particular, the wavefront cannot be a family of surfaces smooth everywhere and equally spaced from one another.

Consider, for example, a field which is the exact solution of the parabolic equation (3.2.4) [and even of Helmholtz's equation (3.2.1)]:

$$E(R) \equiv e^{ikz}E(r, z) = \exp(ikz + i\alpha)(C_x x + i C_y y), \tag{3.5.2}$$

where C_x and C_y are arbitrary complex constants. The amplitude of this field at the z axis (i.e, at $x = y = 0$) is zero. By a suitable choice of the reference point on the z axis and orientation of the rectangular axes x, y, we can make C_x and C_y real-valued. Then, if a path is traced around the z axis in a plane $z = \mathrm{const}$, the phase of the field (3.5.2) varies by $+2\pi$ (if $C_x C_y > 0$) or by -2π (if $C_x C_y < 0$). It is easy to check this if $C_x = +C_y$ or $C_x = -C_y$, since $E(R)$ from (3.5.2) is then proportional to $\exp(+i\phi)$ or to $\exp(-i\phi)$, where $\phi = \arctan(y/x)$ is the azimuthal angle in the path-tracing plane.

This means that the system of wavefront surfaces for the field (3.5.2) on the z axis has a peculiarity similar to a screw dislocation of crystallographic planes in crystals. If we trace a path around the z axis not on the plane $z = $ const, but on the continuous constant-phase surface (i.e., the wavefront), then in making a complete path we pass from a given surface sheet to a neighboring one a distance $-\lambda$ (for $C_x C_y < 0$) or $+\lambda$ (for $C_x C_y > 0$) from the former. It we follow n such paths, we can continuously pass n sheets in the forward or backward direction.

Fields with dislocations of the wavefront were discussed in quantum electronics long ago (without using the term "dislocation"). Thus, for example, an optical resonator possesses transverse-structure modes

$$M_{+1}(r) = re^{i\phi}e^{-r^2/a^2}, \quad \phi = \arctan(y/x), \tag{3.5.3}$$

$$M_{+2}(r) = r^2 e^{2i\phi}e^{-r^2/a^2}, \tag{3.5.4}$$

and similar modes with a complex-conjugate profile, $M_{-n} = M_n^*$. Locally, in the vicinity of the z axis, the field (3.5.3) coincides with (3.5.2) at $C_x = C_y$. The field (3.5.4) has a screw dislocation "multiplicity" equal to 2, i.e., the phase is shifted by $2 \times 2\pi$ on a single path around the axis. It is important, however, that second- and higher-order dislocations are unstable with respect to small field distortions, i.e., they decay into a corresponding number of dislocations of unit force under the action of distortions. In contrast, dislocations of multiplicity ± 1 are stable with respect to small perturbations of the field, which the following considerations will prove.

Due to diffraction, the slowly varying amplitude $E(r, z) = \mathrm{Re}\{E(r, z)\} + i\,\mathrm{Im}\{E(r, z)\}$ is a smooth function of the coordinates (r, z). Hence, the wavefront surface determined by (3.5.1) can have peculiarities only in places where the argument of arcsin is not determined, i.e., where the modulus of the complex amplitude becomes zero, $\sqrt{[\mathrm{Re}\{E(r, z)\}]^2 + [\mathrm{Im}\{E(r, z)\}]^2} = 0$; the imaginary and real parts must equal to zero simultaneously. In a given cross section $z = $ const, the equation $\mathrm{Re}\{E(x, y)\} = 0$ determines a system of lines in the plane x, y (solid lines in Fig. 3.4). The equation $\mathrm{Im}\{E(x, y)\} = 0$ gives another system of lines (dashed lines in Fig. 3.4). The points of intersection of the dashed and solid curves correspond to zero amplitude and, consequently, to dislocations of the wavefront. The converse is valid as well — zero amplitudes should correspond to peculiarities of the wavefront, as otherwise an infinitely rapid phase change at nonzero amplitude would correspond to infinite derivatives $\partial E/\partial r$, which would contradict the wave equation. It is not difficult to understand that for a small disturbance of the complex amplitude $E(r)$ both systems of lines will be slightly deformed, and the points of intersection will not disappear but move to new places.

This reasoning shows that wavefront dislocations may be expected to be present in fields of a strongly inhomogeneous spatial structure, to which speckle fields belong. Indeed, it turns out what wavefront dislocations are

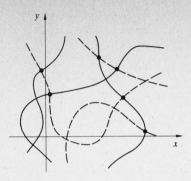

Fig. 3.4. Zero lines of the real part, $\mathrm{Re}\{E(x,y)\} = 0$, (———) and imaginary part, $\mathrm{Im}\{E(x,y)\} = 0$, (– – –) of the amplitude

inherent in speckle fields. Moreover, it becomes possible to calculate precisely the average number of dislocations per unit cross-sectional area of the speckle beam [3.1] and even to determine it experimentally [3.2, 3].

The calculation is based on the fact that zero complex amplitude of the field corresponds to a dislocation. The average number NS of zeros within a certain cross-sectional area S os given by

$$NS = \langle \int \delta(E_1)\,\delta(E_2)\,dE_1\,dE_2 \rangle, \tag{3.5.5}$$

where $E_1 = \mathrm{Re}\{E(x,y)\}$, $E_2 = \mathrm{Im}\{E(x,y)\}$, and $\delta(x)$ is the Dirac delta function. It is not difficult to make sure that (3.5.5) is valid if we take into account that each zero makes a unit contribution to the integral (3.5.5). Changing from variables E_1, E_2 to variables x, y by means of the modulus of a Jacobian

$$|D| = \left| \frac{\partial E_1}{\partial x} \cdot \frac{\partial E_2}{\partial y} - \frac{\partial E_2}{\partial x} \cdot \frac{\partial E_1}{\partial y} \right| \tag{3.5.6}$$

and cancelling S in the assumption of statistical uniformity, we obtain

$$N = \langle \delta(E_1)\,\delta(E_2)\,|D| \rangle. \tag{3.5.7}$$

To evaluate (3.5.7), it is necessary to specify the six-dimensional density of the joint probability distribution of the quantities E_1, E_2, $\partial E_1/\partial x$, $\partial E_2/\partial x$, $\partial E_1/\partial y$, $\partial E_2/\partial y$. We can do this for a speckle field obeying Gaussian statistics, since three complex quantities, i.e., the field $E(x,y)$ and its derivatives $\partial E(x,y)/\partial x$, $\partial E(x,y)/\partial y$, at any given point x, y, form, by virtue of the central limit theorem, a set of joint Gaussian random quantities. The six-dimensional probability density will be of the most suitable form in a system of coordinates for which the corresponding quantities $E_1, \ldots, \partial E_2/\partial y$ are uncorrelated. Therefore, we first calculate the corresponding correlators. We have $\langle E^2 \rangle = \langle E \cdot \partial E/\partial x \rangle = \langle E \cdot \partial E/\partial y \rangle = \langle (\partial E/\partial y)^2 \rangle = \langle (\partial E/\partial y)^2 \rangle = \langle \partial E/\partial x \cdot \partial E/\partial y \rangle = 0$ from the condition of randomness of the absolute phase of the complex field $E = E_1 + iE_2$. Also, for random fields possessing the property of statistical uniformity in space it follows from (5.3.7) that

$$\left\langle E^* \frac{\partial E}{\partial x} \right\rangle = ik\,\bar{\theta}_x I_0\,, \qquad \left\langle E^* \frac{\partial E}{\partial y} \right\rangle = ik\,\bar{\theta}_y I_0\,,$$

$$\left\langle \frac{\partial E^*}{\partial x} \frac{\partial E}{\partial x} \right\rangle = k^2 \overline{\theta_x^2} I_0\,, \qquad \left\langle \frac{\partial E^*}{\partial y} \frac{\partial E}{\partial y} \right\rangle = k^2 \overline{\theta_y^2} I_0\,, \qquad (3.5.8)$$

$$\left\langle \frac{\partial E^*}{\partial x} \frac{\partial E}{\partial y} \right\rangle = k^2 \overline{\theta_x \theta_y} I_0\,,$$

where $I_0 = \langle |E|^2 \rangle$ and the bar denotes averaging over the normalized angular spectrum $j(\theta)$:

$$\overline{F(\theta)} = \int j(\theta) F(\theta)\, d^2\theta\,. \qquad (3.5.9)$$

By choosing the z axis in the central direction of the beam we can make the values of $\bar{\theta}_x$ and $\bar{\theta}_y$ vanish, and by choosing the axis direction in the plane x, y we can reduce the symmetric matrix $\overline{\theta_i \theta_k}$ to the diagonal form so that $\overline{\theta_x \theta_y} = 0$. Then all six random quantities $E_1, \ldots, \partial E_2/\partial y$ will be uncorrelated and, by virtue of the Gaussian statistics, independent. The six-dimensional probability density acquires the form of (3.1.23), and averaging expressions (3.5.6, 7) gives

$$N = \frac{k^2}{2\pi} \sqrt{\overline{\theta_x^2} \cdot \overline{\theta_y^2}}\,. \qquad (3.5.10)$$

If we are using the general system of coordinates, then $\overline{\theta_x^2}$ and $\overline{\theta_y^2}$ are the principal values of the two-dimensional matrix $C_{ik} = \overline{\theta_i \theta_k} - \overline{\theta_i}\,\overline{\theta_k}$, which characterizes the rms scatter in photon directions in the speckle beam. Hence, $N = 2\pi (\det \hat{C})^{1/2}/\lambda^2$.

Since the correlation radius of the speckle field, i.e., the transverse size of speckle, is $r_{\text{cor}} \sim \lambda/\Delta\theta$, it follows from (3.5.10) that the dislocation density approximately coincides with the number of speckles per unit area, that is to say, it is extremely high for typical speckle fields.

This unexpected statement underwent detailed experimental verification [3.2, 3]. In this experiment a speckle field was obtained by transmitting a smooth He − Ne laser beam through a distorting phase plate, a typical OPC experiment. The structure of the speckle-field wavefront was investigated by means of an interferometric scheme, viz., a plane reference wave was directed at a certain angle to the field. The bending of the interference fringes corresponds to the curvature of the wavefront under investigation, while fringe termination or the origination of a new fringe, corresponds to wavefront

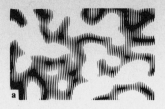

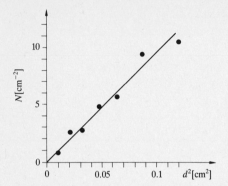

▲

Fig. 3.5a, b. Interferogram of the speckle field shown in Fig. 3.1

◀

Fig. 3.6. Experimental points and theoretical dependence of the dislocation density N [cm^{-2}] on the square of the diaphragm diameter d on a phase plate. The distance from the plate to the registration plane is 9.5 m

dislocations (either positive or negative). Figure 3.5 presents an interferogram of the speckle-field section which was photographed in the absence of the reference wave in Fig. 3.1. It is seen that the number of dislocations is approximately equal to the number of speckles. The interferometric technique also makes it possible to distinguish precisely between places of exact zero amplitude (dislocations) and local minima, see Fig. 3.5a, whereas an ordinary photograph lacks the required dynamic range. Figure 3.5b presents a large-scale interferogram containing two dislocations of opposite signs.

Presented in Fig. 3.6 are experimental points for the dependence of N on the square of the speckle-field divergence. The corresponding theoretical dependence (3.5.10) is denoted by the solid line. It should be emphasized that the *absolute* slope of this line was calculated using by (3.5.10).

Thus, the topological structure of the speckle-field wavefront appears to be rather complicated due to the presence of a great number of irregularly positioned dislocations of different signs. The equiphase surfaces interweave and dislocations describe irregular serpentine trajectories extended along the z axis. The number of positive dislocations is equal, on average, to the number of negative ones. As the wavefront propagates, its smooth section may give rise to a pair of dislocations of opposite signs. The inverse process, i.e., the annihilation of a pair of opposite-sign dislocations, is also permissible.

The presence of wavefront dislocations may place essential topological limitations on the operation of adaptive optical systems which exploit mirrors with flexible surfaces. Indeed, as was mentioned in Chap. 1, a mirror can provide a wave conjugation if its profile exactly coincides with the shape of the wavefront. However, if the wave carries a dislocation, it is impossible to make the mirror follow the wavefront shape without violating the continuity condition.

3.6 Literature

There are a number of excellent books devoted to the theory of random processes and fields as applied to radiophysics and optics [3.4 – 6]. The correlation properties of optical fields are detailed in [3.7 – 9]. The parabolic equation is discussed, for example, in [3.4, 5]. The properties of speckle fields are considered in a number of books [3.10, 11]. The focused "doubly Gaussian" speckle beam is discussed in [3.12, 13] in connection with problems of stimulated light scattering. Wavefront dislocations in fields of various physical natures have been discussed for about ten years [3.14, 15]. Papers [3.1 – 3] are devoted to wavefront dislocations in optical speckle fields.

4. OPC by Backward Stimulated Scattering

In this chapter we shall consider the results of critical experiments on OPC by backward stimulated scattering (OPC-SS) and quantitatively discuss the fundamentals of the physics of OPC-SS, that is to say, the discrimination mechanism. The principal concept of the theory — the specklon — will be introduced and the discrimination conditions against uncorrelated solutions specified for various realizations of SS: in a light guide and in a beam focused with a lens. Section 4.7 gives an account of the major techniques typical of OPC experiments.

4.1 Experimental Discovery of OPC-SS

In the very first experiments on SS, it was observed that the scattered beams were highly directional, i.e., they were either forward-going or backward-going relative to the laser beam direction. Such a property was quite naturally interpreted as the consequence of the geometrical conditions and preferential amplification of those rays of the scattered field which passed along the most intense part of the pump focal region, see Fig. 1.11.

The problem of relating the complicated structure of the incident wavefront to that of the scattered wavefront was raised for the first time by Zel'dovich et al. in 1972 [4.1]. A schematic of their experiment is shown in Fig. 4.1. Radiation from a single-mode pulsed ruby laser was employed as a pump for SBS. The radiation, which had a rather high directivity, passed through a diaphragm D and its divergence was registered by a system C_1. The angular distribution of the pump beam in the far field zone, see Fig. 4.2a ("left eye"), had a divergence of 0.14×1.3 mrad^2, i.e., in the vertical direction the pump beam corresponded to a diffraction-quality plane wave.

The pump radiation was purposely distorted by a phase plate A (aberrator) prepared by etching polished glass in hydrofluoric acid. Behind the phase plate the radiation divergence became as high as 3.5 mrad (i.e., approximately a factor of 100 higher in solid angle), and the field in the far field zone had a large number of irregular maxima and minima (speckle structure) due to the random interference of contributions of various points of the phase plate A.

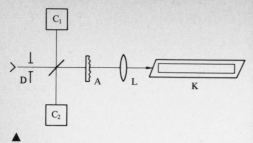

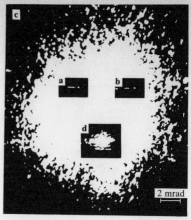

Fig. 4.1. Schematic of the first experiment on OPC by stimulated scattering

Fig. 4.2a – d. Angular distributions of beams in the far field zone on the same angular scale: **(a)** incident beam, before the phase plate; **(b)** conjugated beam, corrected by a phase plate; **(c)** beam distorted by phase plate on two passes and obtained when the cell is replaced by an ordinary plane mirror; **(d)** backscattered in the experiment without a phase plate

When the latter was followed by a mirror (not shown in Fig. 4.1), the divergence of the reflected radiation was additionally increased by a factor of approximately 2 and the beam remained structurally irregular to the same extent. The angular distribution of the radiation which passed through the aberrator, was reflected from a plane mirror, and passed through the aberrator once more thereafter, is shown in Fig. 4.2c ("round hairy physiognomy").

In the basic experiment, the mirror was replaced by a cell K, approximately 100 cm long, with compressed methane (125 atm), within which backward stimulated Brillouin scattering (SBS) developed. The cell contained a light guide, on whose input face was an image, 4×4 mm^2 in cross section, produced by the lens L with a reduction of $1:1.5$, of the illuminated part of the plate A. The total divergence of the radiation in the medium consequently amounted to $\Delta \theta_L$(FWHM) $\approx 8 \times 10^{-3}$ rad. At the cell input this divergence consisted of an irregular portion $\Delta \theta_L \approx 1.5 \times 3.5 \approx 5$ mrad caused by the phase plate, and a regular spherical portion. Due to the interference of irregularly phased angular components, the pump field in the light guide possessed a speckle structure with intensity fluctuations $\Delta I/I \sim 1$. At the far end of the cell the entire divergence acquired an irregular nature owing to reflections from the light guide walls.

The far-field-zone intensity distribution for the scattered wave transmitted through the plate was registered by a system C$_2$, see Fig. 4.2b ("right eye"). The most striking result of the experiment was the fact that the transmission through the plate did not cause any increase in the divergence of the scattered wave but, on the contrary, it caused the divergence to be corrected for to the extent of the original diffraction quality, with even the details of the pump angular structure being repeated. In other words, the SBS-reflected wave turned out to be phase conjugated with respect to the pump wave.

The energy reflectivity in this experiment amounted to 0.26. Measurement of the ratio of the brightness of the scattered wave spot to that of the laser wave spot yielded the same value, which meant that practically all the scattered wave energy was concentrated into the "reconstructed" beam. In modern terms, the "conjugation fraction" was equal to 100%, to an accuracy of $\sim 15\%$.

As has been mentioned, the aberrator A was the cause of spatial inhomogeneities (small-scale speckle structure) of the pump intensity in the scattering volume. To ascertain the effect of such inhomogeneities on SBS, an experiment was carried out without using the phase plate. A photograph of the far-field-zone scattered radiation for this experiment is shown in Fig. 4.2d ("mouth"). The scattered radiation exhibited an extremely high directivity, $\sim 2 \times 10^{-3}$ rad (all angular distributions are shown in Fig. 4.2 on the same scale), but it did not follow the laser pump field structure.

Thus, the presence of strong inhomogeneities in the pump field is the necessary condition for high-quality conjugation in the case of SBS in a light guide.

4.2 Discrimination Conditions Against Non-Conjugated Waves in Light Guides

The discrimination mechanism for self-conjugation by backward SS has been described in Sect. 1.2.4. It is based on the difference in the rates of spatial amplification of the phase-conjugated scattered wave and other scattered waves. Consider the reason for this difference in detail. Suppose that the depletion of the monochromatic pump wave $E_L(r, z) \exp(-i\omega_L t - ik_L z)$ due to convertion into the scattered wave $E_S(r, z) \exp(-i\omega_S t + ik_S z)$ can be neglected. Then the propagation and diffraction of the inhomogeneous field $E_L(r, z)$ can be described by a parabolic wave equation (see Sect. 3.2):

$$\frac{\partial E_L}{\partial z} + \frac{i}{2k} \nabla_\perp^2 E_L = 0 . \tag{4.2.1}$$

Here we consider that the laser wave propagates in the $-z$ direction in order that the backward scattered field should propagate in the positive z direction, see Fig. 4.3. The scattered field $E_S(r, z)$ is amplified in the field of the spatially

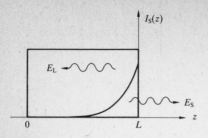

$I_S(z)$

Fig. 4.3. Selection of a system of coordinates in the backward SS problem

inhomogeneous pump $E_L(r, z)$. It is very important that, according to the basic properties of SS (Sect. 2.3), the gain $g(r, z)$ at a fixed point (r, z) is proportional to the local pump intensity $|E_L(r, z)|^2$; the amplitude gain $(1/2)\,g$ [cm^{-1}] is given by $(1/2)\,g(r, z) = (1/2)\,G\,|E_L(r, z)|^2$. The propagation and diffraction of the scattered wave undergoing amplification are then described by the parabolic equation as

$$\frac{\partial E_S}{\partial z} - \frac{i}{2k}\,\nabla_\perp^2 E_S = \frac{1}{2}\,G\,|E_L(r, z)|^2 E_S \,. \tag{4.2.2}$$

Here we suppose that $k_L \approx k_S \approx k$ for SBS. This equation [assuming the pump $E_L(r, z)$ is non-depleted] is linear in the field $E_S(r, z)$. Therefore, the amplification of various spatial configurations of the Stokes wave E_S excited by spontaneous sources near $z = 0$ can be considered independently.

From (4.2.2) we can obtain for the amplification rate of the scattered wave power $P_S(z)$ integrated over the cross section:

$$\frac{dP_S}{dz} = g(z)\,P_S(z)\,, \qquad P_S(z) = \int |E_S(r, z)|^2 d^2 r \,, \tag{4.2.3a}$$

$$g(z) = G[\int |E_L(r, z)|^2 |E_S(r, z)|^2 d^2 r] \cdot [\int |E_S(r, z)|^2 d^2 r]^{-1} \,. \tag{4.2.3b}$$

To derive (4.2.3), one should differentiate $E_S(r, z)$ under the integral in $P_S(z)$ and then use the derivatives from (4.2.2). It is important that the term of the form $\int (E_S^* \nabla_\perp^2 E_S - E_S \nabla_\perp^2 E_S^*)\,d^2 r$ can be reduced, by integration by parts, to the contribution from the transverse boundaries of the scattering region, it consequently becomes zero. This conclusion is quite obvious since diffraction of the radiation only redistributes the intensity over the cross section without varying the total power.

Most investigations of OPC-SS, including the first experiments [4.1, 2], have been carried out for cases when the pump radiation $E_L(r, z)$ possesses a developed speckle structure in the interaction volume, which makes high-quality phase conjugation possible. Speckle inhomogeneities of the pump intensity $|E_L(r, z)|^2$ give rise to a strong inhomogeneity of the profile of the local gain $g(r, z)$, both in the transverse (r) and longitudinal (z) coordinates.

As a result, various spatial configurations $|E_S(r)|^2$ of the scattered field differing in the amount of overlap (4.2.3 b) with the gain profile $G|E_L(r)|^2$ possess different values of the effective gain $g(z)$. Thus, for instance, if $E_S(r) = \text{const} \cdot E_L^*(r)$ in a given cross section, then the maxima of the scattered wave will be exactly correlated with the amplification maxima.

The overlap integral can be easily found in the interesting case when there are many speckle inhomogeneities of the pump over the beam cross section, and integration in (4.2.3 b) can be replaced by ensemble averaging. Using the results of Sect. 3.1, we can obtain for $g(z)$

$$g_{\text{con}}(z) = G\langle I_L^2\rangle/\langle I\rangle = 2g_0, \qquad g_0 = G\langle I_L\rangle, \tag{4.2.4}$$

where $I_L(r, z) = |E_L(r, z)|^2$. If the scattered wave has inhomogeneities uncorrelated with pump inhomogeneities, the averaging in (4.2.3 b) yields

$$g_{\text{uncor}}(z) = g_0, \tag{4.2.5}$$

i.e., the value of g is a factor of 2 lower than that for the phase-conjugated component.

It should be noted that there exist configurations $E_S(r)$ of the scattered field whose overlap integral in a given cross section is even greater than for the conjugated pump wave. For example, such is the case with a field $E_S(r)$ wholly localized within the limits of the most intense pump speckle. However, in the neighboring cross sections, at $z \gtrsim \lambda/\Delta\,\theta_L^2$, the higher match becomes a complete mismatch due to the diffraction of both fields, E_L and E_S. The only configuration $E_S(r, z)$ which remains matched, in spite of diffraction, to the profile of the gain proportional to $|E_L(r, z)|^2$ is the conjugated configuration $E_S(r, z) \propto E_L^*(r, z)$. Configurations uncorrelated with the pump also have a constant overlap integral (with respect to z) and remain uncorrelated over the whole volume.

We now estimate under what conditions the discrimination against non-conjugated (uncorrelated) waves will lead to the phase-conjugated wave being dominant at the system output. With this end in view, we represent the output intensity in the form

$$I_S(z = L) = \alpha \exp(2g_0 L) + \beta \exp(g_0 L), \tag{4.2.6}$$

where the first term corresponds to the phase-conjugated wave and the second one, to uncorrelated waves; α and β ared the intensities of the corresponding priming waves which arise owing to spontaneous scattering in the vicinity of the window of the SS cell at $z = 0$. The condition for OPC discrimination acquires the form $\beta \ll \alpha \exp(g_0 L)$. If we take into consideration the fact that the condition $\exp(2g_0 L) \approx \exp(30)$ usually holds true for the total gain of the strongest component at the SS threshold, then the discrimination condition is $\beta \ll \alpha \exp(15)$.

To estimate the ratio of α to β, we take into account the fact that spontaneous scattering of the pump by incoherent thermal phonons is distributed, on average, with the same intensity in all directions. Therefore, the intensity of an uncorrelated priming wave can be written

$$\beta = \text{const} \cdot \Delta \Theta_{\text{uncor}}, \tag{4.2.7}$$

where $\Delta \Theta_{\text{uncor}}$ is the solid angle in which uncorrelated waves are effectively amplified. The estimation of α for the phase-conjugated component yields an expression similar to (4.2.7) in which $\Delta \Theta_{\text{uncor}}$ is replaced by the diffraction solid angle $\Delta \Theta' = \lambda^2/S$, where S is the cross-sectional area of the light guide. Indeed, the number of oscillators with different transverse indices for a monochromatic field propagating through a surface area $S \, [\text{cm}^2]$ in a solid angle $\Delta \Theta_{\text{uncor}}$ is equal to $\Delta N = S \Delta \Theta_{\text{uncor}}/\lambda^2$ (for a fixed polarization state). One field oscillator with a given transverse configuration corresponds to the phase-conjugated component. Finally, the discrimination condition against non-conjugated waves takes the form

$$\Delta \Theta_{\text{uncor}}/\Delta \Theta' = \Delta \Theta_{\text{uncor}} D^2/\lambda^2 \ll \exp(15) \approx 3 \times 10^6. \tag{4.2.8}$$

Suppose that $D = 0.3$ cm, $S = D^2 = 0.1$ cm^2, $\lambda = 10^{-4}$ cm; then $\Delta \Theta' = \lambda^2/D^2 = 10^{-7}$ sterad. The value of $\Delta \Theta_{\text{uncor}}$ can be determined by various factors, such as losses and depolarization at the light guide walls, deviation of the Stokes ray from the pump beam (if a light guide is absent), cutting of the wings of the Stokes beam by the diaphragm of the detection system, etc. At $D = 0.3$ cm and $\lambda = 10^{-4}$ cm the condition for discrimination against uncorrelated waves limits the operating solid angle of the system so that $\Delta \Theta_{\text{uncor}} \lesssim 3 \times 10^{-1}$ sterad, i.e., the limitation is rather weak. If, however, higher-energy beams are to be conjugated, additional limitations caused by possible optical destruction make it necessary to use light guides of a larger diameter. Thus, for example, at $D = 3$ cm and $\lambda = 10^{-4}$ cm at discrimination condition against uncorrelated noise is of the form $\Delta \Theta_{\text{uncor}} \lesssim 3 \times 10^{-3}$ sterad, this corresponding to a linear angle $\Delta \theta \sim 5 \times 10^{-2}$ rad. The real angle within which uncorrelated waves are effectively amplified in the light guide is, as a rule, considerably larger, and, hence, the discrimination condition would not be satisfied.

If only the quality of the phase-conjugated signal, whose divergence is usually considerably smaller than the above value, is of interest to us, then undesirable uncorrelated noise can be cut off with the aid of diaphragms. However, we should take into account that failing to discriminate against the total amplified noise waves results in noise which, even without getting into the operating angle of the laser system, takes possession of a considerable portion of the energy scattered from the pump into the Stokes field. As a result, even by "cleaning" the phase-conjugated signal with the aid of diaphragms we obtain a low coefficient of reflection into the phase-conjugated wave. Because of this, the conjugation of high-power beams requires the use of additional techniques (Sect. 5.7).

In the above estimate of discrimination we supposed the SS conditions to be steady-state. Because the ratio of the rise index of the phase-conjugated wave to that of the uncorrelated wave is $\sqrt{2}$ instead of 2 (Sect. 4.3) in the nonsteady-state condition, the condition for suppressing uncorrelated waves turns out to be more severe.

Thus, good discrimination against uncorrelated waves in the light guide requires that its diameter should not be too large ($D \lesssim 1$ cm under typical conditions). Otherwise, the coefficient of reflection into the phase-conjugated wave decreases and the quality OPC becomes worse due to an increase of uncorrelated waves in the Stokes field.

4.3 The Specklon

Up to this point, our reasoning has been based on the hypothesis, not yet proved, that a wave, phase conjugated in a certain cross section, $E_S(r)$ $= \text{const} \cdot E_L^*(r)$, retains its conjugation throughout the whole volume. This hypothesis may not seem obvious due to the fact that the scattered field, as distinct from the pump, has to propagate in a medium with a strongly non-uniform gain.

Consider first a speckle-inhomogeneous pump. The propagation of a wave in optically inhomogeneous media is accompanied by distortion, i.e., in terms of our problem, by the extinction of the field due to the transformation of part of it into other spatial configurations. We want to recall the estimate of the extinction coefficient R_{ex} [cm^{-1}], which characterizes the rate at which the original wave becomes more "noisy" due to distortions. Suppose that the relative disturbance of the complex amplitude of the original wave A along the optical inhomogeneity correlation length l_{cor} is a small quantity ($\delta A/A$). Then the relative rate at which it becomes "noisier" over this length is $|\delta A/A|^2$. As distortions caused by various inhomogeneities separated by the length l_{cor} are added randomly, i.e., by intensity, the relative rate for a length $L > l_{cor}$ is given by $R_{ex}L$, where $R_{ex} \approx |\delta A/A|^2/l_{cor}$. In our case, the role of l_{cor} belongs to the Fresnel length of one speckle inhomogeneity of the pump, $l_{cor} \approx \lambda/2\Delta\theta_L^2$ (Sect. 5.1), and $|\delta A/A|$ is equal to $(\exp(\delta g l_{cor}) - 1) \approx \delta g l_{cor}$. Assuming that $\delta g l_{cor} \sim g_0 l_{cor} \lesssim 1$, the rate of extinction from a given wave to distortions is $R_{ex} \sim g_0^2 l_{cor}$.

Usually the need to have insignificant distortion of a wave during its propagation through an inhomogeneous medium limits the propagation length to $R_{ex}L \lesssim 1$. Let us estimate $R_{ex}L$ in the experiment described in [4.1], which uses a phase plate which allows for the irregular part of the divergence $\Delta\theta_L = 5.3 \times 10^{-3}$ rad. We take $g_0 \approx 15/L \approx 0.15$ cm^{-1} at $L = 100$ cm, and so $l_{cor} = \lambda/2\Delta\theta_L^2 \approx 1.7$ cm, $g_0 l_{cor} \approx 0.25$, $R_{ex} \approx 0.038$ cm^{-1}, and over the entire length L we have $R_{ex}L \approx 3.8$, i.e., there are enormous distortions.

It is a striking fact that the wave which is phase-conjugated relative to the speckle-inhomogeneous pump turns out to be almost noiseless. This fact points to the surprising stability of the phase-conjugated wave $E_S(r, z) \propto E_L^*(r, z)$ amplified in the field of the speckle-inhomogeneous pump matched to it. To emphasize the peculiarity and high stability of such a solution of the wave equation for a speckle-inhomogeneous medium, we introduce a special term – the *specklon*.

Let us prove that the specklon is really stable in the sense of distortions of its structure. For this purpose we represent the field $E_S(r, z)$ in the form of the phase-conjugated component (specklon proper), $s(z)E_L^*(r, z)\langle I_L\rangle^{-1/2}$, and field distortions $n(r, z)$ uncorrelated with the specklon, $\langle n(r, z)E_L(r, z)\rangle = 0$:

$$E_S(r, z) = s(z)\frac{E_L^*(r, z)}{\sqrt{\langle I_L\rangle}} + n(r, z) . \tag{4.3.1}$$

Then the equation for the specklon intensity $S(z) = |s(z)|^2$ can be written, allowing for (4.1.4), as

$$\frac{dS}{dz} = 2g_0 S(z) , \tag{4.3.2}$$

i.e., the conjugated specklon is amplified at the doubled rate $2g_0$. Uncorrelated waves $N(z) = \langle |n|^2\rangle$ are excited due to the extinction of the specklon by inhomogeneities and, at in addition, are amplified in the averaged pump intensity, i.e., at a rate g_0. Thus, we have

$$\frac{dN}{dz} = R_{ex}S(z) + g_0 N(z) . \tag{4.3.3}$$

If we take $S(z) = S_0\exp(2g_0 z)$ according to (4.3.2), then from (4.3.3) we obtain

$$N(L) = N_0\exp(g_0 L) + R_{ex}S(L)/g_0 . \tag{4.3.4}$$

The first term corresponds to the excitation of uncorrelated waves by spontaneous primings. We shall assume that the discrimination condition against it, which was discussed in detail in the previous section, is satisfied, $N_0 \exp(g_0 L)$ $\ll S(L)$. The second term describes distortions generated by the specklon itself, due to amplification inhomogeneities.

An important conclusion can be drawn from (4.3.4): the proportion of such distortions $N(L)/S(L)$ is independent of the length L:

$$N(L)/S(L) = R_{ex}/g_0 l_{cor} . \tag{4.3.5}$$

In other words, relative distortions of the specklon are not accumulated following the simple relation $R_{ex} L$. Owing to a larger amplification rate of the specklon in comparison with that of the uncorrelated distortions, accumulation takes place over the length $L_{eff} = (2g_0 - g_0)^{-1}$, determined by the difference in the gain.

In the experiment in [4.1], using a phase plate, the value of $g_0 l_{cor}$ amounted to about 0.25, i.e., the level of specklon distortions was rather low, $N/S \approx 0.25$. Indeed, for the distortion level in [4.1] an experimental limit $N/S \lesssim 15\%$ was obtained. Let us point out possible reasons for the fact that the quality of OPC turned out to be even better than theoretical estimates suggested it could be. First, the rate of discrimination against non-conjugated waves may be higher [$L_{eff} = (g_{con} - g_0)^{-1}$ being lower] if the Stokes waves are pulled into serpentine active micro-waveguides, see Sect. 5.1.1. Secondly, the phase-plate cross section was mapped onto the light-guide input where the pump speckle structure had not yet developed, and the amplification was uniform. Thirdly, the strong pump saturation effect may be responsible for improving the quality of OPC.

The quality of OPC becomes suddenly worse at $g_0 l_{cor} \gtrsim 1$ since in this case distortions over one speckle-inhomogeneity length are very significant: $\exp(g_0 l_{cor}) - 1 \gg 1$. This is excellently illustrated by the results of [4.1], obtained with the original radiation being focused into a light guide and without using a phase plate (Fig. 4.2 d).

Thus, in the OPC-SS problem, we use the term specklon to mean the solution $E_S(r, z)$ of the wave equation for a medium with optical inhomogeneities caused by the speckle structure of a certain pump field $|E_L(r, z)|^2$. The solution possesses the following properties: 1) Owing to the high correlation of the specklon inhomogeneities with medium inhomogeneities, $E_S \propto E_L^*$, i.e., due to spatial resonance, the specklon has a higher gain, and 2) because of this it is stable with respect to distortions.

It should be noted that in some other cases (OPC-SS of focused beams, OPC-SS in the nonsteady-state condition, OPC-SS of depolarized radiation, etc.) the discrimination degree of the gain $(g_{con} - g_{uncor})/g_{uncor}$ and the specklon distortion level will be estimated in some other way. For example, under conditions of essentially nonstationary SS amplification the rise of the

Stokes wave for a plane pump wave obeys the law $|E_S|^2 \propto \exp(2\sqrt{2Gz\Gamma t I_L})$ (Sect. 2.5). It is clear that in the field of the speckle-inhomogeneous pump $E_L(r, z, t) = \frac{1}{2}[E_L^*(r, z)\exp(-i\omega t - ikz) + \text{c.c.}]$ uncorrelated waves will be amplified following the same law if I_L is the space-averaged intensity. Unlike this, the specklon

$$E_S(r, z, t) = \frac{1}{2}[s(z, t)E_L^*(r, z)\exp(-i\omega_S t + ikz + \text{c.c.}]$$

responsible for the conjugation of the pump structure possesses the doubled overlap integral and, hence, will be amplified according to

$$|s(z, t)|^2 \propto \exp(2\sqrt{4Gz\Gamma t I_L}) \,.$$

This is obtained from the expression for the rise of uncorrelated waves by the substitution $GI_L \to 2GI_L$. As a result, the degree of discrimination against uncorrelated wave amplification under nonsteady-state conditions decreases to $\sqrt{2} \approx 1.4$ in the gain.

However, it is important that the general conclusions about the discrimination mechanism of specklon separation and its stability remain in force. The concept of the specklon is also useful for solving the problem of light diffraction in volume holograms of speckle fields where inhomogeneities (often phase ones) of the medium are recorded by the speckle structure of the object field.

4.4 Structure of Uncorrelated Waves

As has been mentioned, the physical mechanism of OPC-SS is based on discriminating the phase-conjugated component above the background of the Stokes waves uncorrelated with the pump. That is why the study of the amplification of uncorrelated waves, their structure and angular distribution is of great interest. Detailed investigations into these questions have been carried out in experiments on stimulated scattering of a speckle-inhomogeneous beam in a light guide of circular cross section [4.3 – 5]. Amplified uncorrelated noise was investigated in experiments on OPC-SS of focused speckle-inhomogeneous beams in [4.6, 7]. We shall now discuss one of these experiments in some detail.

Amplified uncorrelated radiation was registered photographically with a high spatial resolution for the first time in the experiments described in [4.3, 4], the absolute brightness being measured. The speckle-inhomogeneous structure of an uncorrelated component was registered with a high contrast. The angular size of speckles was about 0.2 mrad, this corresponding to the diffraction divergence for the face of a light guide 5 mm in diameter. The high contrast of the speckle structure meant that the scattered radiation was of sufficiently short duration so that the random phases of different uncorrelated components were not shifted within the pulse time τ_p, $\tau_p \lesssim 5\Gamma^{-1}$, where Γ^{-1} is the lifetime of the hypersound grating. The diffraction angular size of speckles testified to the fact that noise waves were, on average, uniformly distributed over the light guide face.

The noise intensity per unit solid angle was measured both photographically and calorimetrically. In the experiments with a light guide, the intensity reflection of the pump per unit solid angle of noise waves near the central direction amounted to $\varrho_n(\theta = 0) \approx 1.5$ sterad^{-1}, and hence the fraction within the diffraction solid angle was $\varrho_n \Delta\Theta' \approx 5 \times 10^{-8}$. The value of $\varrho_n(\theta = 0)$ can be estimated theoretically proceeding from the following consideration. For spontaneous scattering we would have $\varrho_n(\theta = 0) = l\, dR/d\Theta$, where l is the length of the scattering region and $dR/d\Theta$ is the differential (with respect to solid angle) coefficient of extinction in the Stokes component. Under stimulated-scattering conditions this value be multiplied by $\exp(g_0 L) \approx \exp(15) \approx 3 \times 10^6$, and the length l, over which the spontaneous priming of non-conjugated noise is accumulated, should be replaced by the reciprocal of the amplification length, $l = g_0^{-1} = L/15$. Assuming $dR/d\Theta \approx 10^{-7}$ cm^{-1} sterad^{-1} and $L = 50$ cm, we obtain $\varrho_n \approx (dR/d\Theta)L \exp(15)/15 \sim 1$ sterad^{-1}, this value being in good agreement with the experimental figure.

As the noise observation angle θ, measured from the axis, is increased, the noise intensity decreases. For a round light guide operating on total internal reflection (TIR) from an interface (for instance, carbon disulfide — glass or carbon disulfide — quartz) this decrease is due to the depolarizing effect of reflection by the walls. In actual fact, after TIR of the rays directed close to the axis, the polarization of the reflected wave coincides with that of the incident wave; in particular, the pump radiation remains polarized over the entire length of the light guide. However, in the case of uncorrelated waves making large angles with the axis, TIR varies their polarization, thus decreasing the amplification efficiency in the pump field. The angular distribution of noise brightness is described in this case by an empirical formula, $\varrho_n(\theta) = \varrho_n(0) \exp[-(n-1)L\theta^2/D]$, where θ [rad] is measured from the axis in the medium, L is the length of the light guide, D is its diameter, and n is the relative refractive index of carbon disulfide and the envelope (glass, quartz). The distribution $\varrho_n(\theta)$ was practically independent of the pump divergence, which was varied between 10 and 40 mrad (in air), and of the pump power in the saturation mode.

The effective solid angle of the noise, which can be determined from $\Delta\Theta_n = [\varrho_n(0)]^{-1} \cdot \int \varrho_n(\theta) 2\pi\theta\, d\theta$, was equal to $\Delta\Theta_n \approx \pi D/L(n-1) \approx 0.2$ sterad at $L = 50$ cm, $D = 3$ mm, $n-1 \approx 0.1$. Under typical conditions the total reflection into the uncorrelated noise Stokes component was about 10% of the pump energy, whereas about 50% of the pump energy was reflected into the OPC component. For other types of light guides (using Fresnel reflection of strongly inclined rays, or with a rectangular cross section, etc.) the actual angular dependence $\varrho_n(\theta)/\varrho_n(0)$ and the value of $\Delta\Theta_n$ may be different and determined by other physical processes. However, the operating solid angle of most laser devices is considerably less than $\Delta\Theta_n$. That is why, in actually fact, a considerably smaller proportion of noise gets into the operating aperture of the device, i.e., the quality of OPC turns out to be rather high. It should be noted that increasing the diameter of the light guide leads to an increased proportion of scattered-radiation energy corresponding to uncorrelated noise. That is why the coefficient of reflection into the phase-conjugated wave will decrease.

In conclusion, we mention [4.7] in which uncorrelated and OPC components were investigated for SS of a focused beam. An interesting result was obtained, which has not yet been completely explained, namely, the frequency spectrum of the OPC component was narrow whereas that of uncorrelated waves was considerably wider.

4.5 Experimental Investigation of Discrimination in a Light Guide

Though the discrimination mechanism in OPC-SS is considered generally accepted, for a long time there was not direct experimental comparison of the amplification efficiency of various configurations of the Stokes wave by a speckle-modulated pump.

The intensities of uncorrelated and OPC components in backward SBS have been measured in the case of excitation by *spontaneous noise* in [4.8]. Subsequently direct measurements were made [4.8] of the gain coefficients of *specially prepared* uncorrelated and OPC configurations. The results of both experiments confirm the adopted theoretical ideas fairly well. Let us consider these experiments.

In [4.8] OPC-SBS was excited by ruby laser radiation in a light guide of circular cross-section filled with liquid carbon disulfide. The energy transferred to the phase-conjugated component was measured with a calorimeter. Uncorrelated components propagating in a very wide solid angle ($\Delta\Theta \sim 10^{-1} \times 10^{-1}$ rad^2) were registered photographically and calorimetrically and subjected to spectral analysis. With a low pump level the OPC component was absent and the scattered radiation was observed over the wide solid angle. The spectral distribution of the radiation and the dependence of its intensity upon the exciting intensity corresponded to the spontaneous process. When the pump intensity was increased, SS was effectively excited, and the discrete OPC component acquired quite a noticeable intensity. Intensity measurements made it possible to calculate the ratio g_{con}/g_{uncor} for OPC and uncorrelated waves. It was found that $g_{con}/g_{uncor} = 2.43 \pm 0.07$, assuming that the intensity of a spontaneous priming wave exciting the OPC component is the same as that of an arbitrary uncorrelated discrete mode of the light guide. This is indicative of the fact that under the experimental conditions in [4.8] the intensity of the input priming wave was likely to be noticeably larger than $I_L (\lambda/D)^2 (L/30) \, dR/d\Theta$.

A possible explanation of this fact may be that the amplification rate of the phase-conjugated component was higher due to partial pulling into the cores of active micro-waveguides. An additional contribution to the intensity of the OPC component may be attributed to the following effect. Spontaneous priming in some of the most intense speckles at the far end of the cell are strongly amplified and, thus, acquire a larger projection on the OPC component. On the other hand, if the pump intensity averaged over speckle structure in some section of the light guide has a local spike not wholly occupying the cross section, then only a certain portion of all uncorrelated modes, with suitable coordinates and ray directions, survives in that place. This effect manifests itself pronouncedly in the case of a focused pump beam (see below). Pilipetsky et al. in [4.8] supposed that partial coherence of spontaneous priming waves may be the reason for a raised noise discrimination, and they obtained $g_{con}/g_{uncor} = 2.04 \pm 0.06$.

In [4.9] a scheme was used with an additional "differential cell" into which a moderate-intensity speckle-inhomogeneous pump field was directed. Spontaneous noise failed to excite SS in this cell, but a noticeable amplification of specially prepared counterpropagating Stokes waves sent to the cell was observed. The light guide was filled with liquid carbon disulfide. Its rectangular cross section allowed one to keep the polarization of the interacting waves constant. Provision was made for the statistical homogeneity of the pump speckle field in the interaction volume, and the smallness of gain over the length of one serpent bend ($g_0 l_{cor} \lesssim 0.1$) and the amplification linearity.

A specially prepared signal was introduced into the light guide in the opposite direction to the pump. It was frequency shifted so that SBS amplification in the pump field took place. Two kinds of signal were used: a signal exactly conjugating the pump transverse structure and a signal uncorrelated with the pump. In the first case, the OPC signal was generated by thresholdless SBS conjugation of the pump, which was transmitted through the differential cell into a second cell filled with carbon disulfide. In the second case, the signal was generated by SBS of an auxiliary plane wave in an additional cell filled with CS_2, the auxiliary wave being inclined relative to the pump central direction. The dependence of the logarithm of the Stokes wave total gain upon the pump power density proved to be linear. At a given pump power the relative amplitude gain of the OPC wave exceeded that for an uncorrelated wave by a factor of 2.0 ± 0.2, in accordance with theory.

4.6 OPC-SS in Focused Beams

A light guide makes it possible to maintain a high pump power density over a large interaction length L. That is why the SS threshold determined by the relation $2G|E_L|^2 L \approx 30$ turns out to be relatively low. A second important advantage of using a light guide lies in the good miscibility of various angular components of the pump. The presence of several reflections from the light guide walls ensures OPC with a completely coherent envelope.

At the same time, the simplicity of the scheme in which the pump is focused into a nonlinear medium without a light guide attracts experimentalists (whereas the mathematical complexity of the problem frightens theorists). It is for this reason that a scheme employing OPC-SS in a focused pump beam is used in many experiments. Without going into the details of a rather complicated theory (see Chap. 5) we shall estimate the basic parameters.

4.6.1 Speckle Beam

In a focused pump speckle beam the intensity has a smooth nonuniformity in the transverse cross section even after it has been averaged over the speckle structure (Sect. 3.4). This results in two essential differences between OPC in this case and OPC in a light guide: *the relative* discrimination against uncorrelated components becomes lower, but *their number* is sharply reduced.

Let us recall the basic geometrical parameters of a focused speckle beam. If a beam of diameter D_0 is focused into a medium by a lens of focal length F, the angle of convergence of the rays at the waist is $\theta_0 \approx D_0/F$. If, in addition, the divergence θ_1 of the original beam is considerably worse than the diffraction divergence, $\theta_1 \gg \lambda/D_0$, the waist radius at the focus is $a_0 \approx F\theta_1$, and its length $\Delta z_0 \sim 2a_0/\theta_0$. The dimensions of speckle inhomogeneities in the vicinity of the waist are: $\Delta r_{cor} \approx 1/k\theta_0 = a_0/\xi$, $\Delta z_{cor} \approx 1/k\theta_0^2 = \Delta z_0/\xi$. Here $\xi = ka_0\theta_0 \gg 1$ is the parameter characterizing by how many times the divergence of the speckle beam is greather than the diffraction limit.

The total gain of the specklon near the SS threshold over the effective length $L = \pi a_0/\theta_0$ amounts to $\sim 2GI_L(0)\pi a_0/\theta_0 \sim 30$. Therefore, the fraction of intensity (4.3.5) corresponding to the serpentine distortions accompanying the specklon is $N/S \approx R_{ex}/g \sim 5/\xi$. Thus, high-quality OPC of the speckle beam can be provided if $\xi \gg 5$.

To estimate the discrimination degree, suppose that the mean intensity profile of the pump is $I_L \propto \exp[-r^2/a_L^2(z)]$ and of the Stokes wave is $I_S \propto \exp[-r^2/a_S^2(z)]$. It is quite natural to suppose that $a_S(z)$ is always less than $a_L(z)$, both for the phase-conjugated and uncorrelated waves, this being confirmed by calculation (Chap. 5). In other words, the Stokes wave is compressed towards the axis, i.e., into a high-gain region. Then, for the phase-conjugated component we obtain, from the overlap integral (4.2.3), the effective gain in a given cross section:

$$g_{eff}(z) = GI_0 \cdot 2[1+a_S^2(z)/a_L^2(z)]^{-1}, \tag{4.6.1}$$

where the factor 2 is connected with spatial resonance of the phase-conjugated speckle wave. The same formula, but without the factor 2 and with $a_S^{unc}(z)$ replacing $a_S(z)$, is valid for uncorrelated waves. As calculations show, the envelope of the OPC component is approximately a factor of $a_L(z)/a_S(z) \approx (GI_0 a_0/2\theta_0)^{1/2}$ narrower than that of the pump, where I_0 and a_0 are the intensity and radius of the pump beam at the waist. As a result, the effective gain of the phase-conjugated wave is

$$g_{con} \approx 2GI_0(1-2\theta_0/GI_0 a_0)[1+a^2(z)/a_0^2]^{-1}. \tag{4.6.2}$$

Integrating g_{con} over z we obtain

$$2\pi \frac{a_0}{\theta_0} GI_0(1-2\theta_0/GI_0 a_0) \approx 30, \tag{4.6.3}$$

from which $GI_0 a_0/\theta_0 \approx 6.7$.

As distinct from the above, the size of the envelope for uncorrelated waves is given by $(a_{S_{p,q}}^{uncor})^2 = a_L^2(p+q+1)/\xi$. Here p, q are the transverse indices of higher Hermitian-Gaussian modes; $p = q = 0$ for the principal axial Gaussian mode. Then we have for the uncorrelated modes

$$\int_{-\infty}^{\infty} g_{p,q}^{uncor}(z)\,dz = \pi \frac{a_0}{\theta_0} GI_0[1-(p+q+1)/\xi]. \tag{4.6.4}$$

For the mode with the maximum gain, $p = q = 0$, (4.6.4) gives 21 $(1-1/\xi)$. Thus, at $\xi \gg 1$ the ratio of discrimination against the uncorrelated components is $g_{con}/g_{uncor} \approx 30/21 \approx 1.4$, i.e., noticeably less than the Gaussian factor 2 for a light guide. The reason for a decrease in the discrimination lies in the fact that, in order to support spatial resonance, the specklon has to be localized on the periphery of the beam cross section where the gain is low, whereas uncorrelated waves tend to the near-axis portion of the pump beam where the gain is at a maximum.

However, it is for a focused speckle beam that the number of effectively excited uncorrelated modes sharply decreases:

$$\Delta N = p_{max} q_{max} \approx (\xi/21)^2 . \tag{4.6.5}$$

Here p_{max} and q_{max} are specified from (4.6.4) by the condition that the modes corresponding to them are amplified over the whole length by a factor of e less than the central uncorrelated mode. Let us explain the reason for the limited number in (4.6.5). The total number of uncorrelated modes "located" in the pump profile coincides with the number ξ^2 of independent speckles in the pump cross section. However, due to a high gain in the active light guide formed by the pump, only components localized within the transverse length $\Delta r \sim a_0/\sqrt{21}$ and propagating within the angle $\Delta \theta \sim \theta_0/\sqrt{21}$ are effectively amplified (compare with the similar narrowing effect of the Stokes-component spectrum in SS, Sect. 2.5).

Finally, the condition for discrimination against uncorrelated components for a focused speckle beam is

$$\xi \lesssim 21 \exp[(30-21)/2] \approx 2 \times 10^3 . \tag{4.6.6}$$

Detailed experimental investigations of the uncorrelated wave discrimination in the OPC-SBS of a pump speckle beam are described in [4.10]. The value of ξ, the amount by which the beam divergence exceeds the diffraction limit, was varied in the range $10^2 \lesssim \xi \lesssim 10^3$. The pump power $I_0 a_0^2$ required to reach the SS threshold increased linearly with ξ, since the amplification region length $L \propto \xi$ and the intensity [W/cm^2] in the waist was proportional to ξ^{-2}. The fraction of energy corresponding to the exactly phase-conjugated component was measured by the diaphragm method with the use of a phase plate, see Sect. 4.7. The fraction was a maximum, about 80%, in the range $10^2 \lesssim \xi \lesssim 5 \times 10^2$, whereas at $\xi \gtrsim 10^3$ it sharply diminished in accord with the estimate given by (4.6.6). Unfortunately, no experiments were reported in [4.10] for $\xi \lesssim 10^2$. The experiments in [4.11, 12] also showed that in the non-steady-state condition, the requirement on discrimination against uncorrelated noise places a still stronger restriction on ξ, namely, $\xi \lesssim 30$ is necessary at $\tau_p \sim \Gamma^{-1}$.

Up to this point, we have considered a speckle beam for which the focal waist is wholly inside a nonlinear medium, see Fig. 4.4a. If the medium length is considerably less than the waist length (Fig. 4.4b), the situation becomes less favorable for OPC. First, uncorrelated modes are not angle selected and their number increases. Secondly, the quality of OPC worsens due to the fact that the envelope $f(r)$ in the relation $E_S(r) = f(r) E_L^*(r)$ is strongly compressed

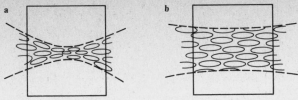

Fig. 4.4. (a) Focused speckle beam located wholly inside a nonlinear medium — a favorable OPC geometry; **(b)** speckle beam in the case of weak focusing — the waist dimension is much larger than the medium's length — an unfavorable OPC geometry

towards the speckle beam axis; moreover, the envelope becomes incoherent within in the cross section.

4.6.2 Ideal Beam

We consider a second limiting case in which backward SS is also accompanied by phase conjugation: a beam with an ideally plane wavefront is converted into a spherical wavefront due to focusing by a lens. Owing to diffraction, the field is not contracted to a point at the focus but to a focal waist, i.e., to a region of size $\Delta r_1 \times \Delta z \sim (\lambda/\theta_0) \times (\lambda/\theta_0^2)$, where $\theta_0 = D/F$ is the convergence angle of rays at the focus. Thus, for instance, if the input beam is ideally Gaussian, then at $\xi = 1$, using (3.4.11), we obtain

$$|E_L(r,z))|^2 = I_0 \frac{a_L^2}{a^2(z)} \exp(-r^2/a^2(z)] , \qquad a^2(z) = a_L^2 + z^2/k^2 a_L^2 , \quad (4.6.7)$$

where I_0 is the intensity at the waist center, a_L is the minimum radius of the waist and $\theta_L = (k a_L)^{-1}$ is the diffraction divergence corresponding to that radius. If the focal waist is wholly inside the nonlinear medium of length L, i.e., if $L \gg k a_0^2$, then among all the Gaussian Stokes beams the most effective amplitude gain, estimated by (4.2.3), is observed for a beam with the same parameter a_L (Fig. 4.5). Indeed, if we suppose that $I_S(r,z)$ is given by (4.6.7) with a_L substituted by a_S, the integral of the effective gain will be

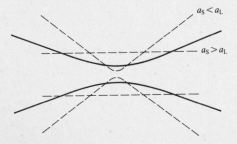

$a_S < a_L$

$a_S > a_L$

Fig. 4.5. Backward SS in the field of an ideally focused pump beam. Stokes beams with $a_S > a_L$ and $a_S < a_L$ are poorly overlapped with the pump

$$\int_{-L/2}^{L/2} g_{\text{eff}}(z)\,dz = \pi G I_0 k a_L^2 \frac{a_L a_S}{a_L^2 + a_S^2} \cdot \frac{2}{\pi} \arctan\left(\frac{L}{2k a_L a_S}\right). \tag{4.6.8}$$

At $L \gg k a_L a_S$ the value of arctan is $\pi/2$, and the integral of the effective gain is a maximum at $a_L = a_S$. We may say that the radial wings of Stokes waves with $a_S > a_L$ are poorly amplified, and the Stokes waves with $a_S < a_L$ quickly leave the amplification region due to diffraction (Fig. 4.5). It is not difficult to check that the higher Hermitian-Gaussian modes of such a beam are also amplified worse. For example, for the mode $E_S \propto x \exp(-r^2/2a_S^2)$, at $a_L = a_S$ the gain decreases twofold; for $E_S \propto xy \exp(-r^2/2a_S^2)$ it decreases fourfold, etc. In other words, only those rays of the Stokes field which pass through a cylinder of diameter $\sim 2a_L$ and length $2k a_L^2$ are efficiently amplified. These rays have a divergence $\theta_S = a_L/k a_L^2 = (k a_L)^{-1}$ which corresponds to the diffraction limit. Thus, in this case OPC is caused by discrimination due to the simple geometry of the amplifying region in the focused pump beam of diffraction quality.

If the length of the medium is not large, $L \ll k a_L^2$, we have for Gaussian beams

$$\int_{-L/2}^{L/2} g_{\text{eff}}(z)\,dz = G I_0 L (1 + a_S^2/a_L^2)^{-1}. \tag{4.6.9}$$

In this case both Gaussian beams with $L/2k a_L \lesssim a_S \lesssim a_L$ and Hermitian-Gaussian modes with higher indices are amplified equally well. As a result, OPC is absent in this case. Thus, the realization of OPC of ideal beams in a homogeneous medium without a light guide requires the following discrimination condition against non-conjugated waves:

$$L \gg k a_L^2 = 1/k \theta_L^2, \tag{4.6.10}$$

i.e., the focal waist should be wholly inside the medium.

It should be noted that accounting for non-uniform amplification in the calculation Stokes beam diffraction leads to deviations from the exact OPC relation $E_S(r) = \text{const} \cdot E_L^*(r)$. If, however, the focal waist is well inside the medium, these deviations are small and consist, mainly, in a slight contraction towards the beam axis of the amplitude distribution of the Stokes wave, compared to that of the pump. This effect is quite understandable in view of the higher gain in the vicinity of the axis.

4.6.3 Weakly Distorted Beam

It is not always worthwhile conjugating beams with an ideal wavefront (either plane or spherical) using methods of nonlinear optics, since in such cases, a plane mirror, corner retroreflector, spherical mirror or some other linear optical elements are suitable. However, OPC of beams with aberration is of special interest.

The case of moderate aberration cannot, unfortunately, be described analytically either in terms of speckle fields or in the framework of the ideal beam. Experiments show that for a beam with moderate aberration focused into a medium without using a phase plate, the quality of OPC is worse than in both limiting cases.

For weakly distorted beams, say $\xi \approx 2.5$, there are about $\xi^2 \approx 6$ random interference maxima in the beam cross section. The largest of them possesses an intensity $\sim \ln \xi^2 \approx 2$ times the mean value and carries an energy $\sim \xi^{-2} \ln \xi^2 \approx 30\%$ of the energy of the whole pump beam. Even if SS develops only from this spot with maximum intensity, the fraction of OPC may amount to $\gtrsim 30\%$. The diffraction interaction of the fields scattered in various spots may lead to an additional rise in the OPC fraction.

Now let us sum up what has been said in this section. To obtain good quality OPC of a weakly distorted beam (with a divergence close to the diffraction limit, $\xi \sim 1$) at a power near the SS threshold, the beam should be focused so that the waist be located completely inside the medium. For beams with noticeable distortions, a phase plate is desirable to increase the divergence up to $\xi = \theta / \theta_{\mathrm{dif}} \gg 5$, it still being necessary to place the waist inside the medium. Finally, one should provide $\xi \lesssim 10^3$ for discrimination against uncorrelated noise; in the nonstationary case ξ should be even smaller.

Unfortunately, there is no theory of OPC-SS of focused beams which properly takes into account saturation (almost always encountered in experiments). With saturation, the central portion of the pump beam undergoes the largest depletion. Due to this, the initial nonuniformity of the gain envelope becomes smoothed, and as a result, the compression of the Stokes beam towards the axis is not so noticeable and the OPC quality increases.

4.7 Registration Methods and Quality Estimation of OPC

In this section we shall briefly describe experimental methods which are employed for registration of OPC on the qualitative level and for quantitative measurements of OPC accuracy. These methods are equally applicable to any OPC procedure.

Preliminary judgements about the possible presence of phase conjugation in an experiment are based on comparing the photographs of the transverse intensity distribution of the incident and reflected beams in a certain section. This section can correspond either to the far field zone of the beams or to their near fields. A particularly pleasing impression is produced by pairs of pictures on which the images bear a strong visual resemblance to one another. Care should be exercised, however, in interpreting such simple photographs. Consider some examples.

Suppose that in the near field zone the incident wave has an intensity distribution $I_0(r)$ and a plane wavefront, $E_{inc}(r) = [I_0(r)]^{1/2}$, and the reflected wave has the same intensity distribution $I_0(r)$ and an arbitrary non-uniform phase distribution, $E_{ref}(r) = [I_0(r)]^{1/2} \exp[i\psi(r)]$. It is obvious that in this case OPC is not observed, though the photographs of the near field zone are identical.

Another example corresponds to good conjugation, even though a sharp distinction can be viewed on the photographs of the near field zones. Let the incident wave be plane with a constant amplitude, $E_{inc}(r) = E_0$, and the reflected wave consist of 85% (in energy) of the exactly conjugated field E_0^* and 15% of a second plane wave inclined to the first:

$$E_{ref}(r) = E_0^* [\sqrt{0.85} + \exp(iq \cdot r)\sqrt{0.15}] . \qquad (4.7.1)$$

In this case the intensity distribution of the incident wave is uniform and that of the reflected wave is of the form

$$|E_{ref}|^2 = |E_0|^2 [1 + 0.72 \cos(q \cdot r)] . \qquad (4.7.2)$$

It can be seen that with good quality phase conjugation (85%) corresponding to maximum measurement accuracy of this parameter in most experiments, the photographs of the beams in the near zone differ substantially: the intensity of the reflected wave varies with a contrast ratio of $(1+0.72)/(1-0.72) \approx 6(!)$.

One more example is related to the comparison of far-field-zone photographs. Let the incident wave correspond to an ideal beam of diameter D with diffraction divergence $\theta_{dif} = \lambda/D \sim 10^{-4}$ rad at $D \sim 1$ cm, and let it produce a spot of size $F\theta_{dif}$ at the lens focus (i.e., in the far field zone). Suppose that only 10% of the total energy of the reflected wave belongs to the exactly conjugated wave of diffraction quality while 90% falls on uncorrelated waves with a fairly small divergence $\theta_{un} \sim 3 \times 10^{-3}$ rad. Then on a photoplate at the lens focus, the energy density produced even by the weak phase-conjugated component will be a factor of $(0.1/0.9) \times (\theta_{un}/\theta_{dif})^2 \approx 100$ higher than that produced by uncorrelated waves. This means that, with normal registration of the bright conjugated central portion, the non-conjugated radiation carrying, in our particular case, the main portion of the energy will not manifest itself on the photographic material, which has a limited dynamic range.

We now consider quantitative methods of measuring OPC accuracy. The phase plate method suggested by Ragulsky [4.13] has been widely practised.

A phase plate (aberrator) is usually a glass plate etched in hydrofluoric acid to obtain random inhomogeneities in thickness. On passing through such a plate, a wave of light acquires a non-uniform phase $2\pi(n-1)\delta h(r)/\lambda$, where n is the glass refractive index, λ is the wavelength in air, and $\delta h(r)$ is the local thickness variation. If the characteristic transverse size of inhomogeneities is $a_\perp$, then, according to geometrical optics, the local deviation angle is $\delta\theta(r) = (n-1)\nabla h(r) \sim (n-1)\delta h/a_\perp$. The value of $\langle|\delta h|\rangle$ is usually chosen such that the non-uniform phase shift is about 2π or somewhat larger. Otherwise, i.e., if $|\delta h|(n-1) \lesssim \lambda$, the regular (non-deviated) wave may keep a considerable amplitude on passing through the plate. At $|\delta h|(n-1) \sim \lambda$ the divergence introduced by the phase plate is of the order of $\delta\theta \sim \lambda/a_\perp$. Phase plates typically used introduce into the beam an irregular ("grey") divergence $\delta\theta \sim 10^{-3}$ to 10^{-2} rad. To create a speckle-inhomogeneous pump in the interaction volume in the process of OPC-SS is the first purpose of the phase plate.

Its other purpose is to analyze the quality of phase conjugation. Suppose that a plane pump wave with a diffraction-quality divergence is incident on the phase plate, and that the reflected field contains the exactly conjugated component and uncorrelated components. To determine the fraction of OPC, these components should be spatially separated. On its backward passage through the phase plate the phase-conjugated fraction of the radiation is converted into a plane wave, and, therefore, a bright kernel of diffraction quality is produced in the far field zone. Uncorrelated waves, on their passage through the plate, yield a beam whose divergence is $\gtrsim \delta\theta$.

Elements with phase inhomogeneities have been used in holographic studies of OPC for a long time. However, it is only the technique of illuminating a phase plate by a diffraction-quality plane wave that enables the localization of the exactly conjugated component and, therefore, quantitative measurements of its characteristics. The energy of the exactly conjugated component can be measured in the far field zone behind the phase plate if a diaphragm of a size corresponding to the beam diffraction divergence is installed there. To advantage of this technique is that the experimental results can be simply interpreted, whereas the necessity of placing a diaphragm of relatively small size exactly within the conjugated laser beam is a drawback. To reduce the sensitivity of the scheme to fluctuations of the incident beam propagation direction, one has to increase the diaphragm diameter, which reduces the measurement accuracy.

Zel'dovich et al. [4.1] used the mirror wedge technique; this method is now widely used for OPC quality measurements. This technique compares the energies of the laser pump and the reflected wave, as well as the brightness of the pump radiation and that of the scattered wave in the far field zone. The energies are measured in the same units by two calorimeters; it is essential that

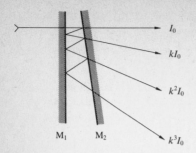

I_0

kI_0

k^2I_0

M_1 M_2 k^3I_0

Fig. 4.6. Scheme of the Ragulsky mirror wedge formed by semi-reflecting mirrors M_1 and M_2; $k = R_1 R_2$, where R_1 and R_2 are the reflectivities of the mirrors

the calorimeter for the reflected wave should collect light propagating within a sufficiently wide solid angle. The brightness of the pump beam with diffraction divergence and that of the scattered wave reconstrcuted by a phase plate are measured with the aid of photographic material. To overcome the limited dynamic range of the latter, either beam is split into a fan of spatially similar beams whose intensities are successively reduced by the same factor, twofold reduction per step being the most convenient. This splitting is performed by a mirror wedge formed by two semi-reflecting plane mirrors (Fig. 4.6) or by a totally reflecting mirror and a semi-reflecting one. As a result, among various beams of the fan there will be several beams whose energy lies within the dynamic range of the photomaterial. Thus, it becomes possible to quantitatively compare both the brightness and the angular distribution for any beams during a single laser pulse. If the same photoplate is used in the experiment, variations in the photomaterial properties and development conditions will not affect the measurement accuracy. Moreover, this experimental scheme automatically provides the blackening curve of a photomaterial.

The wave reconstructed by a phase plate contains a diffraction-quality bright central spot in the far field zone. If the scattered wave consisted only of the conjugated component (with respect to the exciting wave), then the ratio of the brightnesses of the scattered and original waves, B_S/B_L, at the centers of the diffraction spots would coincide with the ratio W_S/W_L of the energies registered by the calorimeters. Any distortion of the scattered field relative to the conjugating field leads to a decrease in B_S/B_L in comparison with W_S/W_L. The fraction of the energy of the reflected wave corresponding to the exactly conjugated component is

$$H = (W_L/W_S) \cdot (B_S/B_L). \qquad (4.7.3)$$

Besides the exact-conjugation fraction, the energy fraction falling into a given solid angle (which may be larger than the diffraction limit) is often of great practical interest. Such a characteristic is especially convenient if the original beam being conjugated is not of diffraction quality. A diaphragm of a suitable size placed in the far field zone is usually employed to measure this parameter. In this case comparison is made between the energy of the whole reflected wave and that of the portion transmitted through the diaphragm.

4.8 Literature

In a number of early works devoted to experiments on different kinds of SS the high directivity of the scattered radiation was noted [4.14 – 17]. The angular divergence of the scattered wave in those experiments was often of the same order of magnitude as that of the pump wave. However, the problem of relating the wavefront of the exciting field to that of the scattered field was not raised.

The problem of relating the wavefront structures of the exciting and scattered fields was raised by Ragulsky in 1971. He proposed and, in collaboration with coworkers in the Lebedev Physical Institute in Moscow, carried out an experiment in which the wavefront reversal in stimulated back scattering was discovered. Immediately afterwards there appeared the theoretical interpretation in terms of discrimination of non-conjugated configurations, proposed by Zel'dovich. All these results were published by *Zel'dovich* et al. [4.1] in 1972. In [4.2] *Nosach* et al. succeeded in OPC-SBS compensation of distortions in a double-pass scheme with reconstruction of the amplified beam to ideal diffraction quality.

The theory of SS for an ideal focused pump beam is considered in [4.18, 19]. The latter also deals with OPC-SS for a pump consisting of two interfering plane waves. Discrimination of uncorrelated waves for a speckle-inhomogeneous pump in a light guide is studied theoretically in [4.20 – 22] and in a number of subsequent publications; see also the Literature to Chap. 5. Conditions for the existence of the conjugating solution are discussed in [4.19 – 23]. The calculation of the coefficient of specklon extinction caused by inhomogeneities of the local gain in the field of the pump speckle wave is given in [4.23, 24]. The picture of a stable specklon is introduced in [4.25]. The work [4.26 – 28] was the first to experimentally realize OPC-SBS in a speckle beam focused into a medium rather than in a light guide. Experimental investigations into OPC-SBS of undistorted and weakly distorted beams focused into a medium were carried out in [4.11, 12, 29]. Experimental studies of the discrimination mechanism of OPC-SS are discussed in [4.3 – 6, 8, 9], and [4.13] deals with methods of registration and quality estimation of OPC.

A theoretical consideration of OPC-SBS of a focused speckle beam is given in [4.6, 10 – 12, 27 – 32]; [4.6] also contains the results of detailed experiments (see also [4.7, 11, 12, 29]).

5. Specific Features of OPC-SS

In this chapter the method of OPC-SS is discussed at greater length. We consider the formal theory of the specklon, the behavior of its envelope, and the theory of OPC-SS in a focused pump speckle beam. An account is given of theory and experiment for OPC of a depolarized pump and a non-monochromatic pump, and the peculiarities of a scheme with a strong regular component are considered as well. All this, in conjuction with the material of Chap. 4, presents the basic information necessary for mastering this particular method of OPC.

5.1 Theory of the Specklon

In this section we shall determine the values of the parameters of interacting speckle beams at which the concept of the specklon may be used. Neglecting the effect of the counterpropagating Stokes wave E_S on the pump E_L we may write the following system of equations describing the propagation of these two waves:

$$\frac{\partial E_L}{\partial z} + \frac{i}{2k_L} \nabla^2_\perp E_L(r, z) = 0 , \tag{5.1.1}$$

$$\frac{\partial E_S}{\partial z} - \frac{i}{2k_S} \nabla^2_\perp E_S(r, z) = \frac{1}{2} G |E_L(r, z)|^2 E_S(r, z) . \tag{5.1.2}$$

Section 5.1.1 will deal with the calculation of specklon distortions connected with speckle inhomogeneity of the gain. In this case we can assume, with sufficient accuracy, that $k_L = k_S$ in (5.1.1, 2) for SBS. In Sect. 5.1.2 we shall discuss distortions caused by the difference in $k_L - k_S$ of the mutually conjugated pump and Stokes waves and by the difference in their central propagation directions. The equation for the diffusion of the amplitude of the smooth envelope $f(r, z)$ in the relation $E_S(r, z) = f(r, z) E_L^*(r, z)$ will be derived in Sect. 5.1.3.

5.1.1 "Serpentine" Distortions

From Sect. 4.3 we know that the phase-conjugated wave is of the form $E_S(r, z)$ $\approx$ const $\cdot E_L^*(r, z)$ and $|E_{S\,conj}|^2 \propto \exp(2g_0 z)$, where g_0 [cm^{-1}] is the intensity gain averaged over speckle inhomogeneities, $g_0 = G\langle|E_L|^2\rangle$. However, as is seen from (5.1.2), such a form for the Stokes wave is not an exact solution of (5.1.2). In Sect. 4.3 a crude estimate was made of the rate R_{ex} [cm^{-1}] of extinction of the specklon by distortions. Here we shall calculate R_{ex}.

The pump intensity $|E_L(r, z)|^2$ has above-average values within the limits of individual speckles which, in passing from one cross section $z =$ const to another, flow into each other and describe serpentine trajectories with speckle-structure inhomogeneity scales. Owing to the higher gain there is a tendency for the Stokes wave to be pulled into the cores of serpentine active micro-waveguides. As a result, exact phase conjugation is perturbed, i.e., distortions are generated. These are called "serpentine" distortions, see [5.1]. On the other hand, diffraction does not permit the Stokes field to be strongly localized in the cores, thus maintaining the spatial resonance of the mutually conjugated waves.

To describe the competition of these two processes, let us seek the solution, allowing for distortions, in the form (4.3.1),

$$E_S(r, z) = (I_L)^{-1/2} s(z) E_L^*(r, z) + n(r, z) , \tag{5.1.3}$$

where $I_L = \langle|E_L|^2\rangle$, and require, for the sake of definitness, that $n(r, z)$ should be orthogonal (uncorrelated) to $E_L^*(r, z)$, i.e., $\langle n E_L\rangle = 0$.

Substituting (5.1.3) into (5.1.2) and separating, on the right side, the components having projections on E_L^* and those orthogonal to E_L^* we obtain

$$\frac{ds}{dz} = g_0 s(z) + \frac{G}{2\sqrt{I_L}} \langle|E_L|^2 E_L n\rangle , \tag{5.1.4}$$

$$\frac{\partial n}{\partial z} - \frac{i}{2k} \nabla_\perp^2 n(r, z) - \frac{1}{2} g_0 n - \frac{1}{2} [g(r, z) - g_0] n$$

$$= \frac{G}{2\sqrt{I_L}} s(z) E_L^* (|E_L|^2 - 2I_L) . \tag{5.1.5}$$

In (5.1.4, 5) the integration over the cross section (projecting) is substituted by averaging over the ensemble of speckle-field realizations $E_L(r, z)$. This system is the exact consequence of the original equation (5.1.2). Equation (5.1.5) is linear with respect to $n(r, z)$ but its right side is proportional to $s(z)$. Thus, its solution can be represented as a sum of noise excited by uncorrelated sources and the result of specklon extinction. The first kind of noise is described by

(5.1.5) without the right side. We suppose that its intensity is negligible due to discrimination.

To estimate the extinction distortions of the specklon, we can use perturbation theory because the parameter $g_0 l_{cor} = g_0 \lambda / \Delta \theta_L^2$ characterizing the fraction of distortions generated on the amplification length is small. Then the term $\propto [g(r, z) - g_0] n(r, z)$ on the left side describing the angular redistribution of distortions due to secondary extinction may be omitted, and for $n(r, z)$ we obtain

$$n(r, z) = \frac{G}{2\sqrt{I_L}} \int_{-\infty}^{z} dz' \int d^2r' \, \Gamma(z - z', r - r')$$
$$\times s(z') E_L^*(r', z')[|E_L(r', z')|^2 - 2I_L] . \tag{5.1.6}$$

Here the Green's function Γ of a parabolic equation is introduced for the field propagating in the $+z$ direction:

$$\Gamma(z, r) = \frac{k^2}{(2\pi)^2} \int d^2\theta \exp(ik\theta \cdot r - ik\theta^2 z/2)$$

$$\equiv \frac{k}{2\pi i z} \exp(ikr^2/2z) . \tag{5.1.7}$$

Let us set up the expression for the ensemble-averaged distortion intensity $\langle |n(r, z)|^2 \rangle$. To do this, one must deal with the following value, which should be integrated over $dz' dz'' d^2r' d^2r''$:

$$\langle E_L^*(r', z')[|E_L(r', z')|^2 - 2I_L] E_L(r'', z'')[|E_L(r'', z'')|^2 - 2I_L] \rangle$$
$$= 2I_L^3 \gamma(r'' - r', z'' - z') |\gamma(r'' - r', z'' - z')|^2 . \tag{5.1.8}$$

Here $\gamma(r'' - r', z'' - z')$ is the normalized function which characterizes the correlation of the complex amplitudes of the pump speckle field at points (r'', z'') and (r', z'). In deriving (5.1.8) we used the expression for the higher moments of fields obeying complex Gaussian statistics in terms of the correlator γ, see Sect. 3.1. According to the Van Zittert-Zernike theorem (Sect. 3.3), taking into account the fact that the pump propagates along $-z$, we have

$$\gamma(r, z) \equiv I_L^{-1} \langle E_L^*(0,0) E_L(r, z) \rangle = \int j(\theta) \exp(ik\theta \cdot r + ik\theta^2 z/2) d^2\theta , \tag{5.1.9}$$

where $j(\theta)$ is the normalized energy distribution over the angular spectrum of the "grey" divergence of the pump speckle field, i.e., the total divergence minus the regular spherical portion. Substituting (5.1.7 − 9) into the six-dimensional integral gives, for the distortion intensity,

$$\langle |n|^2 \rangle = R_{ex} \int_{-\infty}^{z} \exp[g_0(z-z_1)]\,|s(z_1)|^2 dz_1 .$$ (5.1.10)

In obtaining (5.1.10) we used new variables $(z'+z'')/2 = z_1$, $z''-z' = z_2$, and the fact that the specklon amplitude $s(z)$ varies slowly over the correlation length $l_{cor} = \Delta z_2 \sim \lambda/\Delta\theta_L^2$. This allows us to integrate over z_2 from $-\infty$ to $+\infty$, factoring $s(z')s(z'')\exp[g_0(z'+z'')]$ outside $\int dz_2$. Finally, for the extinction coefficient R_{ex} [cm^{-1}], we have

$$R_{ex} = \frac{\pi g_0^2}{k\,\Delta\theta_{eff}^2}, \quad \frac{1}{\Delta\theta_{eff}^2} = \iiint d^2\theta_1 d^2\theta_2 d^2\theta_3 j(\theta_1) j(\theta_2) j(\theta_3)$$
$$\times \delta^{(1)}[(\theta_2-\theta_1)\cdot(\theta_2-\theta_3)] .$$ (5.1.11)

Here $\delta^{(1)}(x)$ is the one-dimensional Dirac δ function whose argument in (5.1.11) is the scalar product of the vectors $\theta_2-\theta_1$ and $\theta_2-\theta_3$. Its order of magnitude $1/\Delta\theta_{eff}^2$ coincides with the reciprocal of the square of the angular width $\Delta\theta_L^2$ of the "grey" divergence of the pump. Thus, for example, for the Gaussian distribution $j(\theta) \propto \exp(-\theta^2/\theta_0^2)$,

$$\frac{1}{\Delta\theta_{eff}^2} = \frac{1}{2\theta_0^2} = \frac{1.38}{\beta_{1/2}^2} ,$$ (5.1.12)

which is expressed in terms of $\beta_{1/2}$ = FWHM, the total angular width at half-maximum intensity, $\beta_{1/2} = 2\theta_0\sqrt{\ln 2}$. For an angular spectrum in the form of a cut parabola, $j(\theta) \propto (1-\theta^2/\theta_0^2)$ at $|\theta| < \theta_0$, calculations give $1/\Delta\theta_{eff}^2 = 1.7/\beta_{1/2}^2 = 8\cdot(3\pi\theta_0^2)^{-1}$.

We now discuss the form of (5.1.11). Individual "gratings" of gain disturbance of the form

$$\tilde{E}_L^*(\theta_1)\tilde{E}_L(\theta_2)\exp[ik(\theta_2-\theta_1)\cdot r + ik(\theta_2^2-\theta_1^2)z/2]$$

are present in the medium; here $\tilde{E}_L(\theta)$ is the amplitude of an angular component of the pump. The scattering of the Stokes wave $\tilde{E}_S(\theta_S)\exp(ik\theta_S\cdot r - ik\theta_S^2\cdot z/2)$ by such a grating gives rise to the polarization wave

$$\tilde{P}(\theta) \propto \tilde{E}_L^*(\theta_1)\tilde{E}_L(\theta_2)\tilde{E}_S(\theta_S)\exp[ik(\theta_S+\theta_2-\theta_1)\cdot r - ik(\theta_S^2-\theta_2^2+\theta_1^2)z/2] ,$$

where $\theta = \theta_S+\theta_2-\theta_1$. For a noise wave to be efficiently excited by this polarization, the dispersion equation should be satisfied, and hence

$$(\theta_S+\theta_2-\theta_1)^2 - (\theta_S^2-\theta_2^2+\theta_1^2) \equiv 2(\theta_2-\theta_1)\cdot(\theta_2-\theta_S) = 0 .$$ (5.1.13)

For a given pair of vectors $\theta_1 = (\theta_{1x},\theta_{1y})$, $\theta_2 = (\theta_{2x},\theta_{2y})$ in the plane of the angular coordinates θ_x, θ_y, possible positions of θ_S and θ satisfying (5.1.13)

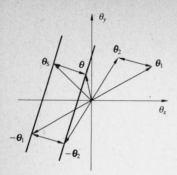

Fig. 5.1. The thick lines denote the loci of those pairs of points θ_S and θ, in the space (θ_x, θ_y) of the Stokes wave angular components, which are efficiently scattered into one another by the gain grating recorded by the pump components $\bar{E}_L(\theta_1)$ and $\bar{E}_L(\theta_2)$

are shown in Fig. 5.1 by thick solid lines. The negative quantities $(-\theta_1)$ and $(-\theta_2)$ indicate that the pump propagates in opposition to the Stokes wave. Finally, allowing for the proportionality $\bar{E}_S(\theta_S) \propto \bar{E}_L^*(-\theta_S)$, supposing $\theta_S = -\theta_3$, and averaging the squares of the respective amplitudes, $|\bar{E}_L(\theta_1)|^2 \propto j(\theta_1)$, etc., we obtain the expression under the integrals in (5.1.11).

It is easy to understand the kind of angular distribution of specklon serpentine distortions in terms of scattering of the individual angular components. To estimate the distortions, we should know the correlator $\langle n^*(r', z) n(r'', z) \rangle$ and its Fourier transform. The angular distribution of the distortions is

$$\frac{1}{R_{ex}} \frac{dR_{ex}(\theta)}{d\Theta} = \Delta\theta_{eff}^2 \iint j(\theta_1) j(\theta_2) j(\theta_2 - \theta_1 - \theta)$$
$$\times \delta^{(1)}[(\theta_2 - \theta_1) \cdot (\theta + \theta_1)] \, d^2\theta_1 d^2\theta_2 . \qquad (5.1.14\text{a})$$

Using (5.1.14a) and the construction in Fig. 5.1 one can see that the angular width of the noise approximately coincides with the width of the angular spectrum $j(\theta)$ of the pump. It is for this reason that serpentine distortions cannot be eliminated by simple diaphragms.

The presence of serpentine distortions has a reaction effect on the specklon gain. Substituting $n(r, z)$ from (5.1.6) into (5.1.4) and averaging, we obtain for the specklon gain

$$\frac{d}{dz}|s(z)|^2 = g_{con}|s(z)|^2, \qquad g_{con} = 2g_0 + R_{ex} . \qquad (5.1.14\text{b})$$

Unlike extinction in the usual nonlinear media, which leads to the attenuation of the propagating image, in our particular case extinction into serpentine distortions increases the amplification rate of the specklon image. This fact is in full agreement with the picture of the specklon field being pulled into the cores of active serpentine micro-waveguides, this raising the total gain. The fraction of serpentine distortions $\langle |n|^2 \rangle / |s|^2$ accompanying the Stokes specklon is stabilized at the level

$$\langle |n|^2 \rangle / |s|^2 = R_{ex}/g_0 = \pi g_0 / k \Delta \theta_{eff}^2$$

and does not depend on the cell length L.

5.1.2 Spectral and Angular Distortions

The OPC-SS process has been observed not only for SBS but for a number of other kinds of stimulated scattering as well. In this section our interest will be centered around stimulated Raman scattering (SRS), which is notable for a relatively large frequency shift, $\alpha = (\omega_L - \omega_S)/\omega_S \sim 0.1$ to 0.3. At first sight, OPC-SS with such a large frequency shift seems impossible. Indeed, even fields $E_L^*(r, z)$ and $E_S(r, z)$ of differing frequencies ω_L and ω_S, respectively, which coincide at a certain cross section $z = $ const, become completely mismatched over the distance $\Delta z \sim \lambda / \alpha \Delta \theta_L^2$ due to the relative difference in diffraction.

In the parabolic approximation of (5.1.1, 2) at $G = 0$, this difference consists in changing the z-coordinate scale of one field relative to the other by a factor $k_L/k_S = 1 + \alpha$. That is why, over a sufficiently large interaction length L, i.e., at $L \gtrsim \Delta z$, the diffraction presents an obstacle to the coincidence of the speckle structures of the free-propagating fields. At the same time, matching over the entire length is required for the discrimination mechanism of OPC-SS to work.

This contradiction is eliminated by the fact that the Stokes wave propagates in a medium with an essentially non-uniform gain. If the nonuniformity is pronounced enough, the medium imposes the propagation law dictated by the pump on the field E_S. As a result, the capture of the Stokes wave into spatial resonance with the gain profile takes place. Let us determine the conditions necessary for such a capture. The OPC solution in spatial resonance is amplified e times in comparison with non-resonant configurations over the discrimination length $l_{dis} = (2g_0 - g_0)^{-1} = g_0^{-1}$. If mismatching due to a frequency shift is small over this length, i.e., if $l_{dis} \lesssim \Delta z$, the conjugated configuration of the Stokes wave will continuously adjust its structure to match the pump structure.

Keeping in mind the small parameter $\alpha(\Delta \theta_L)^2 / \lambda g_0$, we proceed to a quantitative consideration of this question which, like the problem of serpentine noise in the previous section, can be solved both in coordinate space and by expanding the wave amplitude into a Fourier series (into plane waves). Here we use the second method in order to acquaint the reader with the application of the Fourier expansion to OPC-SS problems. The solution of (5.1.1) for the pump in the Fourier representation is of the form

$$E_L(r, z) = \sum_\theta \tilde{E}_L(\theta) \exp(ik_L r \cdot \theta + ik_L \theta^2 z/2), \tag{5.1.15}$$

where θ is the angle of inclination to the z axis. We are seeking the Stokes wave solutions correlated with pump inhomogeneities *in the entire volume*, and we therefore represent $E_S(r, z)$ as

$$E_S(r, z) = \sum_\theta \tilde{E}_S(\theta, z) \exp(i k_L \theta \cdot r - i k_L \theta^2 z/2) , \tag{5.1.16}$$

i.e., in the form of an expansion with respect to the elementary waves of the conjugated pump. It should be stressed that at $k_L - k_S \neq 0$ the exponential factor in (5.1.16), where k_L appears, fails to satisfy (5.1.2) without the right-hand side. The identity transformation of (5.1.2) using (5.1.16) gives

$$\frac{\partial \tilde{E}_S(\theta, z)}{\partial z} + i k \alpha \theta^2 \tilde{E}_S(\theta, z)/2 = \frac{G}{2} \sum_{\theta_1, \theta_2, \theta_3} \tilde{E}_L^*(\theta_1) \tilde{E}_L(\theta_2) \tilde{E}_S(\theta_3, z)$$

$$\times \delta^{(2)}(\theta_3 + \theta_2 - \theta_1 - \theta)$$

$$\times \exp[i k(\theta^2 + \theta_2^2 - \theta_1^2 - \theta_3^2)z/2] , \tag{5.1.17}$$

where $\alpha = (k_L - k_S)/k_S$ and k_L is denoted as k for simplicity.

The right side of (5.1.17) describes the scattering of the angular component $\tilde{E}_S(\theta_3)$ by spatial gratings of the gain $\tilde{E}_L^*(\theta_1) \tilde{E}_L(\theta_2)$, see Sect. 5.1.1. Of particular interest are those terms which produce a coherent contribution which does not become zero after averaging over random phases[1] of the pump angular components $\tilde{E}_L(\theta)$. These terms are of two types. The first includes terms with $\theta_1 = \theta_2$, $\theta = \theta_3$ which describe the amplification in an average-intensity field. Secondly, if the phase-conjugated component is present in the Stokes wave, a coherent contribution is also made by terms with $\theta_2 = -\theta_3$, $\theta_1 = -\theta$. The rest of the terms make no coherent contribution to dE_S/dz and are the source of serpentine distortions of the specklon on the right-hand side of (5.1.5). The conditions under which these distortions are small were discussed in Sect. 5.1.1. When these conditions are satisfied, we may neglect the noncoherent terms and obtain

$$\frac{\partial \tilde{E}_S(\theta, z)}{\partial z} + \frac{1}{2}(i \alpha k \theta^2 - g_0) \tilde{E}_S(\theta, z) = \frac{1}{2} G D(z) \tilde{E}_L^*(-\theta) , \tag{5.1.18}$$

$$D(z) = \langle E_L E_S \rangle = \sum_\theta \tilde{E}_L(-\theta) \tilde{E}_S(\theta, z) . \tag{5.1.19}$$

In connection with this, we should like to recall that the equation $\tilde{A}(\theta) = B^*(-\theta)$ for angular components follows from the relation

---

1 The phases of various angular components of the speckle field are independent by virtue of the supposition about statistical uniformity over the cross section: $\langle \tilde{E}_L^*(\theta_1) \tilde{E}(\theta_2) \rangle \propto j(\theta_1)$ $\times \delta^{(2)}(\theta_1 - \theta_2)$

$A(r) = B^*(r)$. With the help of Fourier analysis, we see that (5.1.19) corresponds to the overlap integral $\int E_L E_S d^2 r$ and, in the case of fields which are statistically uniform over the cross section, is equivalent to the ensemble average. When we seek the solution in the form of (5.1.16), where k_L appears in the exponent, it results in an "inconvenient" term $\propto \alpha \theta^2$ on the left-hand side of (5.1.18). This term appears because the elementary exponential waves from (5.1.16) do not satisfy the free diffraction equation for the Stokes field. However, the advantage is that (5.1.18) contains coefficients independent of z.

The solution

$$\bar{E}_S^0(\theta, z) = \frac{G D_0(z)}{2\mu + i\alpha k \theta^2 - g_0} \bar{E}_L^*(-\theta), \quad D_0(z) = D_0(0) e^{\mu z} \tag{5.1.20a}$$

corresponds to the phase-conjugated wave.

At $\alpha \to 0$ it becomes the exact OPC wave with $\mu = g_0$. Multiplying (5.1.20a) by $\bar{E}_L(-\theta)$ and summing over θ (i.e., ensemble averaging), we obtain, at $D_0 \neq 0$,

$$\sum_\theta G |\bar{E}_L(-\theta)|^2 (2\mu - g_0 + i\alpha k \theta^2)^{-1} = 1, \tag{5.1.20b}$$

from which μ can be found. The solution of (5.1.20b) can be represented as an expansion in terms of powers of the small parameter α:

$$\mu = g_0 - i\alpha k \overline{\theta^2}/2 - 0.5 g_0^{-1}(\alpha k)^2 [\overline{(\theta^2)^2} - (\overline{\theta^2})^2] + O(\alpha^3). \tag{5.1.21}$$

The bar denotes averaging over the angular spectrum of the pump intensity, e.g., $\overline{\theta^2} = \int (\theta_x^2 + \theta_y^2) j(\theta) d^2\theta$.

As can be seen from (5.1.20a) this solution does not exactly conjugate the pump. Actually, at $\alpha k (\Delta\theta_L)^2/g_0 \ll 1$, it follows from (5.1.20a, 21) that

$$E_S(r, z) \approx \text{const} \cdot \exp(\mu z) \left[E_L^*(r, z) + i\frac{\alpha}{k g_0} \nabla_\perp^2 E_L^* \right] \tag{5.1.22}$$

$$\approx \text{const} \cdot \exp(\mu z) E_L^* \left(r, z - \frac{\alpha}{g_0} \right) + O(\alpha^3).$$

To characterize the inaccuracy of OPC quantitatively, it is convenient to introduce the conjugation fraction H characterizing the degree of overlap of the fields E_L^* and E_S:

$$H = \frac{|\int E_L(r, z) E_S(r, z) d^2 r|^2}{\int |E_L|^2 d^2 r \cdot \int |E_S|^2 d^2 r}, \quad 0 \leqslant H \leqslant 1. \tag{5.1.23}$$

When $H = 0$, there is a complete absence of correlation, whereas $H = 1$ at $E_S(r) = \text{const} \cdot E_L^*(r)$. The OPC fraction for the specklon of type (5.1.22) differs slightly from unity:

$$H \simeq 1 - \frac{1}{2}\left(\frac{\alpha k}{|g_0|}\right)^2 [\overline{(\theta^2)^2} - (\overline{\theta^2})^2] . \tag{5.1.24}$$

A decrease in the intensity gain $2\,\mathrm{Re}\{\mu\}$ from (5.1.21) compared to the doubled value $2g_0$ is caused just by the incomplete correlation of the interacting fields. It is interesting to note that at real values of g_0, the small distinction between E_L^* and E_S consists in a spatial shift along z by an amount $\delta z \simeq \alpha / g_0$.

This shows that for a pump which is statistically uniform in the interaction volume, the phase-conjugation condition for SRS is of the form

$$\alpha k [\overline{\theta^4} - (\overline{\theta^2})^2]^{1/2} \lesssim g_0 \lesssim k \Delta \theta_{\mathrm{eff}}^2 , \tag{5.1.25}$$

where the second inequality is required for small serpentine distortions.

Up to this point, we have dealt with the relation $E_S(r, z) = sE_L^*(r, z)$ for slowly varying amplitudes. We recall that the total complex amplitude differs from the slowly varying one in that it contains the factor $\exp(-ik_L z)$ for the pump and the factor $\exp(ik_S z)$ for the scattered wave. It is quite obvious that, with allowance for the quickly varying factors, the relation $E_S = s_1 E_L^*$ does not hold true, since s_1 is not constant but depends on z as $\exp[-i(k_L - k_S)z]$. Nevertheless, the transverse structure of the field can be conjugated.

Let us now proceed to another problem of considerable practical interest, the problem of OPC-SS with an angular shift, known as turning. Consider a situation in which it is necessary to turn the field as a whole while still reproducing the transverse structure:

$$E_S(r, z) = E_L^*(r, z) \exp(ik\psi \cdot r - ik\psi^2 z/2) . \tag{5.1.26}$$

For simplicity we have assumed $|k_L| = |k_S| = k$. The vector $\psi = e_x \psi_x + e_y \psi_y$ lies in the plane normal to the central propagation direction e_z and characterizes both the amount and the direction of the angular shift. If E_L is the single plane wave $E_L = \text{const}$, then a field of the form of (5.1.26) satisfies the parabolic equation (5.1.2) at $G = 0$. If, however, the speckle field E_L is formed by angular components with widths of the order of $\Delta\theta_L$, (5.1.26) is valid in free space only a limited interval Δz. In the parabolic approximation such a mismatch is a simple shift of an amount $\Delta r = \psi \cdot z$ in the transverse plane. This results in the limitation $\Delta z \sim a/|\psi| = \lambda/\Delta\theta_L|\psi|$, where $a = \lambda/\Delta\theta_L$ is the transverse size of speckle inhomogeneities. If the turned OPC wave propagates in the presence of the speckle-inhomogeneous pump, the capture of the Stokes wave into spatial resonance may occur at a sufficiently high gain. Calculations similar to those in the OPC-SRS problem result in an approximate OPC solution:

$$E_S(r, z) = \text{const} \cdot \exp(\mu z) E_L^*(r + \psi/g_0, z) \exp(ik\,\psi \cdot r) + O(\psi^2), \quad (5.1.27)$$

$$\mu(\psi) = g_0 - ik(\bar{\theta} \cdot \psi) - \frac{2k^2}{g_0} [\overline{(\psi \cdot \theta)^2} - (\psi \cdot \bar{\theta})^2] - ik\,\psi^2/2. \quad (5.1.28)$$

The specklon with $\psi = 0$, i.e., the configuration exactly conjugating the pump (without tilting), possesses the highest gain. As in SRS, distortions in (5.1.27) consisting in a transverse shift $\Delta r = \psi/g_0$ give rise to a slight mismatch of the scattered and exciting fields and, as a consequence, to a decrease in the gain. In this case, for OPC of the specklon the condition

$$k[\overline{(\psi \cdot \theta)^2} - (\psi \cdot \bar{\theta})^2]^{1/2} \lesssim g_0 \lesssim k\Delta\theta_{\text{eff}}^2 \quad (5.1.29)$$

must hold. If the second inequality in (5.1.29) is well satisfied, i.e., $g_0/k\Delta\theta_L^2 \ll 1$, the largest permissible angle at which the specklon may still exist is given by $\psi \approx \Delta\theta_L \cdot (g_0/k\Delta\theta_L^2) \ll \Delta\theta_L$. Thus, a solution correlated with the pump, i.e., a specklon, proves to be sufficiently stable with respect to both a frequency and an angular shift.

5.1.3 Specklon Envelope Equation

Let the phase-conjugated wave be of the form

$$E_S(r, z) = f(r, z) E_L^*(r, z), \quad (5.1.30)$$

where $f(r, z)$ is the envelope, smooth on the scale of the speckle structure $E_L(r, z)$, see Fig. 5.2. Locally, such a field is amplified, as before, better than waves uncorrelated with the pump. However, inaccurate constancy of the

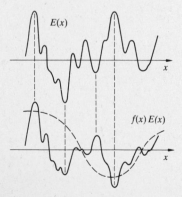

Fig. 5.2. Speckle-inhomogeneous field $E(x)$ and the same field with a smooth envelope, $f(x)E(x)$; the envelope $f(x)$ is shown by the dashed line

spatial dependence of the envelope results, owing to diffraction, in small-structure distortions which affect the amplification rate. The equation for the envelope $f(r, z)$ can be obtained both in the coordinate and Fourier representations. If we write

$$f(r, z) = \int d^2 \psi \tilde{f}(\psi, z) \exp(ik\psi \cdot r), \qquad (5.1.31)$$

then the equation for $f(r, z)$ is almost completely known. From (5.1.27, 28) it follows that

$$\frac{\partial \tilde{f}}{\partial z} = \mu(\psi) \tilde{f}(\psi, z), \qquad (5.1.32)$$

$$\mu(\psi) = g_0 + ik\psi_j \bar{\theta}_j + \psi_j \psi_m [-i\delta_{jm} k/2 - 2k^2 (\overline{\theta_j \theta_m} - \bar{\theta}_j \cdot \bar{\theta}_m)/g_0]. \qquad (5.1.33)$$

It should be noted that the identity $\psi_j \exp(ik\psi \cdot r) \equiv (-i/k)(\partial/\partial x_j) \times \exp(ik\psi \cdot r)$ is valid. As the operations (5.1.31, 32) are linear with respect to f, the use of this identity makes it possible to obtain the equation for $f(r, z)$ in coordinate space by the transformation $\mu(\bar{\psi}) \to \mu(-ik^{-1}\partial/\partial r)$:

$$\frac{\partial f(r, z)}{\partial z} = \left\{ g_0 + \bar{\theta}_j \frac{\partial}{\partial x_j} + \left[\frac{i}{2k} \delta_{jm} + \frac{2}{g_0} (\overline{\theta_j \theta_m} - \bar{\theta}_j \cdot \bar{\theta}_m) \right] \frac{\partial^2}{\partial x_j \partial x_m} \right\} f(r, z). \qquad (5.1.34)$$

To discuss the structure of this equation, let us direct the z axis along the central propagation direction of the pump so that $\bar{\theta}_j = 0$. Also, we let the pump angular spectrum be axially symmetric, so that $\overline{\theta_j \theta_m} = \theta_0^2 \delta_{jm}/2$. Thus, for instance, $\theta_0^2 = \theta_1^2$ for the Gaussian angular spectrum of the pump $j(l) \propto \exp(-\theta^2/\theta_1^2)$. Then (5.1.34) takes the form of the equation of transverse diffusion,

$$\frac{\partial f}{\partial z} - D\nabla_\perp^2 f = g_0 f(r, z), \qquad D = \frac{\theta_0^2}{g_0} + \frac{i}{2k}, \qquad (5.1.35)$$

with the z coordinate playing the role of time and g_0 being the uniform gain of the envelope. The diffusion coefficient has an imaginary part which is small in comparison with its real part, $|\text{Im}\{D\}/\text{Re}\{D\}| = g_0/2k\theta_0^2 \ll 1$, provided serpentine distortions are weak.

We shall apply the above equations to transverse coherentization of a specklon excited by spontaneous noise, and OPC-SS in a focused beam. Here we discuss the limits of applicability of (5.1.34, 35) to the specklon envelope. As has been noted in Sect. 5.1.2, the stable existence of a specklon with a sufficiently small amount of noise requires that the inclination ψ of its envelope should be $|\psi| \lesssim \Delta\theta_L \cdot R_{ex}/g_0 \sim g_0/k\Delta\theta_L$. This condition means that the size

Δr_f of the transverse inhomogeneity of the envelope should not be too small, $\Delta r_f \sim (k\psi)^{-1} \gtrsim (k\Delta\theta_L)^{-1} \cdot (g_0/R_{ex})$, i.e., it should be at least g_0/R_{ex} times larger than the speckle-inhomogeneity size.

5.2 Specklon Phase Fluctuations

Both in laser and SS-active media stimulated radiation begins to develop from the level of spontaneous noise. In the case of SBS this spontaneous noise corresponds to the scattering of the exciting field by hypersonic fluctuation waves. Due to the exponential nature of the amplification, which is $\propto \exp(gz)$, a relatively short layer of the medium located at the beginning of the amplification region ($\Delta l \sim g^{-1}$) gives the main contribution to the amplified Stokes field. The characteristic correlation time of priming fluctuations is $\tau_S \sim \Gamma^{-1}$, where Γ is the scattering linewidth; τ_S coincides with the attenuation time of the hypersonic phonon. Under typical conditions, $\tau_S \sim 10^{-8} - 10^{-9}$ s.

The excitation of SS by spontaneous noise determines a number of important features of OPC-SS. Thus, for instance, the possibilities for OPC extremely weak signals are limited by the level of the amplified spontaneous uncorrelated noise. Below we shall discuss the consequences of excitation of such a character, i.e., fluctuations of the complex amplitude of OPC specklons.

5.2.1 Temporal Phase Fluctuations

The specklon gain factor under linear conditions is given by $\exp[2g(\omega)z]$ where $g(\omega)$ follows, in the main, the contour of the spontaneous scattering line of width Γ. That is why (Sect. 2.5) the correlation time of fluctuations of the amplified radiation increases and appears to be of the order of $\tau_{cor} \sim \Gamma^{-1}$ $\times (2gL)^{1/2} \sim 5/\Gamma$, where $2gL \approx 30$. A five fold reduction in the SS linewidth, compared to the spontaneous contour width, corresponds to this correlation time.

If the Stokes pulse length T_S exceeds $\tau_{cor} \sim 5/\Gamma$, the complex amplitude of the Stokes wave can experience $N \sim T_S/\tau_{cor}$ fluctuations. If, moreover, the effects of SS saturation cannot be neglected, fluctuations of the amplitude modulus are almost completely suppressed, while phase fluctuations remain unchanged.

Specklon phase fluctuations under SS were first reported in [5.2 – 4]. Detailed quantitative investigations of temporal fluctuations of the specklon complex amplitude were described in [5.5 – 7], where the results presented are in good agreement, though recording techniques differ. As an example, we shall dwell on the work of Vasil'ev et al. [5.6]. The laser radiation with a plane wavefront was split by a semi-reflecting mirror into two beams of equal intensity and then directed through two similar phase plates into two cells filled with the same substance. The OPC-scattered radiation was reconstructed by the plates to form plane waves, which were then superimposed onto the semi-reflecting mirror to form a single plane wave with a fluctuating amplitude. The time behavior of the intensity of this wave was registered by a photodetector and displayed on the screen of an oscilloscope.

The OPC-SS process proceeded independently in each cell, and phase difference fluctuations of the reconstructed specklons caused interference intensity fluctuations which were registered by the photodetector. The correlation time of the fluctuations agreed, within experimental accuracy, with the theoretical estimate $5/\Gamma$, where Γ is the spontaneous Brillouin scattering linewidth for the substance employed.

5.2.2 Transverse Coherentization

Of all kinds of original configuration of the scttered field, those for which local maxima coincide with maxima of the pump speckle structure are amplified best. However, it is not difficult to understand that this condition is satisfied not only by the exactly conjugated wave but also by waves like (5.1.30) with envelopes that are smooth on the speckle-structure scale, Fig. 5.2. The OPC component of the field excited by spontaneous sources originally possesses, most probably, an envelope with random spatial fluctuations rather than a smooth one. However, as it propagates, the envelope becomes smoother in shape and its transverse scale increases, that is to say, *transverse coherentization of the envelope* occurs. To describe this process quantitatively, it is convenient to use (5.1.35) for the OPC-specklon envelope. Its solution is of the form

$$f(r, z) = \exp(g_0 z)(4\pi Dz)^{-1}\int d^2r' f(r', z = 0) \exp[-(r - r')^2/4Dz] . \quad (5.2.1)$$

It can be seen that even if the original envelope has considerable random spatial fluctuations, the characteristic transverse scale of correlation increases up to $\Delta r_\perp \sim (Dz)^{1/2}$. The coherentization can be regarded as the result of the preferential amplification of specklons with small inclination angles: $|\psi| \lesssim (2k^2\Delta\theta_L^2 z/g_0)^{-1/2}$. We can write, more accurately,

$$\langle f^*(r_1, z)f(r_2, z)\rangle \approx \text{const} \cdot \exp[2g_0 z - (r_1 - r_2)^2/8Dz] , \quad (5.2.2)$$

where it has been assumed in the integration in (5.2.1) that the initial correlation radius of the envelope is small compared to $(8Dz)^{1/2}$. Then the correlation radius Δr_{cor} at the $\exp(-1)$ level of the correlator maximum is given by

$$\Delta r_{cor} = 2\sqrt{2Dz} = \theta_0 z \frac{4}{\sqrt{2g_0 z}} \approx 0.73\ \theta_0 z \ . \tag{5.2.3}$$

Here we make use of the fact that coherentization occurs only within an exponential amplification length corresponding to a total relative gain of $2g_0 z \approx 30$. Thus, for a speckle beam of diameter D the condition required for OPC a completely coherent envelope, excited by spontaneous noise, is $r_{cor} \approx 0.73\ \theta_0 z > d$. For SS in a light guide, this means that most of the pump rays should experience at least one reflection from the walls.

5.3 Theory of OPC-SS in Focused Speckle Beams

We shall consider a "doubly Gaussian" beam (Sect. 3.4) as a model of the speckle-inhomogeneous focused pump beam. In this case, even the pump intensity averaged over the speckle structure has a non-uniform profile:

$$\langle I_L(r,z) \rangle = I_0 (1 + z^2/z_0^2)^{-1} \exp[-r^2/a^2(z)] \ ,$$
$$a^2(z) = a_0^2 + \theta_0^2 z^2 = a_0^2 (1 + z^2/z_0^2) \ , \quad \xi = k a_0 \theta_0 \ , \tag{5.3.1}$$

where $a(z)$ is the transverse size of the beam in terms of the HWe^{-1}M criterion, the origin of coordinates is placed at the center of the waist, θ_0 is the divergence of the beam at the waist in terms of the same criterion, and $z_0 = a_0/\theta_0$ is the waist length (HWHM). The parameter $\xi \gg 1$ characterizes the excess of the divergence over its diffraction limit, $\theta_{dif} = (k a_0)^{-1}$, and approximately equals the ratio of $a(z)$ to the transverse size $\Delta r_\perp$ of speckles.

As long as the main gain $g(r,z) = G\langle I_L(r,z)\rangle$ is strongly non-uniform, the specklon will be localized mainly near the beam axis where the gain is a maximum. Accordingly, OPC with a non-uniform envelope may be expected, $E_S(r,z) = f(r,z) E_L^*(r,z)$. We should take the equation for the envelope $f(r,z)$ in the form of (5.1.34), which contains the quantities $\overline{\theta_j}$, $\overline{\theta_j \theta_m}$ characterizing the mean values of the pump angular spectrum variables. Using the Van Zittert $-$ Zernike theorem we may write

$$\overline{\theta_j} = (ik)^{-1} \frac{\partial}{\partial (r_1 - r_2)_j} \gamma(r_2, r_1) \Big|_{r_2 = r_1} \ ,$$
$$\overline{\theta_j \theta_m} = (ik)^{-2} \frac{\partial^2}{\partial (r_1 - r_2)_j \partial (r_1 - r_2)_m} \gamma(r_2, r_1) \Big|_{r_2 = r_1} \ , \tag{5.3.2}$$

where $\gamma(r_2, r_1) = \langle E^*(r_2) E(r_1) \rangle [\langle |E(r_1)|^2 \rangle \langle |E(r_2)|^2 \rangle]^{-1/2}$. For a doubly Gaussian beam, $\overline{\theta_j}$ and $\overline{\theta_j \theta_m}$ are continuous functions of the coordinates. Calculations using (3.4.9) give

$$\bar{\theta}_j(r, z) = -\frac{z^2}{z^2 + z_0^2} \cdot \frac{x_j}{z}, \qquad \overline{\theta_j \theta_m} = \bar{\theta}_j \cdot \bar{\theta}_m + \frac{1}{2} \delta_{jm} \theta_0^2 (1 + z^2/z_0^2)^{-1}.$$
$$(5.3.3)$$

Since the envelope $f(r, z)$ will be localized near the beam axis within a distance considerably less than $a(z)$, the expression for $g_0(r, z)$ in (5.1.35) may be replaced by $GI_0[1 - r^2/a^2(z)](1 + z^2/z_0^2)^{-1}$. For that same reason, we should take the value of the diffusion tensor D_{jm} near the axis, $D_{jm}(r, z)$ $\approx \delta_{jm} \theta_0^2 / GI_0 \approx \delta_{jm} D$, the imaginary part of D_{jm} being neglected because we assume that the serpentine noise is low. With all the above approximations, which are valid at $\xi \gg 1$, we may write the envelope equation as

$$\frac{\partial f}{\partial z} + \frac{z}{z^2 + z_0^2} (r \cdot \nabla) f - D \nabla_\perp^2 f - GI_0[1 - r^2/a^2(z)](1 + z^2/z_0^2)^{-1} f(r, z) = 0.$$
$$(5.3.4)$$

Direct substitution shows that functions of the form

$$f(r, z) = \exp[\Gamma_{mn}(z) - r^2/2b^2(z)] \, H_m(x/b(z)) \, H_n(y/b(z)) \qquad (5.3.5)$$

satisfy (5.3.4). Here $H_m(x)$ is the mth Hermite polynomial and $\Gamma_{mn}(z)$ is the index in the exponent of the wave amplitude,

$$\Gamma_{mn}(z) = [GI_0 z_0 - 2(m + n + 1)] \arctan(z/z_0). \qquad (5.3.6)$$

The quantity $b(z)$ gives the effective radius within which the envelope for the solution with $m = n = 0$ is localized:

$$b^2(z) = a^2(z)/(GI_0 z_0). \qquad (5.3.7)$$

If the focal waist is wholly inside the nonlinear medium, $z_1 < z < z_2$, i.e., if $|z_1|, |z_2| \gg z_0$, the total amplitude gain may be obtained from (5.3.6) by substituting π for arctan and multiplying it by 2 (for intensity):

$$2[\Gamma_{mn}(z_2) - \Gamma_{mn}(z_1)] \approx 2\pi[GI_0 z_0 - 2(m + n + 1)]. \qquad (5.3.8)$$

Above the SS threshold, for the specklon with indices equal to zero we must have $2\pi(GI_0 z_0 - 2) \approx 30$, which $GI_0 z_0 \approx 6.8$. It can be easily seen that the gain of the next two solutions, with indices $(1, 0)$ and $(0, 1)$, proves to be weaker by a factor of $\exp(2\pi) \approx 500$ (with respect to intensity), and their contributions may be ignored. This applies even more to solutions with higher indices. This means that for OPC-SS of a speckle beam with its waist wholly inside the medium, the conjugation near the SS threshold will occur with a completely coherent envelope.

Discrimination against higher solutions will be noticeably less, amounting to $\exp\{2(m + n) \cdot [\arctan(z_2/z_0) - \arctan(z_1/z_0)]\}$ if the waist is not wholly inside the nonlinear medium.

For non-conjugated waves $E_S(r, z)$ we may simply average the gain to give

$$\left\{ \frac{\partial}{\partial z} - \frac{i}{2k}\nabla_\perp^2 - \frac{1}{2}GI_0[1 - r^2/a^2(z)](1 + z^2/z_0^2)^{-1} \right\} E_s(r, z) = 0 . \quad (5.3.9)$$

By making a direct substitution, we can check that (5.3.9) has the follwing solutions:

$$E_{pq}(r, z) = (1 + z^2/z_0^2)^{-1/2}\exp\left[\tilde{\Gamma}_{pq}(z) - \frac{r^2}{2c^2(z)} \right] H_p\left(\frac{x}{d(z)} \right) H_q\left(\frac{y}{d(z)} \right) ,$$

$$(5.3.10)$$

$$c^2(z) = \xi^{-1}a^2(z)(\sqrt{1 - iGI_0z/\xi} - iz/z_0)^{-1} , \qquad (5.3.11)$$

$$d^2(z) = \xi^{-1}a^2(z)(1 - iGI_0z_0/\xi)^{-1} , \qquad (5.3.12)$$

$$\tilde{\Gamma}_{pq}(z) = \{\tfrac{1}{2}GI_0z_0[1 - (p+q+1)/\xi] - i\xi(p+q+1)\}\arctan(z/z_0) . \quad (5.3.13)$$

At $GI_0 \to 0$ these modes go over into functions known from the theory of resonators with spherical mirrors. If $\xi \gg 1$, i.e., if the pump divergence is much larger than its diffraction limit, then for the mode with $p = q = 0$ the gain is practically equal to a geometric-optics integral of the amplitude gain along the axis. Incidentally, when the serpentine distortion is small, we can show $GI_0z_0/\xi \ll 1$.

We now discuss discrimination against non-conjugated waves. The ratio of the amplitude gains is

$$\Gamma_{00}/\tilde{\Gamma}_{00} = 2(GI_0z_0 - 1) \cdot (GI_0z_0 - 1/\xi)^{-1}$$

$$\approx 2 - \frac{4}{GI_0z_0} \approx 1.42 \qquad (5.3.14)$$

because of the condition $2\pi(GI_0z_0 - 2) \approx 30$. The intensity of the non-conjugated solution with $p = q = 0$ is localized near the axis, within $\Delta r \sim a_0/\sqrt{\xi} \ll a_0$. Therefore, the gain of this solution is determined by the pump intensity along the beam axis. There are two reasons for a decrease (as compared to the simple Gaussian twofold) in the degree of discrimination. Firstly, the conjugated solution with $m = n = 0$ is localized within a somewhat larger transverse size $\Delta r \sim a_0(1 + GI_0z_0)^{-1/2}$ compared to the non-conjugated one with $p = q = 0$. Due to the decrease in gain with radius, this fact leads to a decrease in the total gain of the specklon. Secondly, because the envelope $f(r, z)$ is not constant, small-structure distortions result and, consequently, a smaller amplitude gain compared to the doubled mean value. This fact is clearly illustrated by the case of inclined (tilted) specklons, see (5.1.27, 28). In terms of mathematics, it manifests itself in the fact that (5.3.4) is of the form of an amplitude diffusion equation with a real-valued coefficient D. These two contributions are approximately equal. In order that the reader's attention should not be overburdened with details, we ventured, in Sect. 4.6, to limit ourselves to an account of the first contribution only, and its strength was purposely doubled to obtain the correct result.

5.4 OPC of Depolarized Radiation

The polarization properties of OPC-SS are of considerable interest. The relation

$$E_S(r, z) = \text{const} \cdot E_L^*(r, z) \tag{5.4.1}$$

corresponds to exact spatial-polarization conjugation. On the other hand, it is known (Sect. 2.8) that in backward SBS with a circularly polarized pump $e_L \propto (e_x + i e_y)/\sqrt{2}$ the Stokes wave has the same polarization unit vector $e_S = e_L$, rather that the complex conjugate. In other words, as far as polarization is concerned, an SBS mirror works as a conventional one rather than a conjugating one.

To discuss the question quantitatively, let us write a system of equations for the vectorial fields $E_L(r, z)$ and $E_S(r, z)$. From the transversality of these fields it follows that $e_z \cdot E_L = e_z \cdot E_S = 0$, and, since SBS is a scalar-type scattering, we have (Sect. 2.8)

$$\left(\frac{\partial}{\partial z} - \frac{i}{2k} \nabla_\perp^2 \right) E_L^*(r, z) = 0 ,$$

$$\left(\frac{\partial}{\partial z} - \frac{i}{2k} \nabla_\perp^2 \right) E_S(r, z) = \frac{1}{2} G(E_L^* E_S) E_L . \tag{5.4.2}$$

In this section we shall be interested in OPC of depolarized pump beams. This problem is, first and foremost, of considerable practical interest in view of the need to compensate inhomogeneities of anisotropic media, such as rods of solid-state amplifiers. It is of no less importance that there is a detailed theory concerning this problem and detailed experiments aimed at the verification of the theory. Using this problem as an example, it is possible to demonstrate vividly that the physics of SS interactions of speckle fields is now understood rather well.

5.4.1 Theory of OPC-SS of Depolarized Radiation

We use the term depolarization to mean inhomogeneity of the polarization vector over the beam cross section and not temporal inconstancy of this vector which is the usual definition. Any transverse field $E_L(r)$ can be resolved into components along an arbitrary basis of mutually orthogonal $(e_1 \cdot e_2^* = 0)$ complex unit vectors e_1 and e_2:

$$E_L(r, z) = e_1 E_1(r, z) + e_2 E_2(r, z) . \tag{5.4.3}$$

It is convenient to choose e_1 and e_2 so that the speckle structures of fields E_1 and E_2 are uncorrelated. Then the degree of polarization p can be specified by

$$\langle (E_L \cdot E_L^*) \rangle = I_L , \quad \langle E_1^*(r, z) E_2(r, z) \rangle = 0 ,$$

$$\langle |E_1|^2 \rangle = I_L(1 + p)/2 , \quad \langle |E_2|^2 \rangle = I_L(1 - p)/2 . \tag{5.4.4}$$

In the general case, where the unit vectors e_1 and e_2 are complex at $0 < p < 1$, we have partial elliptical polarization of a monochromatic wave. The scattered field can be represented as a resolution along the same unit vectors e_1 and e_2 (but not along their conjugates):

$$E_S(r, z) = S_1(r, z) e_1 + S_2(r, z) e_2 . \tag{5.4.5}$$

Substituting (5.4.3, 5) into (5.4.2) gives

$$\left(\frac{\partial}{\partial z} - \frac{i}{2k} \nabla_\perp^2 \right) S_1 = \frac{1}{2} G E_1 (E_1^* S_1 + E_2^* S_2) ,$$

$$\left(\frac{\partial}{\partial z} - \frac{i}{2k} \nabla_\perp^2 \right) S_2 = \frac{1}{2} G E_2 (E_1^* S_1 + E_2^* S_2) . \tag{5.4.6}$$

We shall analyze the right side of (5.4.6) in the usual manner. Since the nonuniformity in gain weakly distorts the propagating fields providing the low serpentine distortion condition is satisfied, then the only terms on the right side of (5.4.6) which lead to effective excitation of waves are those which satisfy the wave equation for a homogeneous medium. Therefore, the first equation in (5.4.6) contains, in addition to the space-averaged amplification of the wave S_1 of the type $\langle |E_1|^2 \rangle S_1$, the coherent excitation of the conjugated components $E_1^*(r) \langle E_1 S_1 \rangle + E_2^* \langle E_1 S_2 \rangle$. Terms of the type $\langle E_1 E_2^* \rangle S_2$ are absent owing to the choice of polarization unit vectors, in agreement with (5.4.4) $- \langle E_1 E_2^* \rangle = 0$. Terms of the type $\langle E_1^* S_1 + E_2^* S_2 \rangle E_1(r, z)$ do not excite, on average, the wave S_1, since the configuration $E_1(r, z)$ (without conjugation) describes diffraction in the direction opposite to the Stokes wave. Applying the same procedure to the second equation in (5.4.6) we obtain

$$\left(\frac{\partial}{\partial z} - \frac{i}{2k}\nabla_\perp^2\right)S_1 = \frac{1}{2}G(\langle|E_1|^2\rangle S_1 + \langle E_1 S_1\rangle E_1^* + \langle E_1 S_2\rangle E_2^*),\quad (5.4.7a)$$

$$\left(\frac{\partial}{\partial z} - \frac{i}{2k}\nabla_\perp^2\right)S_2 = \frac{1}{2}G(\langle S_1 E_2\rangle E_1^* + \langle|E_2|^2\rangle S_2 + \langle E_2 S_2\rangle E_2^*),\quad (5.4.7b)$$

Missing terms on the right side of (5.4.7) are sources of serpentine distortions; their contributions may be neglected if the condition $G|E_L|^2/k\Delta\theta_L^2 \ll 1$ is satisfied.

Consider first Stokes waves which are spatially uncorrelated with both pump components. For such waves $\langle S_1 E_1\rangle = \langle S_2 E_1\rangle = \langle S_1 E_2\rangle = \langle S_2 E_2\rangle = 0$, and, as a result, the z dependence of the form $e \cdot \exp(\mu z)$ is governed by μ:

$$e = e_1, \mu = \tfrac{1}{4}g_0(1+p),$$
$$e = e_2, \mu = \tfrac{1}{4}g_0(1-p),\qquad\qquad\qquad\qquad (5.4.8)$$

where $g_0 = GI_L$. In other words, a given component $e_{1,2}$ of an uncorrelated Stokes wave is sensitive only to the space-averaged intensity of the same polarization component of the pump.

Among correlated solutions, the highest gain is observed for the specklon

$$E_S(r,z) = M_1(r,z) = e_1 E_1^*(r,z)\exp[g_0(1+p)z/2],\qquad (5.4.9)$$

which is the solution of (5.4.7). There exists an analogous correlated solution of the same equations for the second polarization component,

$$E_S(r,z) = M_2(r,z) = e_2 E_2^*(r,z)\exp[g_0(1-p)z/2].\qquad (5.4.10)$$

All the solutions (5.4.8 – 10) have a spatially constant polarization unit vector (e_1 or e_2) and do not interact with a "foreign" polarization component of the pump. Indeed, the first equation in (5.4.7) contains a nontrivial term $E_2^*\langle S_2 E_1\rangle$ responsible for the interaction of different polarization components. This term relates to the coherent scattering of the pump component $e_1 E_1(r,z)$ by that part of the scalar hypersonic grating of $\delta\varepsilon$ which is recorded by the interference of "foreign" polarization components: $e_2^* E_2^*(r,z)$ with $e_2 S_2(r,z)$. A similar term $E_1^*\langle S_1 E_2\rangle$ appears in the second equation of (5.4.7) as well. However, these components failed to work for solutions (5.4.8 – 10), since for them $\langle S_1 E_2\rangle = \langle S_2 E_1\rangle = 0$.

There exist two more correlated solutions of (5.4.7) for which all these properties do not hold true. In these solutions a given polarization component of the Stokes field appears to be correlated with the spatial structure of the orthogonal pump component:

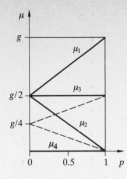

Fig. 5.3. Dependence on the pump polarization degree p of gains increments of different eigensolutions for the Stokes wave. (———) for M_1-M_4, (– – –) are uncorrelated waves

$$E_S(r,z) = M_3(r,z) = [E_1^*(1-p)e_2 + E_2^*(1+p)e_1] \exp(\tfrac{1}{2}g_0 z), \qquad (5.4.11)$$

$$E_S(r,z) = M_4(r,z) = E_1^* e_2 - E_2^* e_1) \exp(0 \cdot z) = [E_L^* \times e_z]. \qquad (5.4.12)$$

An interesting property of the "intricate" solution (5.4.11) is the fact that its amplitude gain is independent of the degree of polarization p of the pump at a given total pump intensity $I_L = \langle (E_L \cdot E_L^*) \rangle$. For this solution the lessening effect of depolarization is exactly compensated by an increase in the amplitude gain due to spatial correlation. The wave (5.4.12) is strictly orthogonal to the pump at all points in space and, therefore, fails to record some gratings and be amplified. The dependence of the amplitude gains for the solution found upon the degree of polarization p of the pump at $I_L = $ const is illustrated in Fig. 5.3.

5.4.2 Experimental Results

A schematic of the experiment of *Blaschuk* et al. [5.2, 3] aimed at verifying the theoretical considerations above is illustrated in Fig. 5.4. The scheme in this experiment includes almost all elements usually employed in OPC-SBS investigations as well as special-purpose polarization devices.

A single-mode neodymium laser beam is transmitted through a Faraday cell F and a polarizer P at whose output a linearly polarized beam is obtained with a controllable direction of polarization. The beam then passes through a diaphragm A, phase depolarizing plate DP, and lens L_1 to enter the cell. The back-scattered beam passes DP and is incident on the elements of the registration system. The angular spectra of the two polarization components are registered at the focal plane of a lens L_2 separately, having been separated by a calcite birefringent prism W_1. The energy of the two polarization components is measured by calorimeters C_1 and C_2; plate B is positioned at the Brewster angle. Photocells S_1 and S_2 provide registration of the SS threshold via a "loop" scheme, see Sect. 2.5.5; a photocell S_3 and calorimeter C_3 register the temporal variation and energy of a laser pulse; W_2 is a Ragulsky wedge.

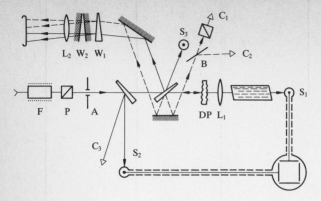

Fig. 5.4. Schematic diagram of the experiment on OPC-SS of depolarized radiation [5.2, 3]. See text for description

This scheme was capable of separating and registering the energy of each of the fields (5.4.9 – 12). The most important element of the scheme was the special phase plate which inserted into the wave both phase and polarization-state inhomogeneities and served as an OPC quality analyzer. The depolarizer was a calcite plate cut parallel to the optic axis and etched in nitric acid. The etching produced small pits with an average depth of $\sim 8 \, \mu m$ and $\sim 250 \, \mu m$ across, this being sufficient to insert a path difference $\geq \lambda$ for orthogonal polarizations ($n_o - n_e = 0.16$). The plate was placed into an immersion oil of refractive index $n \approx (n_o + n_e)/2$. Due to this, the divergences of the beams with orthogonal polarizations were the same at the outlet of the depolarizer and were much smaller than without immersion.

To control the degree of pump polarization within the nonlinear medium, the original linearly polarized plane wave from the laser was directed onto the depolarizer after passing a Faraday cell, which turned the wave polarization vector through a contrallable angle α with respect to the depolarizer axis. The degree of beam polarization in the cell was $p = |\cos 2\alpha|$. The backward scattered radiation passed through the depolarizer and was directed to a system for recording the energy and angular distribution. Presented in Fig. 5.5 are the experimental energy fractions H_1 und H_2 for specklons M_1 and M_2, respectively, as functions of the degree of polarization p of the pump. The dependence of the reciprocal of the pump threshold power on p is shown in Fig. 5.6.

In the case of completely polarized radiation ($p = 1$) only the single OPC specklon M_1 is excited. Practically all the backwardly scattered energy ($95\% \pm 10\%$) is transferred to it. Thus, uncorrelated waves appear to be almost fully suppressed by the discrimination mechanism. With decreasing p, uncorrelated solutions have, as before, an amplitude gain half as large as that for the specklon M_1 and thus are absent in the scattered radiation at all values

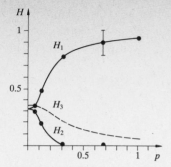

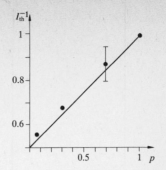

Fig. 5.5. Experimental dependence of the energy fractions H_1, H_2, and $H_3 = 1 - H_1 - H_2$ in specklons M_1, M_2, and solution M_3 on the degree of pump polarization at the input

Fig. 5.6. Experimental dependence of the reciprocal of the SBS threshold power (in relative units, determined by the "loop" scheme) on the degree of pump polarization p at the input

of p. The same is true for the solution M_4 for which the gain is zero. At $p \lesssim 0.5$ the presence of the structure M_3 with a fraction $H_3 = 1 - H_1 - H_2$ becomes noticeable, and at $p \lesssim 0.2$, M_2 is also present. If the pump is completely depolarized within the medium ($\alpha = 45°$), the energy is shared equally, on average, among all three solutions, i.e., $H_1 \approx H_2 \approx H_3 \approx 1/3$. This experimental result is in good agreement with the theoretical prediction that μ_1, μ_2, μ_3 are equal in the case of complete depolarization.

Special measurements showed that phases of different specklons excited by spontaneous sources are uncorrelated and vary randomly from pulse to pulse (for details refer to [5.2, 3]).

In the experiment, the waves M_2 and M_3 arise at a larger degree of polarization p than would follow from the simple formula $\exp(g_i L)$ with g_i given by (5.4.9 – 12). This discrepancy is probably due to SS saturation effects: a stronger polarization component $e_1 E_1$ is depleted more strongly due to the transfer of energy to the specklon M_1, and, accordingly, the effective degree of polarization in the interaction volume decreases.

A great advantage of the loop scheme used in [5.2, 3] for recording a pulse shape, see Sect. 2.5, consists in the possibility of recording the threshold of SS where saturation effects do not manifest themselves. The experimental data on the reciprocal threshold power fit the theoretical dependence $g_1 \propto (1 + p)$ well for the specklon M_1, see Fig. 5.6.

Thus, the experimental results obtained in [5.2, 3] support in detail the specklon picture of OPC-SS. An important conclusion following from the theory of SS of depolarized beams and confirmed by experiment is the phenomenon of nonlinear polarization selection, whereby a specklon whose polarization coincides with the polarization of a more intense component of the exciting radiation has the highest amplitude gain. Because of the exponential dependence of the output intensity on the gain $[\exp(gz) \sim \exp(30)]$, this specklon dominates in the scattered radiation, which, accordingly, appears (at $p \neq 0$) to be completely polarized. One more example of nonlinear selection of this kind will be discussed in the next section.

A second conclusion is the fact that in backward SBS of depolarized radiation complete spatial-polarization OPC cannot be achieved. This is connected both with the nonlinear selection of a stronger component and with the absence of the required phase relations among specklons excited by spontaneous noise.

A scheme was realized in [5.8] in which the original radiation was split into two beams with orthogonal linear polarizations by means of a birefringent prism. Then, one of the beams was transmitted through an element turning its polarization through an angle of 90°, brought together with the other beam, and directed to a nonlinear medium. In this case, the ordinary OPC-SS developed in the field of linearly polarized radiation. Inasmuch as all elements were reversible, the reflected wave was capable, on its backward passage, of conjugating small-structure inhomogeneities of the polarizations state.

5.5 Nonlinear Selection of Non-Monochromatic Radiation

In the case of non-monochromatic exciting radiation, the ratio of the spectral width $\Delta\omega$ to the half-width Γ of the spontaneous scattering line is an important parameter. If the pump linewidth is much less than Γ, then SS may be described in terms of the interaction of monochromatic waves. Let a pump $E_L(t)$ consist of two monochromatic components with frequencies ω_1 and ω_2, respectively, such that $\omega_1 - \omega_2 = \Delta\omega$, and the Stokes radiation consist of frequency-shifted harmonics $\omega_1 - \Omega$, $\omega_2 - \Omega$:

$$E_L(t) = \exp(-i\omega_1 t)[E_{1_L} + e^{i\Delta\omega t}E_{2_L}] ,$$
$$E_S(t) = \exp[-i(\omega_1 - \Omega)t][E_{1_S} + e^{i\Delta\omega t}E_{2_S}] . \tag{5.5.1}$$

A shortened equation for the build-up of a hypersonic wave $\delta\varrho(t) = \tilde{\varrho}(t) \times \exp(i\Omega t) + \tilde{\varrho}^*(t)\exp(-i\Omega t)$ can be written as

$$\frac{d\tilde{\varrho}}{dt} + \Gamma\tilde{\varrho} = -iAE_L^* E_S e^{-i\Omega t}$$
$$\equiv -iA(E_{1_L}^* + e^{-i\Delta\omega t}E_{2_L}^*) \cdot (E_{1_S} + e^{i\Delta\omega t}E_{2_S}) . \tag{5.5.2}$$

From its solution,

$$\tilde{\varrho}(t) = -\mathrm{i}\,\frac{A}{\Gamma}\left[\,(E_{1_\mathrm{L}}^* E_{1_\mathrm{S}} + E_{2_\mathrm{L}}^* E_{2_\mathrm{S}}) + \frac{E_{1_\mathrm{L}}^* E_{2_\mathrm{S}}}{1 + \mathrm{i}\,\varDelta\,\omega/\Gamma}\,\mathrm{e}^{\mathrm{i}\varDelta\omega t}\right.$$

$$\left. + \frac{E_{2_\mathrm{L}}^* E_{1_\mathrm{S}}}{1 - \mathrm{i}\,\varDelta\,\omega/\Gamma}\,\mathrm{e}^{-\mathrm{i}\varDelta\omega t}\right], \tag{5.5.3}$$

it follows that, at $\varDelta\omega \gg \Gamma$, each spectral component of the pump effectively records a density disturbance while interfering with the corresponding mono-chromatic Stokes component. Thus, at $\varDelta\omega \gg \Gamma$, the situation is analogous to the polarization problem, in which a density disturbance was excited only by the interference of waves with coinciding polarizations. Unlike the polarization problem, where the number N of independent temporal components equals 2, this number may be far larger, $N \sim \varDelta\omega/\Gamma$, in the temporal problem.

Continuing this analogy, we note that the non-monochromatic field of the factorized form

$$E_\mathrm{L}(r, z, t) = e_\mathrm{L}(t)E_\mathrm{L}(r, z)\,, \tag{5.5.4}$$

together with the Stokes wave

$$E_\mathrm{S}(r, z, t) = c \cdot e_\mathrm{L}(t)E_\mathrm{L}^*(r, z)\,, \tag{5.5.5}$$

in which the *temporal* structure *is reproduced without conjugation* and *the spatial* structure *is phase-conjugated*, produce an efficient excitation of the resonantly responding hypersonic wave

$$\delta\varrho(r, z, t) = -\mathrm{i}A\,c\,[E_\mathrm{L}^*(r, z)]^2 \int_{-\infty}^{t} |e_\mathrm{L}(t')|^2 \exp[-\Gamma(t - t')]\,dt'\,. \tag{5.5.6}$$

The scattering of the pump by this density disturbance gives rise to the ampli-fication of the Stokes wave (5.5.5) with a doubled gain. It should be under-lined that any rapid oscillations of the phase of the function $e_\mathrm{L}(t)$ have no influence, according to (5.5.6), on the efficiency of excitation of the Stokes wave given by (5.5.5).

An arbitrary non-factorized pump field $E_\mathrm{L}(r, t)$, incoherent both in space and time, may be always represented as an infinite series expansion

$$E_\mathrm{L}(r, t) = \sum_i \alpha_i e_i(t)\varepsilon_i(r)\,, \tag{5.5.7}$$

where the $e_i(t)$ are functions normalized and orthogonal over an interval $T \lesssim \Gamma^{-1}$, such that $\varDelta\omega \cdot T \gg 1$. The uniqueness of such an expansion, as in the polarization problem, is achieved by requiring the structures $\varepsilon_i(r)$ to be orthogonal in space. These structures may be considered normalized if the rest of the factors are included into the coefficients α_i. Such an expansion is called the Karunen-Loev expansion.

Let us seek the solution for the Stokes field at $\Delta\omega \gg \Gamma$, as in the polarization problem, in the form

$$E_S(r, z, t) = \sum_i f_{ik}(z) e_i(t) \varepsilon_k^*(r, z) . \qquad (5.5.8)$$

Using the biorthogonality of the expansion basis, neglecting corrections of the order of $\Gamma/\Delta\omega$, and considering $\varepsilon_k(r, z)$ to be speckle fields obeying Gaussian statistics, we obtain a system of equations[2]

$$\frac{df_{ik}}{dz} = \frac{1}{2} G(|\alpha_i|^2 f_{ik} + \alpha_i \alpha_k^* f_{ki}) . \qquad (5.5.9)$$

We denote the number of terms in expansion (5.5.7) by N. Then f_{ik} is an $N \times N$ matrix, and the system (5.5.9) has N^2 linearly independent solutions. All N^2 eigensolutions [i.e., in the form $\exp(\mu z)$] of the system can be found and written explicity. First, there are N solutions of the form

$$f_{ik} = \delta_{im} \delta_{km}, \quad \mu_m = G|\alpha_m|^2, \quad m = 1, \dots, N . \qquad (5.5.10)$$

Then, there are $N(N-1)/2$ "intricate" solutions with the matrix

$$f_{ii} = 0, \quad f_{ik} = \frac{\alpha_i}{\alpha_k} f_{ki} \text{ for } i \neq k, \quad \mu_{ik} \equiv \mu_{ki} = \frac{1}{2} G(|\alpha_i|^2 + |\alpha_k|^2) , \qquad (5.5.11)$$

and the same number of non-amplified solutions with the matrix

$$f_{ii} = 0, \quad f_{ik} = -\left(\frac{\alpha_k}{\alpha_i}\right)^* f_{ki} \text{ for } i \neq k, \quad \mu = 0 . \qquad (5.5.12)$$

The case of $N = 2$ identically coincides, both in terms of the system of equations and system of modes, with the polarization problem, see Sect. 5.4.

On the grounds of the solutions found, we can conclude that, at $\Delta\omega \gg \Gamma$ and near the SS threshold, the process of nonlinear selection should lead to the factorized OPC component $e_{i_0}(t) \varepsilon_{i_0}^*(r)$ represented in the pump with the greatest weight $|\alpha_{i_0}|^2$ being predominant in the scattered field. The importance of this conclusion can be illustrated by the problem of OPC of a weak signal entering a powerful amplifier. The latter amplifies the power of a signal directed to a nonlinear medium and simultaneously adds to the signal the amplified spontaneous radiation whose total power may considerably exceed the power of the amplified signal. Nevertheless, the latter retains its factorized

---

2 Summation with respect to repeated indices is not to be carried out in the formulas of this section

form, and noise is distributed among many components of the expansion (5.5.7). Therefore, because of nonlinear selection it is possible to find conditions under which the noise is not reflected and the signal experiences spatial OPC.

Unfortunately, the selection efficiency is high only if pump saturation is neglected. In the framework of the model considered, each factorized Stokes specklon of the form (5.5.10) takes energy only from the respective pump component. That is why, under conditions of noticeable saturation, the distribution of $|\alpha_i|^2$ in the bulk of the medium becomes equalized and the selection efficiency diminishes. Such a tendency for $N = 2$ in the polarization problem was registered experimentally, see Sect. 5.4.

Another limiting case of OPC-SS is also of interest, namely, when the spectral width $\Delta\omega$ of the spatially incoherent pump (5.5.7) is less than or of the order of the linewidth Γ. It is obvious that at $\Delta\omega \ll 1$ (say, $\Delta\omega < \Gamma/30$) the OPC-SS process proceeds quasi-statically, with the phase conjugation of the pump transverse configuration occurring at every instant:

$$E_S(r, t) \propto E_L^*(r, t) . \tag{5.5.13}$$

The nontrivial results of recent theoretical and experimental model investigations [5.9, 10] show that such quasi-static OPC is maintained up to $\Delta\omega \sim \Gamma$.

5.6 Effect of Saturation in OPC-SS

In the vast majority of experiments the SS process proceeds with a noticeable depletion of the pump due to energy transfer to the Stokes wave. Indeed, the pump intensity range within which SS can be observed, before the onset of saturation, is rather narrow: $20 \lesssim G|E_L|^2z \lesssim 30$. Also, reaching a high reflectivity in OPC-SS automatically means that the SS process occurs with deep saturation condition. Saturation at backward SS was theoretically considered, but irrespective of OPC, in Sect. 2.4.

The system of wave equations taking depletion into account is of the form (2.1.1, 2), the former being modified as

$$\frac{\partial E_L}{\partial z} + \frac{i}{2k} \nabla_\perp^2 E_L(r, z) = \frac{1}{2} G|E_S(r, z)|^2 E_L(r, z) . \tag{5.6.1}$$

At the input of the medium ($z = L$) the pump field has a developed speckle structure. In the course of pump depletion by an *inhomogeneous* Stokes wave along z, both the average *intensity* and the *structure* of the pump *vary*. Hence, we cannot consider that the pump speckle structure diffracts in accord with the laws of a homogeneous medium. The Stokes wave is excited by spontaneous noise and, therefore, obeys Gaussian statistics. Due to its interaction with the pump, the Stokes wave varies its structure in addition to being amplified. It is essential for further consideration that it is quite reasonable to regard the fields as retaining Gaussian statistics in the bulk of the volume, providing serpentine distortions are low. Then, methods of specklon theory may be applied to the fields, and (2.1.2, 5.6.1) can be represented as

$$\frac{dI_S}{dz} = \frac{dI_L}{dz} = GI_L I_S(1+H) , \qquad (5.6.2)$$

$$\frac{dH}{dz} = G(I_L + I_S)H(1-H) , \qquad (5.6.3)$$

$$I_{L,S}(z) = \langle |E_{L,S}(r, z)|^2 \rangle , \qquad H(z) = \frac{|\langle E_L E_S \rangle|^2}{I_L I_S} . \qquad (5.6.4)$$

We note that $H(z)$ coincides with the fraction of OPC, understood in the usual sense, but at the output section $z = L$. At all other sections $H(z)$ is the coefficient of overlap between the Stokes wave and the pump field actually present at a given section rather than with the original pump field. At $H = 1$, (5.5.8) describes SS of mutually conjugated fields, and the effective coefficient of interaction is $2G$ (the so-called Gaussian 2). At $H = 0$, spatial resonance waves disappears and energy exchange among uncorrelated waves takes place, i.e., the interaction with the coefficient G. From (5.5.9), it follows that dH/dz becomes zero both at $H = 0$ (fields remain uncorrelated over the entire interaction length) and at $H = 1$ (precise OPC in all cross sections).

As a matter of fact, (5.6.2) represents a system of two equations for I_L and I_S, with the exact integral

$$I_L(z) - I_S(z) = \text{const} = I_L(L) - I_S(L) = I_L(L)(1-R) \qquad (5.6.5)$$

in which R denotes the reflectivity (with respect to intensity). There are then only two equations remaining, for $I_L(z)$ and $H(z)$, say. Introducing the designation

$$y(z) = GI_L(L) \int_z^L [1 + H(z')] dz' , \qquad (5.6.6)$$

we obtain the solution in the form

$$I_L(z) = I_L(L) \cdot (1-R) \cdot \{1 - R \exp[-y(1-R)]\} ,$$
$$I_S(z) = I_L(z) - I_L(L)(1-R) , \qquad (5.6.7)$$

$$H(z)[1 - H(z)]^{-2} = \text{const} \cdot I_L(z) I_S(z) . \tag{5.6.8}$$

The constant in (5.6.8) can be determined from the boundary condition for $H(z = 0)$ which characterizes, at the cross section $z = 0$, the value of the Stokes signal projection on the pump field which reaches that cross section. For a typical situation of OPC due to spontaneous noise, $H(0) \ll 1$. A detailed examination of the behavior of the functions $I_{L,S}(z)$ and $H(z)$ at different values of the dimensionless parameters $G L I_L(L)$ and $H(0)$ requires numerical calculation. The relation between the OPC quality, i.e., $H(L)$, and the reflectivity R is of particular interest. This relation can be derived directly from (5.6.7, 8)

$$H(L) = 1 - b^2/2 - (b^2 + b^4/4)^{1/2} ,$$
$$b = [I_S(0)(1 - R)/I_L(L) R H(0)]^{1/2} , \tag{5.6.9}$$

and we have $H(L) \approx 1 - b$ at $b \ll 1$, $H(L) \approx b^{-2}$ at $b \gg 1$. The condition for high-quality OPC, i.e., for sufficient discrimination against amplified uncorrelated noise, under saturation conditions is $b \ll 1$. For comparison with discrimination under linear conditions (Sect. 4.2) we note that $H(z = 0) \approx (\theta_{\text{dif}}/\theta_n)^2$, where $\theta_{\text{dif}} \sim \lambda/D$ is the diffraction angle and θ_n^2 is the solid angle filled by the noise being amplified. In addition, $I_S(0)$ includes the contribution of the whole of the amplified angular noise spectrum, and the fraction of the pump which is scattered into the exactly conjugated configuration (i.e., into the solid angle θ_{dif}^2) under typical conditions amounts to about $\exp(-30)$. Finally, $I_L(0) \approx I_L^{\text{thr}} \approx I_L(L)(1 - R)$ and the discrimination condition, with account for saturation, can be written as

$$1 - H \approx b \approx \exp(-15) \left(\frac{\theta_n}{\theta_{\text{dif}}}\right)^2 \sqrt{\frac{(1 - R)^2}{R}} \lesssim 1 . \tag{5.6.10}$$

The factor $\exp(-15) \cdot (\theta_n/\theta_{\text{dif}})^2$ is a small parameter determining the discrimination quality under linear conditions (without saturation).

Expression (5.6.9, 10) represent the major result of this section and make it possible to draw the following conclusion. With increasing reflectivity, i.e., passing into a region of a deeper saturation, the degree of noise discrimination increases, and the discrimination condition is facilitated by a factor of $(1 - R)$. Thus, in the framework of this approach, saturation fruitfully contributes to discrimination against noise excited by uncorrelated spontaneous sources.

5.7 OPC-SS with Reference Wave

Chapter 2 is devoted to the SS interaction of ideal waves which have no transverse structure. In contrast, Chaps. 3 – 5 deal with fields with a strongly developed transverse structure, i.e., with speckle fields. In practice we may encounter situations in which the pump field is a coherent superposition of these two kinds of fields.

Transmitting a plane wave through an amplitude mask with an amplitude transmission factor $t(r) = [T(r)]^{1/2} \geqslant 0$ allows noticeable fraction of the energy of the non-diffracted component in the transmitted wave to be retained:

$$\frac{W_{\text{plane}}}{W_{\text{trans}}} = \frac{[\int t(r)\,d^2 r]^2}{S\int t^2(r)\,d^2 r} \tag{5.7.1}$$

where S is the cross-sectional area. In particular, if the mask consists only of places with $t = 0$ or $t = 1$, then

$$\frac{W_{\text{plane}}}{W_{\text{trans}}} = \frac{S_1}{S}, \tag{5.7.2}$$

where S_1 is the area of the transparent portion of the mask.

If the plane wave is distorted with the aid of a phase mask, $t(r) = \exp[i\phi(r)]$, then the transmitted wave also contains a plane wave component at a moderate phase modulation depth. If $\phi(r)$ is a random function obeying Gaussian statistics, then we have

$$\frac{W_{\text{plane}}}{W_{\text{trans}}} = \exp[-\langle(\phi - \langle\phi\rangle)^2\rangle] . \tag{5.7.3}$$

This formula can be easily obtained by using the characteristic function of a random quantity described by Gaussian statistics, see Sect. 3.1. At small values of $\langle\Delta\phi^2\rangle$, (5.7.3), which may be reduced to $W_{\text{plane}}/W_{\text{trans}} \approx 1 - \langle\Delta\phi^2\rangle$, is valid without the supposition that $\delta\phi(r)$ obeys Gaussian statistics.

If the plane wave is transmitted through a medium with weak inhomogeneities, it will also contain a nondistorted portion to which inhomogeneous components will be added. In this section we consider the theoretical and experimental results on SS of such a field.

5.7.1 SS of Radiation with Incomplete Speckle Modulation

Let the pump field consist of a plane wave and a speckle part:

$$E_L(r, z) = E_0 + E_1(r, z) . \tag{5.7.4}$$

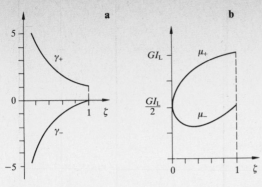

Fig. 5.7. (a) Dependence of coefficients γ_+ and γ_- and (b) of the corresponding amplitude gains μ_+ and μ_- on the parameter ζ for SS in the field of an incomplete speckle-modulation pump

We shall define the energy fraction of the speckle component by the parameter $\zeta = \langle |E_1|^2 \rangle / I_L$, where $I_L = |E_0|^2 + \langle |E_1|^2 \rangle$ is the total intensity of the pump.

Equation (5.1.2) for the Stokes wave is linear. Assuming the serpentine distortions are small, we may consider separately the waves which are uncorrelated with the pump, such that $\langle E_S E_1 \rangle = 0$, and which have no plane wave component conjugated with respect to E_0. For such waves, the amplitude dependence of the gain is given by the factor $\exp(G I_L z / 2)$, where $G I_L$ is the space-averaged gain coefficient. It can be easily obtained by averaging the right side of (5.1.2) over the ensemble of the speckle field $E_1(r)$, providing $E_1(r)$ and $E_S(r)$ are independent.

Let us seek the Stokes waves correlated with the pump in the form

$$E_S(r, z) = a_0(z) E_0^* + a_1(z) E_1^*(r, z) . \tag{5.7.5}$$

Separating, on the right side of (5.1.2), the terms coherently exciting the plane phase-conjugated wave E_0 and the conjugating specklon $E_1^*(r, z)$, we obtain

$$\frac{da_0}{dz} = \frac{1}{2} G I_L (a_0 + \zeta a_1) ,$$

$$\frac{da_1}{dz} = \frac{1}{2} G I_L [(1 - \zeta) a_0 + (1 + \zeta) a_1] . \tag{5.7.6}$$

This system gives two exponential solutions, see Fig. 5.7:

$$E_\pm(r, z) = [E_0^* + \gamma_\pm E_1^*(r, z)] \exp(\mu_\pm z) ,$$

$$\gamma_\pm = \pm \sqrt{\zeta^{-1} - 3/4} - 1/2 , \quad \mu_\pm = \tfrac{1}{2} G I_L [1 + \zeta(1/2 + \sqrt{\zeta^{-1} - 3/4})] . \tag{5.7.7}$$

It is convenient to introduce fractions $\zeta_\pm$ characterizing the conjugated speckle part in the solutions,

$$\zeta_\pm = \zeta |\gamma_\pm|^2 (1 - \zeta + \zeta |\gamma_\pm|^2)^{-1} .$$

We now discuss the solutions obtained. At $\zeta \to 1$ the pump consists almost only of the speckle field; then, for the larger value μ_+ we have $\mu_+ \approx GI_L$ and the solution corresponds to the pure phase-conjugated speckle wave: $\gamma_+ \to 1$, $\zeta_+ \to 1$. The solution with the smaller amplitude gain μ_- describes the plane wave E_0^* which is uncorrelated with the speckle pump, $\gamma_- \approx (\zeta - 1) \to 0$, $\zeta_- \to 0$, and has an average gain, $\mu_- = (1/2)GI_L$. Another limiting case, $\zeta \to 0$, is of greater interest. Here $\mu_\pm \approx \frac{1}{2}GI_L(1 \pm \sqrt{\zeta})$, $\gamma_\pm \approx \pm 1/\sqrt{\zeta}$, $\zeta_+ \approx \zeta_- \approx 1/2$. If $\zeta = 0$, the pump consists of the pure plane wave, and in the presence of the spatially constant gain coefficient all configurations are amplified equally well. However, at very small $\zeta > 0$, $\mu_\pm$ differ sharply from their mean values. This fact can be easily understood in the light of the example from Sect. 4.7. When the energy fraction in the speckle part is small, say $\zeta \sim 0.04$, its interference with the plane wave results in large spatial variations of the gain, $\delta g/g \approx \pm 2\sqrt{\zeta}$, reaching $\pm 40\%$ or 80% on both sides of the mean value. Field maxima in the wave E_+ fall on pump maxima, whereas those in the wave E_- fall on pump minima. The value of γ_+ satisfies the inequality $\gamma_+ > 1$, and at $\zeta \to 0$ we have $\gamma_+ \to \infty$. This means that the fraction of speckle component in the scattered field E_+ is markedly higher than in the pump, i.e., the contrast of the image formed by the interference of E_0 with $E_1(r)$ is higher in the scattered field E_+.

The availability of the regular component E_0 ("reference wave") in the pump causes a decrease in the degree of discrimination, $2\mu_+/GI_L < 2$ and at first sight, this should make the OPC quality deteriorate. However, the scheme with a reference wave may be advantageous in a number of cases, see Sects. 5.7.2, 4. The reference wave can be transmitted via a light guide without being distorted and then conjugated separately (for example, in low-threshold SS of an ideally focused beam). Because of this, the signal coming to the input of the main interaction volume consists of the conjugated reference wave. The latter excites only waves E_+ and E_- containing no uncorrelated components. Therefore, in this case the discrimination condition for suppression of the single solution E_- turns out to be considerably easier, $(\mu_+ - \mu_-)L \gtrsim 1$.

5.7.2 SS with Reference Wave Through Focusing with a Lens

If the field (5.7.4) is focused by a lens into a nonlinear medium, its structure immediately behind the lens will be of the form

$$E_L(r, z) = \exp(-ik_L r^2/2F)[E_0 + E_1(r, z)], \qquad (5.7.8)$$

corresponding to a wave with incomplete spatial modulation. As the field propagates, the component E_0 will be sharply focused to a diffraction-size waist: $a_\perp \sim \lambda D/F$, $a_\parallel \sim \lambda D^2/F^2$, where F is the focal length and D is the beam diameter on the lens, see Fig. 5.8. Unlike the component E_0, the speckle part is focused to a waist ξ times greater in all three coordinates, ξ being the ratio

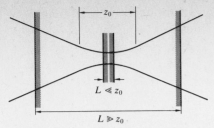

Fig. 5.8. Lens focusing of a pump beam, containing a diffraction-quality non-distorted component and wide-angle components, into an SS-active medium of thickness L. Two possible cases are shown: $L \gg z_0$ and $L \ll z_0$, where z_0 is the waist length of the wide-angle components. At $L \ll z_0$ only the regular component is scattered (nonlinear selection); at $L \gg z_0$ wide-angle components are phase-conjugated and "emphasized" in intensity

of the irregular divergence $\Delta\theta_1$ of the speckle field to the diffraction divergence λ/D of the reference wave. As a result, even if the regular component possesses a relatively small fraction of the input beam energy, $(1 - \zeta) \lesssim 1$, then its intensity within the focal waist will be ξ^2 times larger than the speckle-part intensity due to the sharp focusing. Since the SS threshold is determined by the integral of the intensity extended along the length, the threshold for the regular part of the beam is about a factor of ξ lower than for the speckle part. That is why, in experiments on SS with a reference wave focused by a lens, the Stokes component is usually excited from the bright focal waist.

The geometry of experiments determines various scattering regimes. If a medium is short, $L \ll F^2 \Delta\theta_1/D$ so that its length is less than the length of the focal waist for the speckle part, the fields E_0 and E_1 practically do not overlap in the nonlinear medium. As a result, the regular component is scattered independently. In this case, conjugation of the speckle part is impossible since the discrimination condition against uncorrelated waves is not satisfied. Moreover, the power density in this part of the beam is usually lower than its threshold value and the speckle part of the beam is not scattered at all. The nonlinear selection of the regular spatial component of the signal then takes place. This may be essential when it is necessary to conjugate a highly directional weak signal and avoid the conjugation of parasitic speckle components $E_1(r)$ arising due to the broad-angle superluminescence of the amplifier. The signal selection in such a geometry may be successful even if the noise energy is far larger than the energy of the amplified signal; it is important that the brightness of the signal should be markedly higher than that of the noisy component. We note that such nonlinear selection will also occur if the medium is long but the lens focus is positioned not in the center of the medium but immediately at the cell input window.

Another case is observed if $L \gg F^2 \Delta\theta_1/D$. The focused speckle field is well mixed with the focused regular component at distances z from the focus larger than or of the order of the length $z_0 \sim F^2 \Delta\theta_1/D$ of the speckle-field waist. In the region where both components are mixed, SS proceeds under the regime corresponding to a pump with incomplete spatial modulation. As a result, in this geometry, first, the excitation of the component conjugating the pump speckle part will occur. Second, in the absence of saturation the relative energy of the speckle component will increase (an increase in the image contrast). Third, the component E_1^* is excited not by uncorrelated primings but by the conjugated reference wave scattered in the diffraction-quality waist. Owing to this, the discrimination conditions will be easier to fulfil, $\int(\mu_+ - \mu_-)\,dz \gtrsim 1$, where the integration is extended along the length within which the reference and speckle beams are mixed.

In the intermediate case the conjugation is experienced only by those weakly deviating angular components of the speckle part which move together with the reference wave, i.e., within the angle $\Delta\theta \lesssim LD/F^2$.

A large number of experiments on OPC-SS of plane waves transmitted through amplitude masks have been conducted by *Sokolovskaya* with co-workers [5.11 – 13] since 1976. The SRS component of the scattered radiation was mainly recorded in their experiments, which we shall discuss in Sect. 5.8.

5.7.3 OPC-SS of Under-Threshold Signals

Suppose that it is necessary to conjugate radiation whose intensity is lower than the SS threshold. This can be achieved by mixing the signal to be conjugated with another wave (reference one) whose power exceeds the SS threshold and then by realizing OPC-SS of the total field obtained. In fact, such a situation was considered in Sect. 5.7.2 where an ideal component was taken as the reference wave with over-threshold power. Another scheme for OPC of under-threshold signals making use of the interaction in a light guide can be advantageous in many respects.

If the reference wave possesses a speckle structure, the total field of the reference wave and a signal may be considered as a single pump speckle wave; the task of conjugating it comes down to that considered above.

Nontrivial situations arise if the time behavior of the reference wave differs from that of the signal. *Efimkov* et al. [5.14] directed the under-threshold signal being tested to a nonlinear medium at the moment when the OPC-SS of the reference wave had already reached its steady state. It was shown experimentally and theoretically that if the frequency of the reference wave coincided with that of a signal wave, the reflectivity of the latter reached the steady-state value equal to the reflectivity of the reference wave for a time $\tau \sim \Gamma^{-1}$. Here $\Gamma\,[\mathrm{s}^{-1}]$ is the Lorentz width of the SBS line. In [5.15] the duration of the under-threshold signal was approximately the same as that of the reference wave, and the signal frequency shift was varied. The theoretical and experimental investigations reported here showed that the effective OPC reflection of the signal is observed in a narrow spectral region; the difference between the reference and signal wave frequencies should not exceed the spontaneous scattering linewidth Γ.

The nontriviality of the results obtained in these two experiments consists in the fact that the build-up time of SBS due to spontaneous noise (if the SS threshold is exceeded moderately) is approximately $2gz = 30$ times larger than the value of Γ^{-1}. Also, the spectral width of radiation scattered under SS conditions is approximately $\sqrt{2gz} \approx 5$ times narrower than the linewidth Γ. The real response time of the system was considerably less, $\tau \sim \Gamma^{-1}$. The reason for such a quick response to a weak signal is as follows. It is convenient to examine the process in terms of four-wave mixing. The under-threshold signal E_3 interferes with the conjugated reference wave $E_2 = E_S \propto E_L^*$ and records the traveling hypersonic grating $E_3^* E_2$ localized near the input window of the cell where $E_2 \equiv E_S$ reaches a noticeable intensity. Recording this grating takes only the local build-up time $\tau \sim \Gamma^{-1}$ of the steady-state amplitude of the resonant hypersound.

This method of OPC of weak signals is often referred to as the thresholdless OPC-SS scheme.

5.7.4 OPC-SS of Large-Diameter Beams

If the diameter of a light guide has to be large for some reason, the discrimination condition against uncorrelated waves fails in the case when OPC-SS of the speckle pump is primed by spontaneous noise. To facilitate discrimination, it is quite reasonable to increase the projection of the input signal on the conjugated configuration. With this end in view, *Basov* et al. [5.16] suggested and realized the following solution. A coherent plane wave was mixed with the field which needed to be conjugated, the latter being directed into a light guide at a relatively steep angle. This plane wave passed through the light guide without touching its walls. Then it was focused by a lens into an additional cell containing the same substance, within which backward SBS with phase conjugation developed.

In this situation, the priming signal in the basic cell is not spontaneous noise with a small (of the order of $\lambda^2/D^2 \Delta \Theta_{nc}$, where $\Delta \Theta_{nc}$ is solid angle) projection on the conjugated wave but a plane-wave signal conjugated with respect to the input reference wave. If the power of the reference wave is $(1 - \zeta)$ times the power of the basic speckle wave which is to be conjugated, $|E_0|^2 \cdot (|E_0|^2 + |E_1|^2)^{-1} = 1 - \zeta$, the projection on the conjugating solution E_+ (5.7.7) is of the order of $(1 - \zeta)$ for this way of priming. Even when the reference wave has a small relative intensity, say $(1 - \zeta) \sim 5\%$, the projection of about 5% of the reflected reference wave on the conjugating solution is many times larger than the projection $\lambda^2/D^2 \Delta\Theta_{nc} \lesssim 10^{-6}$ when the excitation is due to spontaneous noise. As a result, the discrimination condition against non-conjugated waves appears to be satisfied. It should be noted, however, that the intensity I_{pr} of the priming reference wave should not be too large, since otherwise the total gain in the basic cell would be small, $2g_0L \approx \ln[I_L/(1-\zeta)I_{pr}]$, and the discrimination would decrease.

5.8 OPC by Other Types of SS

The discrimination mechanism of OPC-SS is based on the coincidence of local maxima in the exciting and amplified fields. This mechanism will work, in principle, for any kind of amplification if the local gain increases with increasing local intensity of the speckle-inhomogeneous pump. Up to now, discrimination in OPC has been studied, in addition to the "classical" case of SBS, for the following amplification mechanisms: stimulated Raman scattering (SRS), stimulated Rayleigh-wing scattering (SRWS), stimulated temperature scattering due to absorption (STS), and amplification at the speckle-inhomogeneous pump of a dye solution under superluminescence.

SRS and dye superluminescence are characterized by a large frequency shift, up to a few times ten per cent. This fact hinders their practical use in laser systems. However, the study of these two mechanisms is of scientific interest. The first experimental investigations of OPC-SRS were performed in [5.11, 17] which were followed by a large number of other publications, e.g., [5.12, 13, 18]. In [5.17], the phase plate technique was used to detect OPC. A specific feature of [5.11 – 13] and other work by *Sokolovskaya* and co-workers is the use of amplitude – (not phase –) distorting masks and the focusing of radiation into a medium without using a light guide. In [5.11 – 13] they describe the reconstruction mechanism of an image carried by the irregular part of the Stokes wave in terms of reading out a volume amplifying hologram by means of a spherical Stokes wave emerging from a diffraction-quality waist. In this case they mean that the hologram is recorded by the interference of the regular part of the pump with its irregular part. The presence of OPC was judged by the reconstruction of the mask image by the wave at the Stokes frequency.

It is interesting to detect and measure quantitatively longitudinal shifts of the image and changes in its transverse scale. Both the shift and scale change are connected with the difference in the laws of propagation for the pump and the Stokes wave in air where they do not interact. The measurement results illustrate a "detachment" of the Stokes wave field from the pump field at the boundary of the nonlinear medium.

OPC-SS of ultrashort (25 ps) pulses through SRWS is reported in [5.13]. This kind of SS has a relatively small frequency shift and a short build-up time, so OPC by such a mechanism can find practical applications, including the problem of depolarized fields.

OPC by STS was observed and studied in [5.19 – 21] and subsequently by many others. STS possesses a very small frequency shift (which is favorable to applications) but, unfortunately, a fairly long build-up time. A striking feature of OPC-STS is that it is accompanied by a noticeable broadening of the scattered wave spectrum. This fact is likely to be caused by phase modulation through thermal variations in the refractive index of the heated medium [5.19].

5.9 Literature

The theory of OPC-SS has been advanced in a large number of publications. The importance of the condition that the "serpentine"-distortion parameter $g/k\Delta\theta^2$ be small in order to maintain the field spatial structure in a non-uniformly amplifying SS-active medium is discussed in [5.22 – 24]. A quantitative estimate of "serpentine" distortions of the specklon is given in [5.1] and, for a number of particular cases, in [5.25 – 27]. Spectral distortions of the specklon are considered in [5.28] with particular cases being discussed in [5.29, 30]. Angular distortions and the equation for the specklon envelope were obtained in [5.31]. Specklon phase fluctuations have been studied experimentally in [5.2 – 7]. The picture of specklon transverse coherentization is studied theoretically in [5.31].

The theory of OPC-SS in focused speckle beams is considered in [5.32 – 35], and that of OPC-SS of spatially depolarized radiation in [5.36]. Experimental investigations of the effect of depolarization on OPC-SS are reported in [5.2 – 4, 8]. Nonlinear selection of non-monochromatic radiation in OPC-SS is considered theoretically in [5.37, 38], while [5.9, 10, 39, 40] are devoted to the study of OPC-SS of spatially incoherent radiation. The effect of saturation on OPC-SS of speckle fields is reported in [5.41 – 43] and the theory of OPC-SS of beams with incomplete spatial modulation is given in [5.44]. OPC-SS of beams transmitted through an amplitude mask with a subsequent focusing by a lens is investigated in detail in [5.11 – 13, 45 – 47] and references therein. The interpretation of OPC-SS of such beams was suggested here in terms of the holographic scattering of the reference Stokes wave produced within the focal waist the non-distorted beam. The method of OPC-SS of an under-threshold signal was suggested and realized in [5.48]; see also [5.14, 15, 49, 50]. The use of the reference wave to increase the OPC-SS quality was suggested and realized in [5.16]. For OPC-SS by other scattering mechanisms (apart from SBS) refer to [5.11 – 13, 17 – 21, 51]. Numerical modeling of OPC-SS is considered in a number of references: see [5.52] and references therein.

6. OPC in Four-Wave Mixing

OPC in four-wave mixing (FWM) has been realized for various third-order optical nonlinearities. Irrespective of the type of third-order nonlinearity of a medium, a phenomenological study of the generation of a phase-conjugated wave in FWM is carried out in much the same way. In this chapter we will present a phenomenological study of this method which enables us to reveal its major features. The features of particular mechanisms of nonlinearity will be discussed in Chap. 7.

Phase-conjugated images in normal static holography have been discussed in earlier work by *Gabor* [6.1, 2] and *Bragg* [6.3]. If a plane reference wave $E_1(R)$ and signal $E_3(R)$ are incident on a hologram while it is being recorded, the disturbance of optical parameters recorded in the material contains, among other terms, an interference term of the form $\delta\varepsilon(R) \propto E_1(R)E_3^*(R)$. Suppose that during reconstruction the hologram is illuminated by a plane wave $E_2(R)$ which is in exact opposition to the recording reference wave $E_1(R)$; this means $E_2(R) = \text{const} \cdot E_1^*(R)$. Then the field $E_{\text{recon}} \propto \delta\varepsilon(R)E_2(R)$ reconstructed by the hologram will contain a part corresponding to the wave which is phase-conjugated with respect to the signal $E_3(R)$: $E_{\text{recon}} = \text{const} \cdot E_1E_2E_3^*(R) \propto |E_1|^2 \cdot E_3^*(R)$, see Fig. 1.5a. Unfortunately, in static holography the generation of a phase-conjugated wave during reading out is separated in time from recording. However for most practical applications, optical phase conjugation in real time is important. OPC by dynamic holography is also referred to as OPC in four-wave mixing (OPC-FWM). At present this method is being studied as extensively as OPC-SBS is.

6.1 Principles of OPC-FWM

Dynamic holography exploits media which vary in their optical properties immediately, with no additional treatment, under the action of incident radiation. A typical OPC-FWM scheme is illustrated in Fig. 1.6. Two plane reference waves $E_1(R)$ and $E_2(R)$ in exact opposition to one another $[E_2(R) \propto E_1^*(R)]$ are introduced into a nonlinear medium simultaneously with the signal to be conjugated $E_3(R)$. We assume that all three waves are of the same frequency and coherent. The following processes take place in the bulk of the medium. The interference of the waves $E_1(R)$ and $E_3(R)$ gives rise to the permittivity holographic grating

$$\delta\varepsilon_{13}(R) = \beta_{13}E_1(R)E_3^*(R) \,. \tag{6.1.1}$$

The scattering of the counterpropagating reference wave $E_2(R)\exp(-i\omega t)$ by this grating produces the phase-conjugated wave, its amplitude E_4 being proportional to the addition to the dielectric displacement $\delta D_4 = \beta_{13}E_1E_2E_3^*(R)\exp(-i\omega t)$. The interference of $E_2(R)$ with $E_3(R)$ also induces the holographic grating

$$\delta\varepsilon_{23}(R) = \beta_{23}E_2(R)E_3^*(R) \,, \tag{6.1.2}$$

and reading out this reflection grating by the wave $E_1\exp(-i\omega t)$ also excites the phase-conjugated wave, $E_4 \propto \delta D_4 = \beta_{23}E_1E_2E_3^*(R)\exp(-i\omega t)$. Moreover, in fast-responding media there is one more contribution; such media respond both to the time-averaged square $|E|^2$ of the real-valued light field $E_{real} = \frac{1}{2}[E\exp(-i\omega t)+E^*\exp(i\omega t)]$ and to oscillations of this square at twice the light frequency, $\delta\varepsilon(t) \propto E^2\exp(-2i\omega t)+E^{*2}\exp(2i\omega t)$. Of considerable interest to us is the spatially homogeneous part of the temporal modulation $\delta\varepsilon(t)$ induced by the reference waves E_1 and E_2:

$$\delta\varepsilon_{12}(t) = \beta_{12}E_1E_2e^{-2i\omega t} \,. \tag{6.1.3}$$

The scattering of the signal field $E_{3\,real} = \frac{1}{2}[E_3(R)\exp(-i\omega t)+E_3^*(R) \times \exp(i\omega t)]$ due to the modulation (6.1.3) also gives rise to a wave which is phase-conjugated with respect to the signal, $E_4 \propto \delta D_4 = \beta_{12}E_1E_2E_3^*(R) \times \exp(-i\omega t)$.

A similar description of all these processes in terms of a third-order polarization of the medium is accepted in nonlinear optics. In a medium with a third-order nonlinearity, the polarization P (i.e., the dipole moment per unit volume) contains an additional term

$$\delta P^{NL} = \chi^{(3)}EEE^*e^{-i\omega t} \tag{6.1.4}$$

caused by the field $E\exp(-i\omega t)$. Here $\chi^{(3)}$ [cm^3/erg] is the third-order susceptibility of the medium. Substituting the sum of the fields $E = (E_1+E_2+E_3) \times \exp(-i\omega t)$ into (6.1.4), we obtain, among other terms, the term

$$P_4 = 2\chi^{(3)}E_1E_2E_3^*(R)e^{-i\omega t} \,, \tag{6.1.5}$$

which is responsible for the generation of the phase-conjugated wave. Since the dielectric displacement $D = \varepsilon E$ is related to the polarization P by

$D = E + 4\pi P$, the phenomenological constant $\chi^{(3)}$ contains the contributions of all three processes:

$$2\chi^{(3)} = (\beta_{13} + \beta_{23} + \beta_{12})/4\pi .$$

Expression (6.1.5) makes it possible to calculate the intensity of the phase-conjugated wave. Let us use a shortened equation for the slowly-varying amplitude E_4 of the phase-conjugated wave field $E_4(z) \exp(i k_4 \cdot R - i\omega t)$ excited by $P_4 \exp(i k_4 \cdot R - i\omega t)$ (a standard method in nonlinear optics [6.4, 5]):

$$2ik\frac{dE_4}{dz} = -\frac{\omega^2}{c^2} 4\pi P_4^{NL} = -\frac{\omega^2}{c^2} 8\pi\chi^{(3)} E_1 E_2 E_3^* . \qquad (6.1.6)$$

Without losing generality, suppose that the wave E_3 and the phase-conjugated wave $E_4(z) \exp(i k_4 \cdot R)$ are plane waves, $k_4 = -k_3$, $|k_3| = k$. Then the slowly varying amplitudes E_1, E_2, E_3 of the plane waves on the right side of (6.1.6) are independent of coordinates. Integrating (6.1.6) with respect to dz yields

$$|E_4|^2/|E_3|^2 = |L\gamma\sqrt{I_1 I_2}|^2 , \qquad \gamma = \frac{4\pi\omega}{c\sqrt{\varepsilon}} \chi^{(3)} . \qquad (6.1.7)$$

Here $I_{1,2} = |E_{1,2}|^2$ are the intensities of the reference waves and L is the path traveled by the signal wave in the volume of interaction with the reference waves. In the case of a purely reactive nonlinearity ($\chi^{(3)} = \chi^{(3)*}$) the constant γ can be expressed in terms of the self-focusing constant ε_2 from the relation $\varepsilon = \varepsilon_0 + \frac{1}{2}\varepsilon_2|E|^2$, namely, $\gamma = \omega\varepsilon_2/2c\sqrt{\varepsilon}$. The constant γ characterizes the rate of change in the nonlinear phase for linearly polarized light, $\phi_{non} = 0.5\gamma Iz$. If the intensity I is expressed in MW/cm^2, then the dimensions of γ are cm/MW. For liquid CS$_2$ at $\lambda = 1.06\,\mu$m, $\gamma = 4.6 \times 10^{-3}$ cm/MW. Thus, for instance, at $L = 10$ cm, $I_1 = I_2 = 10$ MW/cm^2, the coefficient of reflection into the phase-conjugated wave amounts to $|E_4|^2/|E_3|^2 \sim 20\%$. An approximation in which E_4 is proportional to the first power of the factor $\chi^{(3)} E_1 E_2 E_3^*$ is usually called the Born approximation. It essentially allows the retroaction of the wave E_4 upon the original waves E_1, E_2, E_3 to be neglected. The inequality $|E_4|^2 \ll |E_3|^2$ is the necessary condition for the Born approximation to apply.

6.2 Selective Properties of OPC-FWM

In Sect. 6.1 the reflectivity was calculated in the simplest case when all three waves E_1, E_2, E_3 had exactly equal frequencies, the reference waves were exactly counterpropagating, and the OPC efficiency was not high, $|E_4|^2/$

$|E_3|^2 \ll 1$. In this section we consider the case in which the frequencies do not coincide exactly and the reference waves are not in exact opposition to one another. Anticipating the results, we state that the efficiency of the phase conjugation can have a relatively sharp resonance in the three-dimensional space comprising frequency ω_3 and direction k_3 of the signal. This enables one to conjugate selectively only that part of the signal which possesses a given frequency and direction.

Let two reference waves and the signal wave be directed into a layer of thickness l of a medium with nonlinearity $\chi^{(3)}$:

$$E = E_1 \exp(-i\omega_1 t + ik_1 \cdot R) + E_2 \exp(-i\omega_2 t + ik_2 \cdot R)$$
$$+ E_3 \exp(-i\omega_3 t + ik_3 \cdot R), \qquad (6.2.1)$$

and suppose all three to be plane waves. The component of the nonlinear polarization responsible for OPC of the signal E_3 is given by

$$P_4 = 2\chi^{(3)} E_1 E_2 E_3^* \exp(-i\omega_4 t + iQ \cdot R) \qquad (6.2.2)$$

inside the layer $0 < z < l$ and is zero outside it. Here

$$\omega_4 = \omega_1 + \omega_2 - \omega_3, \qquad (6.2.3)$$

$$Q = k_1 + k_2 - k_3. \qquad (6.2.4)$$

Equation (6.2.3) may be interpreted as the law of conservation of energy, $\hbar\omega_1 + \hbar\omega_2 = \hbar\omega_3 + \hbar\omega_4$, in the elementary four-photon process, i.e., conversion of a pair of pump quanta into quanta of the signal and conjugated waves. The conjugated wave is efficiently radiated by the polarization (6.2.2) if its spatial dependence $\propto \exp(iQ \cdot R)$ satisfies the wave equation for the field of frequency ω_4:

$$|Q| \equiv |k_1 + k_2 - k_3| = \frac{\omega_4}{c}\sqrt{\varepsilon}. \qquad (6.2.5)$$

Below we designate $\omega_4\sqrt{\varepsilon}/c$ by k_4.

Equation (6.2.5) is referred to as the spatial synchronism condition, or the phase matching condition. If it is satisfied, the efficiency of OPC-FWM is at a maximum, and (6.1.6, 7) hold true. If the frequencies of the reference waves coincide, $\omega_1 = \omega_2$, and the waves propagate in exact opposition to one another, then $k_1 + k_2 = 0$ and the phase matching condition holds true for a signal wave in any direction, providing its frequency ω_3 coincides with that of the references waves, $\omega_4 = \omega_3 = \omega_1 = \omega_2$, see Fig. 6.1a.

Let us derive the law governing the OPC efficiency when there are deviations from (6.2.5). With this end in view, we should solve the wave equation

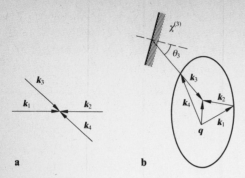

Fig. 6.1. (a) Phase-matching condition in the degenerate case $\omega_1 = \omega_2 = \omega_3 = \omega_4$, $k_1 + k_2 = 0$, $k_3 + k_4 = 0$. **(b)** Phase-matching ellipsoid in FWM

$$\nabla^2 \tilde{E}_4 + k_4^2 \tilde{E}_4(R) = -8\pi \frac{\omega_4^2}{c^2} \chi^{(3)} E_1 E_2 E_3^* \exp(iQ \cdot R) \qquad (6.2.6)$$

for the "fast-varying" amplitude of the field $\tilde{E}_4(R)$. In this case we imply that $\chi^{(3)} = 0$ outside the layer. Since the problem is uniform in the layer plane $(x, y) = r$, the solution should be sought in the form $E_4(R) = E_4(z) \times \exp(iQ_\perp \cdot r)$, where $Q_\perp = (Q_x, Q_y)$ is the transverse component of the vector $k_1 + k_2 - k_3$. Then (6.2.6) can be transformed to

$$\frac{d^2 \tilde{E}_4}{dz^2} + (k_4^2 - Q_\perp^2)\tilde{E}_4 = -\frac{\omega_4^2}{c^2} 8\pi \chi^{(3)}(z) E_1 E_2 E_3^* \exp(iQ_z z) . \qquad (6.2.7)$$

To solve (6.2.7), we use, as usual, the shortened-equation method. Supposing $\tilde{E}_4(z) = E_4(z) \exp(iz\sqrt{k_4^2 - Q_\perp^2})$, where $E_4(z)$ is the slowly varying amplitude, we find that inside the layer

$$2i\sqrt{k_4^2 - Q_\perp^2}\, \frac{dE_4}{dz} = -8\pi(\omega_4^2/c^2)\chi^{(3)} E_1 E_2 E_3^*\, e^{i\Delta kz} , \qquad (6.2.8)$$

$$\Delta k = Q_z - \sqrt{k_4^2 - Q_\perp^2} . \qquad (6.2.9)$$

The expression $\sqrt{k_4^2 - Q_\perp^2}$ can be written $\sqrt{k_4^2 - Q_\perp^2} = k_4 \cos\theta$, where θ is the angle between the propagation direction of E_4 and the normal to the layer (z axis). Integrating (6.2.8) over the layer thickness l yields the reflectivity as

$$\left|\frac{E_4}{E_3}\right|^2 = \left|\frac{\gamma\sqrt{I_1 I_2}(e^{i\Delta kl}-1)}{\Delta k \cos\theta}\right|^2 \equiv |\gamma L\sqrt{I_1 I_2}|^2 \left(\frac{\sin x}{x}\right)^2 ,$$

$$x = \Delta kl/2 . \qquad (6.2.10)$$

We note that at $\Delta k = 0$ (i.e., $x = 0$), (6.2.10) becomes the same as (6.1.7), and the reflectivity is proportional to the square of the path $L = l/\cos\theta$ traveled by the signal in the volume of interaction with the pump.

The phase matching condition (6.2.5) has a simple geometrical interpretation. Let us fix the directions and frequencies of the reference waves, i.e., ω_1, k_1 and ω_2, k_2. At $|Q| = k_4$ (phase matching condition) from (6.2.4) it follows that the vectors k_3, k_4, and $(k_1 + k_2)$ form a closed triangle (Fig. 6.1 b). If the refractive index is frequency independent, the condition (6.2.3) means that the sum $(\omega_1 + \omega_2)\, n/c$ of the lengths of the wave vectors, k_3 and k_4, is constant. This means that the end of the vector k_4 which coincides with the origin of the vector k_3, lies on the ellipsoid of revolution whose focuses are at the extremities of the vector $k_1 + k_2$.

We now consider the parameters of this phase matching ellipsoid in the most interesting case when the reference waves are nearly counterpropagating and their frequencies are close, so that $(k_1 + k_2)/k = \alpha$, where $|\alpha| \ll 1$. In this case, the phase-conjugated wave is nearly in opposition to the signal and its frequency is also close to the signal frequency.

We have to solve two problems. First, it is necessary to relate the frequency ω_3^0 and direction of propagation n_3^0 of the signal, corresponding to the exact phase matching condition, to the parameters of the reference waves. Second, we should find the phase matching bandwidth, i.e., express $x = \Delta k l / 2$ from (6.2.10) in terms of the deviations $\omega_3 - \omega_3^0$ and $n_3 - n_3^0$ of the signal wave parameters from their optimum values.

If $k_1 + k_2 = 0$, which is possible only with exactly counterpropagating reference waves of equal frequencies, then exact phase matching occurs when all four frequencies coincide for any signal direction. If $k_1 + k_2 = k\alpha \neq 0$, transformations yield

$$\omega_3^0 \approx \tfrac{1}{2}(\omega_1 + \omega_2)(1 - \tfrac{1}{2}\alpha \cos \beta^0)\,, \tag{6.2.11}$$

where β^0 is the angle between the vectors k_3^0 and $-\alpha$, Fig. 6.1 b. In other words, at $k_1 + k_2 \neq 0$, there is the possibility of changing the optimum signal frequency ω_3^0 by varying the angle β^0 through manipulating either the signal direction or the parameters of the reference waves (their directions or frequencies). Equation (6.2.11) is the solution of the first of our problems.

To determine the width of the frequency-angular selection of the OPC filter, it is necessary to express $x = \Delta k l / 2$ in terms of frequencies and directions of the interacting waves:

$$\Delta k = k\,[(\omega_4 - \omega_3)/\omega - \alpha \cos \beta]/\cos \theta\,, \tag{6.2.12}$$

where $\omega_4 = \omega_1 + \omega_2 - \omega_3$ and $\omega = (\omega_1 + \omega_2)/2$. Putting Δk equal to zero, we reproduce the condition (6.2.11) of exact phase matching. Further, we should separately consider the spectral width $\Delta \omega_3$ of the filter at the fixed direction n_3 and its angular width $\Delta \beta$ with respect to the fixed frequency ω_3 of the signal. The half-width of the function $\sin^2 x/x^2$ at the first zeros (HWOM) is determined by $x = \Delta k l/2 = \pi$. Then, allowing for (6.2.12), we obtain

$$\Delta\omega_3(\text{HWOM}) = \pi\omega/kL , \tag{6.2.13}$$

$$\Delta\beta_3(\text{HWOM}) = 2\pi(Lk\alpha\sin\beta^0)^{-1} . \tag{6.2.14}$$

Here $L = l/\cos\theta$ is the path of the signal in the interaction region.

It should first be noted that the OPC filter possesses angular selectivity only with respect to one coordinate, i.e., β. If the direction is changed so that it is always on the cone of constant β (i.e., $\beta = \beta_0$), the phase matching condition will not be violated.

Thus, OPC-FWM enables one to realize controllable selection of the phase-conjugated signal with respect to frequency and reception angle. For instance, at $L \sim 3$ cm, $k \sim 10^5$ cm^{-1}, we have $\Delta\omega/\omega \sim 10^{-5}$, i.e., $\Delta\omega/2\pi \sim 3\times10^9$ Hz (see the experiment described in [6.6]). At moderate values $\alpha \sim 3\times10^{-4}$ (obtained when the reference waves are not in exact opposition) and $\sin\beta^0 \sim 0.5$, the angular selectivity amounts to $\Delta\beta_3 \sim 0.1$ rad $\approx 6°$, see the experiment in [6.7]. Control of the center of the signal reception band can be achieved simply changing the direction or frequency of one of the reference waves. We also note that at $\omega_3 \neq (\omega_1 + \omega_2)/2$, the frequency ω_4 of the reflected signal differs from the frequency ω_3 of the incident one. At $k_1 + k_2 = k\alpha \neq 0$, the direction k_4 is turned through the angle $\psi = \alpha\sin\beta^0$ with respect to the direction $(-k_3)$ which is in exact opposition to the signal. All these possibilities – selection with tuning, OPC with a controllable frequency shift, OPC with a controllable additional turning angle – are of significant importance for practical applications.

The selective properties of OPC-FWM considered here are based on the wave effects of phase matching in the interaction volume. Another selection mechanism is also possible, based on the sluggishness or resonant nature of the local nonlinear response of the medium; this will be covered in Sect. 6.6.

6.3 Polarization Properties of OPC-FWM

OPC-FWM has the advantage over OPC-SBS as far as their polarization properties are concerned, since in OPC-FWM it is possible to achieve, in principle, complete spatial polarization. However, this requires a definite choice of the experimental geometry and the polarizations of the reference waves. Consider these questions in detail. We will limit ourselves to FWM in isotropic media. The most general phenomenological form of a third-order polarization proportional to the product of E_1, E_2, E_3^* is

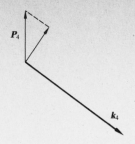

Fig. 6.2. Wave E_4 is excited only by polarization component $P_{4\perp}$ perpendicular to wave vector k_4

$$P_4 = \mu_{13}(E_1 \cdot E_3^*)E_2 + \mu_{23}(E_2 \cdot E_3^*)E_1 + \mu_{12}(E_1 \cdot E_2)E_3^* . \tag{6.3.1}$$

Various physical mechanisms make different contributions to the coefficients μ_{ik}. In this section we shall consider them as given constants.

As electromagnetic waves are transverse, only the projection of the polarization P_4 on the plane normal to the direction n_4 of the phase-conjugated wave makes a contribution (see Fig. 6.2):

$$E_4 \propto P_4 - n_4(n_4 \cdot P_4) . \tag{6.3.2}$$

Equations (6.3.1, 2) show that in the general case the polarization of the phase-conjugated wave E_4 does not coincide either with that of the signal E_3 or with the complex conjugate of E_3. This fact can be associated both with the complicated interplay of μ_{ik} and the polarizations of the reference waves and with the trivial effect of the projection operation (6.3.2). Equations (6.3.1, 2) make it possible to chose the interaction conditions required for reflecting a certain given polarization of the signal into the phase-conjugated wave with the necessary polarization (which does not coincide with that of the signal, generally speaking).

From a theoretical viewpoint, it is interesting to find the condition for exact OPC of the signal, including its polarization state: $E_4 \propto E_3^*$. From a practical viewpoint, such complete spatial-polarization OPC is required, for example, to compensate for distortions arising in solid-state lasers. Consider two schemes realizing complete OPC-FWM.

In one of them (Fig. 6.3a), the reference waves of the pump have the same linear polarization, $E_1 \propto E_2 \propto e_x$. The propagation direction of the signal E_3 coincides with e_x, so that $(E_3^* \cdot E_1) = (E_3^* \cdot E_2) = 0$ (a transverse scheme). As a result, (6.3.1, 2) yield

$$E_4(R) \propto \mu_{12}(E_1 \cdot E_2)E_3^*(R) . \tag{6.3.3}$$

The realization of such a scheme requires a non-scalar nonlinearity; in the experiment described in [6.8] the orientational nonlinearity of CS_2 was used for this purpose.

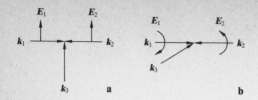

Fig. 6.3a, b. Schemes providing complete spatial-polarization OPC-FWM: (a) transverse, (b) longitudinal

The other scheme, [Fig. 6.3 b], corresponds to nearly colinear propagation of the waves E_1 and E_3 (a longitudinal scheme). Here it is required that the nonlinearity should satisfy the condition $\mu_{13} = \mu_{23}$ and the unit vectors of the polarizations of the reference waves be circular and mutually orthogonal: $E_1 = E_1(e_x + ie_y)/\sqrt{2}$, $E_2 = E_2(e_x - ie_y)/\sqrt{2}$. In this case $(E_1 \cdot E_2) = 0$, the vector (6.3.1) is automatically transverse to the propagation direction and

$$P_4 = \mu_{13} E_1 E_2 E_3^*(R) . \tag{6.3.4}$$

The longitudinal scheme of complete OPC was realized in [6.9].

In the following sections of this chapter we shall assume that the polarizations of all of the four interacting waves are identical and linear.

6.4 Effect of Absorption on OPC-FWM Efficiency

In most cases a nonlinear medium absorbs. Let us suppose that the signal E_3 propagates approximately colinearly with the reference wave E_1 and that the reference wave E_2 is obtained by reflection of E_1 from a mirror whose reflectivity is R. In this case $E_1(z)E_2(z) = E_1^2(0) \exp(-\alpha L)\sqrt{R}$. In the Born approximation, i.e., to first order with respect to the nonlinearity constant, $E_3(z) = E_3(0) \exp(-\alpha z/2)$, and the equation for the reflected signal $E_4(z)$ is

$$\frac{dE_4}{dz} - \frac{\alpha}{2} E_4(z) = i\gamma E_1^2(0)\sqrt{R} \exp(-\alpha L)E_3^*(0)\exp(-\alpha z/2) . \tag{6.4.1}$$

Solving this equation with the boundary condition $E_4(z = L) = 0$ yields

$$\left| \frac{E_4(0)}{E_3(0)} \right|^2 = [\gamma I_1(0)L]^2 R \exp(-2\alpha L)\{[1 - \exp(-\alpha L)]/\alpha L\}^2 . \tag{6.4.2}$$

We shall be interested in the optimization of the conjugation coefficient (6.4.2) in various physical situations. First consider the case in which the linear absorptivity α makes no contribution to the nonlinearity, $\gamma \propto \chi^{(3)}$. With increasing thickness L the OPC reflectivity first increases proportional to L^2, and then at $\alpha L \gtrsim 1$ it decreases as $\exp(-2\alpha L)$. The optimum is achieved at $\alpha L_{\text{opt}} = \ln 2 \approx 0.7$, this corresponding to a 50% linear transmission of the sample $[\exp(-\alpha L_{\text{opt}}) = 0.5]$; in this case $|E_4/E_3|^2 = R(\gamma I_1 \alpha^{-1})^2/16$.

It is of interest to compare this scheme, in which the reference wave E_2 is obtained by reflection of the wave E_1, transmitted through the medium, from a mirror, with that in which reference waves are produced by splitting the original beam of intensity I_1 into two beams, $I_1' = 0.5 I_1$, $I_2' = 0.5 I_1$. In the latter scheme

$$\left|\frac{E_4}{E_3}\right|^2 = \frac{1}{4}(\gamma I_1 L)^2 e^{-\alpha L}\left(\frac{1 - e^{-\alpha L}}{\alpha L}\right)^2. \tag{6.4.3}$$

Then the optimal thickness is $\alpha L_{\text{opt}} = \ln 3 \approx 1.1$, and $|E_4/E_3|^2 \approx (\gamma I_1 \alpha^{-1})^2/27$, i.e., approximately a factor of 1.7 lower than in the scheme using a mirror with $R = 1$.

Finally, in the case of OPC in dyes and other resonant media under conditions far from absorption saturation the nonlinearity constant γ is proportional to the absorptivity α, $\gamma = C\alpha$. In the scheme using a mirror with $R = 1$, we should then optimize the function

$$\left|\frac{E_4}{E_3}\right|^2 = (CI_1)^2 \exp(-2\alpha L)[1 - \exp(-\alpha L)]^2. \tag{6.4.4}$$

Here, the optimum is also achieved at 50% transmission of the cell containing the working medium: $1 - \exp(-\alpha L_{\text{opt}}) = 0.5$, $(\alpha L_{\text{opt}}) \approx 0.7$, $|E_4/E_3|^2 = (\gamma I_1 \alpha^{-1})^2/16$.

6.5 Theory of Coupled Waves

In the previous sections the coefficient of reflection of the signal into the phase-conjugated wave was calculated under the assumption that the amplitudes of the waves E_1, E_2, E_3 were constant in the interaction volume. If the signal E_3 is weak compared to the reference waves E_1 and E_2, the depletion of the latter due to energy transfer to the waves E_3 and E_4 may be neglected in most cases. If, however, calculations using (6.1.7) yield $|E_4|^2/|E_3|^2 \gtrsim 1$, one more process should be taken into account even for non-depleted reference waves. The wave E_4 interacts with the reference waves E_1 and E_2 and generates the phase-conjugated wave E_4^*, i.e., the original signal wave E_3, with an efficiency ~ 1. Thus, it is necessary to solve a system of coupled equations for the amplitudes $E_3(z)$ and $E_4(z)$. We write this system as

$$\frac{dE_4^*(z)}{dz} - i\gamma E_1^* E_2^* e^{-i\Delta kz} E_3(z) = 0, \tag{6.5.1a}$$

$$\frac{dE_3(z)}{dz} - i\gamma E_1 E_2 e^{i\Delta kz} E_4^*(z) = 0. \tag{6.5.1b}$$

Here we consider that the signal $E_3(z)$ enters a medium with a local fast-responding reactive nonlinearity via the plane $z = 0$ in the $+z$ direction. To simplify the formulas, we consider small angles of incidence (between e_z and k_3) so that $\cos\theta_3 \approx 1$. The coefficient γ is introduced in (6.1.7).

We discuss possible reasons for wave detuning, Δk. If the reference waves are not exactly counterpropagating and if their frequencies do not coincide with one another and with ω_3, there exists a contribution to Δk of the form (6.2.12). Another source of wave detuning is connected with the self-action of the reference waves. If two counterpropagating plane waves are in a medium with a local reactive third-order nonlinearity, the behavior of the amplitudes E_1 and E_2 is defined by the Helmholtz nonlinear equation $\nabla^2 E + k^2 E + k\gamma|E|^2 E = 0$. Its approximate solution is $E(z) = E_1 \exp(ik_1 z) + E_2 \exp(ik_2 z)$, where $k_1 = k + \gamma(|E_1|^2 + 2|E_2|^2)/2$, $k_2 = k + \frac{1}{2}\gamma(|E_2|^2 + 2|E_1|^2)$, see Sect. 6.7. Thus, the product $E_1(z)E_2(z)$ acquires the factor $\exp[0.5 i\gamma(|E_2|^2 - |E_1|^2)z]$, i.e., Δk acquires an additional term proportional to the intensity difference of the reference waves. Finally, taking (6.2.12) into account, we have

$$\Delta k = k[(\omega_4 - \omega_3)/\omega - \alpha\cos\beta] - 0.5\gamma(|E_1|^2 - |E_2|^2). \tag{6.5.2}$$

When the wave E_4 is not present, the reference waves affect the propagation of E_3 by changing the wave-vector length by an amount $\delta k_3 = \gamma(|E_1|^2 + |E_2|^2)$. The wave-vector length for E_4 is also changed by the same amount. Therefore, the effect of the reference waves on the signal E_3 and phase-conjugated wave E_4 does not vary either (6.5.2) or (6.5.1).

The boundary conditions for the system (6.5.1) are

$$E_3(z = 0) = E_3(0), \qquad E_4^*(z = L) = 0. \tag{6.5.3}$$

The second condition in (6.5.3) means that the conjugated wave E_4 is not incident on the opposite face of the layer ($z = L$), and thus the wave E_4 arises only inside the layer. If we have $a_3(z) = E_3(z)\exp(-i\Delta kz/2)$, $a_4^*(z) = E_4^*(z)\exp(i\Delta kz/2)$, then (6.5.1) becomes a system with constant coefficients. The final result for $E_4^*(z = 0)$ is

$$E_4^*(z = 0) = E_3(0)(-i e^{i\phi}) \frac{\sin\mu L}{(\mu/\varkappa)\cos\mu L + i(\Delta k/2\varkappa)\sin\mu L}, \tag{6.5.4}$$

$$\mu = \sqrt{\varkappa^2 + (\Delta k/2)^2}\,, \quad \varkappa = |\gamma E_1 E_2|\,, \quad e^{i\phi} = \frac{E_1 E_2}{|E_1 E_2|}\,. \tag{6.5.5}$$

At $\Delta k = 0$ the reflectivity is

$$|E_4(0)/E_3(0)|^2 = \tan^2(\varkappa L)\,, \quad \varkappa = \gamma \sqrt{I_1 I_2}\,. \tag{6.5.6}$$

From (6.5.6) we can draw an important conclusion: at $\varkappa L \gtrsim \pi/4$ OPC-FWM produces a phase-conjugated wave amplified with respect to the incident signal. This amplification is caused by energy transfer from the pair of reference waves E_1 and E_2 to the pair of waves E_3 and E_4. At $\varkappa L = \pi/2$ the reflectivity becomes infinite. This means the possibility of parametric generation of the pair of mutually conjugated waves E_3 and E_3^* even in the absence of the input signal.

The possibility of OPC with high amplification appears extremely attractive. However, it is under these high amplification conditions that the conjugation quality falls catastrophically. First, at $\varkappa L \gtrsim 1$, the self-focusing of the reference waves proceeds intensively, this resulting in distortions of their structure. Second, near the threshold of parametric FWM generation, even weak parasitic illumination of the interaction volume (say, the scattering of reference waves by glaring surfaces or inhomogeneities of the medium) are amplified as effectively as the proper signal.

If the phase matching is not ideal, $\Delta k \neq 0$, then from (6.5.4) we have

$$\left| \frac{E_4(0)}{E_3(0)} \right|^2 = \frac{\sin^2 \mu L}{\cos^2 \mu L + (\Delta k/2\varkappa)^2}\,. \tag{6.5.7}$$

Figure 6.4 presents the dependence of (6.5.7) on the parameter $\varkappa L = \gamma L \sqrt{I_1 I_2}$ for several values of the detuning parameter $\Delta k L$. It is not difficult to obtain the Born approximation (6.2.10) from (6.5.7) at $\Delta k \gg \varkappa$.

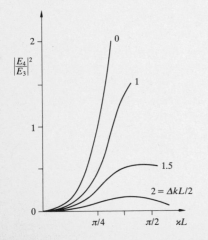

Fig. 6.4. Dependence of the OPC reflectivity $|E_4/E_3|^2$ on the reduced intensity $\varkappa L$ of the reference waves at various detuning amounts $\Delta k L/2$ shown on the curves

At first sight, it follows from (6.5.7) that with unequal reference wave intensities and inexact counterpropagation, the detuning Δk which arises suppresses the harmful effect of parametric FWM generation. In fact, however, if $\varkappa L \gtrsim \pi/2$, there always exist pairs of frequencies and directions, ω_3 and ω_4, k_3 and k_4, for which Δk from (6.5.2) is exactly zero, and parametric generation does occur. However, the development of parametric FWM generation due to noise may require a sufficiently long time exceeding the pulse length of the references waves (Sect. 6.6).

It should be noted that the FWM interaction satisfies the Manley-Row relation, compare with Sect. 2.4; in the absence of absorption the addition to the number of quanta of the field E_3 is equal to the number of quanta of the field E_4:

$$|E_3(z)|^2 = |E_3(z=L)|^2 - |E_4(z)|^2 . \tag{6.5.8}$$

This conservation law is the exact consequence of (6.5.1); the minus sign in (6.5.8), as in Sect. 2.4, is connected with the counterpropagation of the interacting waves. Hence, the amplification of the signal E_3 by the layer is given by

$$\left| \frac{E_3(z=L)}{E_3(z=0)} \right|^2 = 1 + \frac{\sin^2 \mu L}{\cos^2 \mu L + (\Delta k/2\varkappa)^2} . \tag{6.5.9}$$

Equation (6.5.1) relates to the case of non-resonant non-absorbing media. There are many publications devoted to the theory of OPC-FWM in absorbing media and in media with an active resonant nonlinearity. If the nonlinearity is, as before, reactive and fast-responding, but there is linear absorption, the intensities of the reference waves vary within the volume so that $I_1(z) = I_1(0) \exp(-\alpha z)$, $I_2(z) = I_2(L) \exp[-\alpha(L-z)]$, and $I_1(z)I_2(z) = I_1(0)I_2(L) \exp(-\alpha L)$, i.e., $I_1(z)I_2(z)$ is independent of z. Unfortunately, because of the dependence on z of $I_1(z)$ and $I_2(z)$, the phase of the product $E_1(z)E_2(z)$ depends on z in a complicated manner due to self-action effects. This fact makes the analytical solution of the problem difficult. If, in addition, the nonlinearity contains an active component, $\mathrm{Im}\{\chi^{(3)}\} \neq 0$, an exponential dependence is not observed even for the intensities $I_1(z)$ and $I_2(z)$, and then $I_1(z)I_2(z) \neq \mathrm{const}$. Nevertheless, in theoretical considerations [6.10], it is usually supposed that $E_1(z)E_2(z) = \mathrm{const}$ and terms responsible for absorption of the waves E_3 and E_4 are introduced in (6.5.1). Then (6.5.4) can be modified by the trivial change $\Delta k \to \Delta k - i\alpha$, $\varkappa \to \varkappa_0 \exp(-\alpha L/2)$, where $\varkappa_0$ is calculated from the product of the input intensities $I_1(0)$ and $I_2(L)$. When the nonlinearity contains a resonant active component, the constant γ from (6.5.1a) should be replaced by γ^*, γ being considered a complex parameter. In particular, for exact phase matching, $\mathrm{Re}\{\Delta k\} = 0$, and we have

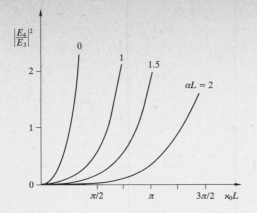

$$\left|\frac{E_4(0)}{E_3(0)}\right|^2 = |\xi + \sqrt{1-\xi^2}\cot(\varkappa L\sqrt{1-\xi^2})|^{-2}, \tag{6.5.10}$$

where $\xi = \alpha/2\varkappa = (\alpha/2\varkappa_0)\exp(\alpha L/2)$, see Fig. 6.5.

Let us discuss the effect of absorption on the threshold of parametric FWM generation. At $0 < \xi < 1$ we can always find a sufficiently large L such that the FWM-generation threshold, determined by (6.5.10) becoming infinite, can be exceeded. For this case $(\varkappa L)_{\mathrm{thr}}\sqrt{1-\xi^2}$ lies in the interval from $\pi/2$ to π. At $\xi > 1$, FWM generation is impossible. In amplifying media, $\alpha < 0$, $\xi < 0$, the threshold value decreases, $(\varkappa L)_{\mathrm{thr}} < \pi/2$, and at $\xi \to -\infty$, $(\varkappa L)_{\mathrm{thr}} \to 0$. The general tendency can be formulated as follows. With weak absorption, $\alpha L \ll 1$, the threshold of generation rises as $(\varkappa_0 L)_{\mathrm{thr}} \approx \pi/2 + \alpha L(\pi^2+4)/2\pi$. With strong absorption, $\alpha L \gg 1$, the threshold rises exponentially:

$$(\varkappa_0 L)_{\mathrm{thr}} \approx (\alpha L/2)\exp(\alpha L/2).$$

A great advantage of the popular model of *Abrams* and *Lind* described above [6.10], employed when considering OPC-FWM in resonant media with the assumption $E_1(z)E_2(z) = \text{const}$, is its simplicity. The results obtained using this model far from the FWM-generation threshold describe various experimental situations fairly well. One of the drawbacks of this model is that the approximation that both the phase and amplitude profile of the product $E_1(z)E_2(z)$ are constant is not rigorous, and ignores the strong effects (at $\varkappa L \sim 1$) of self-action and the interaction of counterpropagating reference waves in a medium with amplitude-phase nonlinearities.

6.6 Nonstationary Effects in OPC-FWM

Nonstationary effects in OPC-FWM can be conventionally separated into three groups. In the first, simplest group, the response of an FWM device to external fields can be regarded as instantaneous on the time scale of changes in interacting fields. In the Born approximation, in a quasi-static case we have

$$E_4(t) = i\gamma L E_1(t) E_2(t) E_3^*(t) . \tag{6.6.1}$$

If all three waves are produced by one and the same source with no time delay, then $I_4(t) = \text{const} \cdot [I_0(t)]^3$. The conjugation of a signal frequency modulation is also of interest. If the reference waves are monochromatic, $E_1(t)E_2(t) = \text{const}$, and the slowly varying amplitude of the signal E_3 has a positive frequency deviation, $E_3(t) \propto \exp(-iat^2)$, $a > 0$, then the reflected field $E_4(t) \propto \exp(iat^2)$ has a negative frequency deviation. This may be of use for the compensation of a dispersion spread of short light pulses (see Chap. 1). Other examples of the formation of the required time behavior of $E_4(t)$ are possible on the basis of the temporal modulation of the amplitudes of the waves $E_1(t)$, $E_2(t)$, and $E_3(t)$.

The second group of nonstationary effects is connected with the sluggishness of the nonlinear response. In the general case, the gratings $\delta\varepsilon_{13} \propto E_1 E_3^*$ and $\delta\varepsilon_{23} \propto E_2 E_3^*$, see (6.1.1, 2), have different relaxation times, $\tau_{13} = \Gamma_{13}^{-1}$ and $\tau_{23} = \Gamma_{23}^{-1}$. Therefore the time behavior of $E_4(t)$ for a thin medium in the Born limit can be expressed as

$$\begin{aligned}
E_4(t) \propto \Big\{ &\varkappa_{12} E_1(t) E_2(t) E_3^*(t) \\
&+ \varkappa_{13} \Gamma_{13} E_2(t) \int_0^t E_1(t') E_3^*(t') \exp[-\Gamma_{13}(t-t')]\, dt' \\
&+ \varkappa_{23} \Gamma_{23} E_1(t) \int_0^t E_2(t') E_3^*(t') \exp[-\Gamma_{23}(t-t')]\, dt' \Big\},
\end{aligned} \tag{6.6.2}$$

where $t = 0$ is the instant when the interacting fields are switched on. At $t \lesssim \Gamma_{i3}^{-1}$, $\varkappa_{12} = 0$, the contribution of a corresponding grating is accumulated $\propto \int^t E_i E_3^*\, dt$, i.e., at equal frequencies of all waves, $|E_4/E_3|^2 \propto I(t) \int_0^t I^2 dt$. In other words, under nonstationary operation, the effective nonlinearity constant $\chi^{(3)}$ increases approximately linearly with the pulse duration. For a time $t \gtrsim \Gamma_{i3}^{-1}$, the contribution of a corresponding grating reaches its steady-state value.

From the viewpoint of the selective properties of the FWM interaction, the effect of nonlinearity sluggishness on the efficiency of OPC of signals with a variable frequency ω_3 is of considerable interest. If the reference waves $E_1(t)$ and $E_2(t)$ are monochromatic, $E_1 \propto \exp(-i\omega_1 t)$, $E_2 \propto \exp(-i\omega_2 t)$, the steady-state value of the response, say for the grating $\delta\varepsilon_{13}$, is given by

$$\delta\varepsilon_{13} = \varkappa_{13}E_1E_3^* \left[1 + i(\omega_3 - \omega_1)/\Gamma_{13}\right]^{-1}. \qquad (6.6.3)$$

Thus, the contribution of this grating to the efficiency of FWM reflection diminishes according to Lorentz's law with increasing $\omega_3 - \omega_1$. Due to this, the nonlinearity sluggishness may be the mechanism causing frequency selection in OPC-FWM. For the contribution of the grating $\delta\varepsilon_{13}$, the center frequency of the OPC filter is $\omega_3^0 = \omega_1$ and its half-width (HWHM) is equal to Γ_{13}; similarly, for the contribution of the grating $\delta\varepsilon_{23}$, $\omega_3^0 = \omega_2$, $\Delta\omega_3 = \Gamma_{23}$.

We have discussed the effect of the sluggishness of the nonlinear response on the selective properties of the FWM interaction. On the other hand, Sect. 6.2 dealt with the effect of the phase matching condition on the selectivity, see (6.2.10, 12). Usually the position of the center of the effective OPC line and the bandwidth are determined by one of the above two mechanisms. In the general case, the profile of the FWM filter line is given by the product of the profile due to each of the two mechanisms:

$$|E_4/E_3|^2 = (L\sqrt{I_1I_2})^2|\gamma(\omega_1 - \omega_3, \omega_2 - \omega_3)|^2 \frac{\sin^2(\Delta kL/2)}{(\Delta kL/2)^2}, \qquad (6.6.4)$$

where Δk is defined by (6.2.12).

Finally, the third group of nonstationary effects is associated with the finiteness of the travel time $\tau = L/v$ of the photon over the interaction region; here v is the propagation velocity. This group may be considered on the basis of the equations for traveling waves:

$$\frac{\partial E_3}{\partial z} + \frac{1}{v}\frac{\partial E_3}{\partial t} = i\gamma E_1E_2E_4^*(z, t), \qquad (6.6.5a)$$

$$\frac{\partial E_4^*}{\partial z} - \frac{1}{v}\frac{\partial E_4^*}{\partial t} = i\gamma E_1^*E_2^*E_3(z, t). \qquad (6.6.5b)$$

If the phase matching condition is satisfied, we may consider that $E_1E_2 = \text{const}$, i.e., is independent of both z and t. The boundary conditions for the system (6.6.5) are specified as

$$E_3(z = 0, t) = E_3(t), \qquad E_4(z = L, t) = 0. \qquad (6.6.6)$$

In the Born approximation ($|\gamma E_1E_2L| \ll 1$) we have $E_3(z, t) \approx E_3(t - z/v)$, and then integrating (6.6.5b) yields

$$E_4^*(z = 0, t) = -\frac{i}{2}\gamma vE_1^*E_2^* \int\limits_0^{2L/v} E_3(t - \tau)d\tau. \qquad (6.6.7)$$

The upper integration limit in (6.6.7) is the time taken by the wave to travel there and back through a medium of thickness L with a velocity v. If the wave

$E_3(t)$ is switched on in a step-like manner at $t = 0$, $E_4^*(t)$ rises linearly at $0 < t < t_0 + 2L/v$ and then becomes stabilized at its steady-state value; $I_4(t)$ rises, at the initial stage as the square of time. Thus, the build-up time in the Born approximation is $\tau_0 = 2L/v$. For instance, $\tau_0 \approx 10^{-9}$ s for $L = 10$ cm and $v = 2 \times 10^{10}$ cm/s.

A more general approach in the case of monochromatic reference waves is to express the signal $E_3(t)$ in terms of a Fourier integral and find the response $\tilde{E}^*(\Omega_4) = r(\Omega_3)\tilde{E}(\Omega_3)$ for each Fourier component of the signal $\tilde{E}(\Omega_3)$ using the theory of coupled waves (Sect. 6.5). In this case, the time behavior of the reflected signal is given by the integral

$$E_4^*(z = 0, t) = \int_{-\infty}^{\infty} r(\Omega_3)\tilde{E}(\Omega_3)\exp(-i\Omega_3 t)\,d\Omega_3. \tag{6.6.8}$$

In the Born approximation $r(\Omega_3) \propto \exp(ix)\sin(x/X)$, where $X = \Omega_3 L/v$, and from (6.6.8) we can reproduce (6.6.7). The major advantage of (6.6.8) consists in the possibility of obtaining the result for high reflectivity when the Born approximation fails to work. The system of equations (6.6.5) is equivalent to (6.5.1) with $\Delta k = 2\Omega_3/v$. Therefore we may use the result for $r(\Omega_3)$ given by (6.5.4).

First consider a temporal rise of the reflected signal in the vicinity of the threshold of parametric FWM generation. We denote the rate of threshold excess by $\eta = (\gamma L\sqrt{I_1 I_2} - \pi/2)/(\pi/2)$; then $\eta > 0$ above the threshold. Using (6.5.1, 6.6.8) for the step-like input signal

$$E_3(t) = \theta(t), \qquad \tilde{E}(\Omega_3) = i[2\pi(\Omega_3 + i\varepsilon)]^{-1}_{\varepsilon \to +0} \tag{6.6.9}$$

at $|\eta| \ll 1$, $t > 0$ we obtain

$$E_4^*(t) = \frac{2i}{\pi\eta}(1 - e^{\Gamma t}), \qquad \Gamma = \frac{\pi^2 v}{4L}\eta. \tag{6.6.10}$$

Under the threshold, $\eta < 0$, a high stationary reflectivity is achieved for the time

$$\tau_{st} = \frac{1}{\Gamma} = \left(\frac{2L}{v}\right)\left(\frac{2}{\pi^2\eta}\right) = \tau_0\frac{1}{\pi}\left|\frac{E_4}{E_3}\right|_{st}. \tag{6.6.11}$$

To an accuracy of a factor $\pi^{-1} \approx 0.3$, this time is equal to $\tau_0 = 2L/v$ multiplied by the amplitude reflectivity $|E_4/E_3|_{st}$ under stationary condition, assuming $|E_4/E_3|_{st} \gg 1$. No stationary condition is observed above the threshold in this approximation; the reflected intensity rises exponentially, $\propto \exp(2\Gamma t)$, Γ being proportional to the rate of threshold excess, see (6.6.10). In real situations the time required to reach the stationary condition is determined by depletion of reference waves when $|E_4(t)|^2 \sim |E_{1,2}|^2$.

6.7 Reference Wave Instability

For a high-quality phase-conjugated wave to be generated in FWM, an optically homogeneous working medium and reference waves uniform in the interaction volume and of a sufficiently high intensity are required. These requirements turn out to be contradictory: smooth high-intensity waves in nonlinear reactive media are unstable. We start by discussing instability in the self-focusing of a unidirectional beam.

6.7.1 Self-Focusing of a Beam as a Whole

The propagation of a unidirectional wave $E(r, z) \exp(ikz - i\omega t)$ in a reactive nonlinear medium is described by the parabolic equation

$$\frac{\partial E}{\partial z} - \frac{i}{2k} \nabla_\perp^2 E = i \frac{\gamma}{2} |E(r, z)|^2 E(r, z) . \qquad (6.7.1)$$

Here $\gamma = \omega^2 \varepsilon_2 / 2kc^2$, $k = \omega \sqrt{\varepsilon_0}/c$. If the wave is plane, (6.7.1) has the strict solution

$$E(r, z) \equiv E(z) = E_0 \exp\left(i \frac{\gamma}{2} |E_0|^2 z \right) . \qquad (6.7.2)$$

If the intensity $|E|^2$ is non-uniform over the cross-sectional area, the equiphase surface (wavefront) becomes distorted due to nonlinearity. At $\gamma > 0$ the wavefront lags behind to a greater extent in place of higher intensity, so the rays normal to the wavefront surface begin to deviate towards places of higher intensity (Fig. 6.6). The shape of the wavefront is determined by an approximate equation $k(z-L) + (\gamma/2)L|E(r)|^2 = $ const. Here we suppose that the incident wave has a planar wavefront and that the input distribution of intensity $|E(r)|^2$ does not vary noticeably over the medium thickness $\Delta z = L$. If we assume $|E(r)|^2 = |E_0|^2(1 - r^2/a^2 + ...)$, the radius of curvature of the wavefront in the beam center is

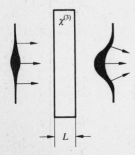

Fig. 6.6. External self-focusing of a beam by a layer of a nonlinear medium

$$\varrho = \frac{ka^2}{\gamma L |E_0|^2} = \frac{a^2}{L} \cdot \frac{2\varepsilon_0}{\varepsilon_2 |E_0|^2}, \tag{6.7.3}$$

and at $\gamma > 0$ this leads to a focusing of the central part of the beam.

Strictly speaking, this formula may be used if $L \ll \varrho$. With increasing thickness L of the medium, focusing leads to an increase in the amplitude at the center, the process of self-focusing is accelerated, and so on. The characteristic distance z_1 at which the beam collapses (to a point, in the framework of geometrical optics) can be estimated by supposing $z_1 \approx \varrho \approx L$:

$$z_1 \sim a(2\varepsilon_0/\varepsilon_2 |E_0|^2)^{1/2}. \tag{6.7.4}$$

Equation (6.7.4) is valid only in the case when within the distance z_1 the spread of the beam due to diffraction is weaker than its self-focusing contraction. At a distance z, the diffraction divergence $\Delta\theta \sim \lambda/a$ leads to an increase in the transverse size of the beam by $\Delta a \sim z \Delta\theta \sim z\lambda/a$, so from the condition $\Delta a \lesssim a$ we obtain $z \lesssim l_{Fr} = a^2/\lambda$. Hence, we may conclude that the self-focusing nonlinearity is stronger than the diffraction spread only if $z_1 \lesssim l_{Fr}$, i.e.,

$$P \approx \frac{nc}{8\pi} |E_0|^2 a^2 \gtrsim \frac{c}{4\pi} \frac{\lambda_0^2 n}{\varepsilon_2} \approx P_{cr}. \tag{6.7.5}$$

Here we have introduced the critical self-focusing power P_{cr} [erg/s].

Let us make some numerical estimates. For CS_2, $\varepsilon_2 \approx 5 \times 10^{-11}$ cm³/erg and at $\lambda_{vac} = 1$ μm, $n = 1.5$ we have $P_{cr} \sim 10^5$ W. It is worth noting that it is the beam power [W] that turns out to be critical but not, say, the power density [W/cm²]. If the beam cross-sectional dimension is increased by a factor of N at the same total power P, then the intensity $|E|^2$ and the nonlinear lens power $\propto \varepsilon_2 |E|^2$ will be decreased by a factor of N^2. However, the Fresnel diffraction length $l_{Fr} = a^2/\lambda$ is increased by the same factor, so the ratio of it to the self-focusing length z_1 remains the same.

We note that the critical self-focusing power is an inherent property for the two-dimensional (with respect to transverse coordinates) problem. For a z-propagation problem with a single transverse dimension, at any power per unit length [W/cm] there exists a self-channeling stable solution.

Returning to the real two-dimensional problem and allowing for diffraction, we may write

$$z_1 \sim (a^2/\lambda)\sqrt{P_{cr}/(P - P_{cr})}. \tag{6.7.6}$$

6.7.2 Plane Wave Instability

In a typical experimental situation a transverse dimension $a \sim 1$ cm of a beam corresponds to a very large Fresnel length, $a^2/\lambda \sim 100$ m. Even if the beam

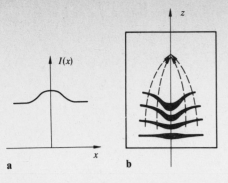

Fig. 6.7. (a) Initial disturbance of the plane wave intensity. **(b)** Wavefront surfaces are curved due to nonlinearity and focus the rays

power is a factor of 10^4 higher than the critical power of the self-focusing length z_1 amounts to about 1 m. That is why within the dimensions of the experimental setup, the self-focusing of the beam as a whole usually produces only phase distortions. At a large threshold excess the effects of small-scale self-focusing due to an exponential rise of disturbances (along z) prove to be essential.

The physical reason for instability arising at $\gamma > 0$ is the following. Let the intensity distribution have a transverse inhomogeneity, such as is shown in Fig. 6.7. Owing to nonlinearity, the wavefront is so distorted that the rays are focused. As a result, the intensity inhomogeneity becomes stronger.

Let us seek the solution of (6.7.1) in the form

$$E(r,z) = E_0 \exp\left(i\frac{\gamma}{2}|E_0|^2 z\right)[1 + u(r,z)] , \tag{6.7.7}$$

where u is a perturbation, $|u| \ll 1$. In a linear (with respect to u) approximation we then obtain from (6.7.1)

$$\frac{\partial u}{\partial z} - \frac{i}{2k}\nabla_\perp^2 u = i\frac{\gamma}{2}|E_0|^2(u + u^*) ,$$

$$\frac{\partial u^*}{\partial z} + \frac{i}{2k}\nabla_\perp^2 u^* = -i\frac{\gamma}{2}|E_0|^2(u + u^*) . \tag{6.7.8}$$

We may seek the solutions of (6.7.8) in the form $u(r,z) = u_0 \cos(q \cdot r + \phi)$ $\times \exp(\mu z)$; for the eigenvalues of the incremental growth we have the system of equations

$$\left(\mu + i\frac{q^2}{2k} - i\frac{\gamma}{2}|E_0|^2\right)u_0 - i\frac{\gamma}{2}|E_0|^2 u_0^* = 0 ,$$

$$i\frac{\gamma}{2}|E_0|^2 u_0 + \left(\mu - i\frac{q^2}{2k} + i\frac{\gamma}{2}|E_0|^2\right)u_0^* = 0 . \tag{6.7.9}$$

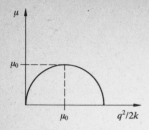

Fig. 6.8. Instability gain μ as a function of the transverse wave vector q^2; $\mu_0 = 0.5\,\gamma\,|E_0|^2$

Putting the determinant of this system equal to zero, we obtain

$$\mu_{1,2} = \pm\frac{1}{2k}\sqrt{q^2(2q_{\mathrm{opt}}^2 - q^2)}\,, \qquad q_{\mathrm{opt}}^2 = k\gamma|E_0|^2\,. \tag{6.7.10}$$

The dependence of the instability gain μ_1 on the transverse wave vector $|q|$ is presented in Fig. 6.8. It is at a maximum at $q = q_{\mathrm{opt}}$:

$$\mu_{\max} = \tfrac{1}{2}\gamma|E_0|^2\,, \tag{6.7.11}$$

which coincides with the rate of nonlinear phase shift in the basic plane wave. The instability itself can be regarded as the parametric decay of a pair of quanta of the basic wave E_0 into a pair of copropagating disturbance quanta with the transverse components $+q$ and $-q$ of the wave vector. The phase matching condition with respect to the z component of the wave vector, $\Delta k_z \approx q^2/k$, limits permissible directions of this decay by means of the relation $\Delta k_z \lesssim (\gamma/2)|E_0|^2$, where $\gamma|E_0|^2$ characterizes the parametric coupling constant [cm^{-1}]. Indeed, at $|q| > q_{\mathrm{opt}}/\sqrt{2}$ disturbances become indifferently stable (intermediate between stable and unstable) since pure imaginary values of $\mu_{1,2}$ correspond to them. The optimum disturbance size $a_{\mathrm{opt}} \sim 2\pi/q_{\mathrm{opt}}$ corresponds to the area a_{opt}^2 through which the power

$$P \approx \frac{nc|E_0|^2}{8\pi}\,a_{\mathrm{opt}}^2 \approx \frac{\lambda_0^2 nc}{4\pi\varepsilon_2} = P_{\mathrm{cr}}\,, \tag{6.7.12}$$

equal to the critical power, is transferred. Thus, a beam of dimension $a \gg a_{\mathrm{opt}}$ with a power many times greater than the critical power tends to break down into filaments, with of the order of the critical power passing through each filament. The development of the input disturbance with $|u_0| \ll 1$ requires a distance $z \sim [\mu(q)]^{-1}\ln(1/|u_0|)$, and for disturbances with the most dangerous period we have $z \sim (\gamma|E_0|^2/2)^{-1}\ln(1/|u_0|)$.

Therefore, for given length z and beam intensity $|E_0|^2$, it is possible, in principle, to avoid small-scale self-focusing. This requires, however, a severe limitation on the amplitude of the initial disturbances:

$$|u_0| \lesssim \exp\left(-\frac{\gamma}{2}|E_0|^2 z\right). \tag{6.7.13}$$

Disturbances usually arise because of the diffraction of the field by the edges of diaphragms or due to the scattering by inhomogeneities of the medium. The small-scale self-focusing is an essential factor limiting the radiation power in amplifiers for short light pulses.

6.7.3 Instability of Counterpropagating Waves

The instability of a unidirectional plane wave is of a convectional nature [6.11], such that initial disturbances develop exponentially in space in the direction of the wave propagation. At $\gamma|E_0|^2 L \lesssim 4$, this instability exhibits itself rather weakly: even the most dangerous disturbances rise by no more than $\exp(2) \approx 7$ times. As distinct from this, a pair of counterpropagating waves already exhibits, at $\gamma|E_0|^2 L \sim 2$, not convectional but *absolute* instability, i.e., when a certain threshold is exceeded a disturbance rises exponentially with time. Such an instability represents parametric generation. To investigate this, let us consider the following field in a nonlinear medium:

$$E(R) = E_+(r, z) e^{ikz} + E_-(r, z) e^{-ikz}. \tag{6.7.14}$$

The nonlinear dielectric displacement $0.5\, \varepsilon_2 |E|^2 E$ for this field is given by

$$D^{NL} = \frac{\varepsilon_2}{2} [(|E_+|^2 + 2|E_-|^2) E_+ e^{ikz} + (|E_-|^2 + 2|E_+|^2) E_- e^{-ikz}$$

$$+ E_+^2 E_-^* e^{3ikz} + E_-^2 E_+^* e^{-3ikz}]. \tag{6.7.15}$$

The terms $\propto \exp(\pm 3ikz)$ do not radiate electromagnetic waves at the frequency ω of the original field. Therefore the shortened equations for the amplitudes $E_+(r, z)$ and $E_-(r, z)$ are

$$\frac{\partial E_+}{\partial z} - \frac{i}{2k} \nabla_\perp^2 E_+ = i\frac{\gamma}{2}(|E_+|^2 + 2|E_-|^2) E_+ ,$$

$$\tag{6.7.16}$$

$$\frac{\partial E_-}{\partial z} + \frac{i}{2k} \nabla_\perp^2 E_- = -i\frac{\gamma}{2}(|E_-|^2 + 2|E_+|^2) E_- .$$

Thus, the wave E_+ experiences an addition to the refractive index from the wave E_- with a doubled strength, and vice versa. This doubling is connected with the superposition of two processes: refraction of the wave $E_+ \exp(ikz)$ by the homogeneous permittivity disturbance $\delta\varepsilon = 0.5\, \varepsilon_2 |E_-|^2$ and reflection of the wave $E_- \exp(-ikz)$ by the permittivity grating $\delta\varepsilon = 0.5\,\varepsilon_2 E_+ E_-^*$ $\times \exp(2ikz)$. Let us seek the solution of (6.7.16) in the form

$$E_+(r,z) = E_1 \exp\left[i\frac{\gamma}{2}(I_1+2I_2)z\right] \cdot [1+u(r,z)]\,,$$

$$E_-(r,z) = E_2 \exp\left[-i\frac{\gamma}{2}(I_2+2I_1)z\right] \cdot [1+v(r,z)]\,, \qquad (6.7.17)$$

where $I_{1,2} = |E_{1,2}|^2$ are the intensities of the plane reference waves. For small u and v, the linearized system (6.7.16) is of the form

$$\frac{\partial u}{\partial z} - \frac{i}{2k}\nabla_\perp^2 u = i\frac{\gamma}{2}I_1(u+u^*)+i\gamma I_2(v+v^*)\,,$$

$$\frac{\partial v}{\partial z} + \frac{i}{2k}\nabla_\perp^2 v = -i\frac{\gamma}{2}I_2(v+v^*)-i\gamma I_1(u+u^*)\,, \qquad (6.7.18)$$

and one more pair of equations for $\partial u^*/\partial z$ and $\partial v^*/\partial z$ obtained from (6.7.18) by complex conjugation should also be included.

At $I_2 \to 0$, the first equation in (6.7.18) transforms into (6.7.8), describing the development of disturbances in a unidirectional wave. However, if the propagation direction of the disturbances

$$u(r,z) = E_3(z)\exp(iq\cdot r - iq^2 z/2k)\,,$$

$$v(r,z) = E_4(z)\exp(-iq\cdot r + iq^2 z/2k)$$

makes a sufficiently large angle θ with the z axis, i.e., if $|\theta| = |q|/k \gg \sqrt{\gamma I/k}$ $\approx \sqrt{\varepsilon_2 I/\varepsilon_0}$, then the term $u^* \propto v \propto \exp(-iq\cdot r + iq^2 z/2k)$ experiences a strong phase mismatch with respect to the terms proportional to u and v^*. Therefore, omitting the terms with u^* and v in (6.7.18), we obtain a pair of coupled equations for the mutually conjugated waves E_3 and E_4^* equivalent to (6.5.1).

To solve (6.7.18), we use new variables $p = u+u^*$, $h = u-u^*$, $m = v+v^*$, $n = v-v^*$ and assume that

$$p(r,z) = p\cos(q\cdot r + \phi)\exp(\mu z)\,, \qquad (6.7.19)$$

with similar expressions for h, m, n. Then (6.7.18), together with its conjugate, can be written as

$$\mu p + i\frac{q^2}{2k}h = 0\,, \qquad \mu h + i\left(\frac{q^2}{2k} - \gamma I_1\right)p - 2i\gamma\sqrt{I_1 I_2}\,m = 0\,,$$

$$\mu m - i\frac{q^2}{2k}n = 0, \quad \mu n - i\left(\frac{q^2}{2k} - \gamma I_2\right)m + 2i\gamma\sqrt{I_1 I_2}\,p = 0. \tag{6.7.20}$$

Putting the determinant of (6.7.20) equal to zero yields a fourth-degree equation for the eigenvalues of μ. For the simplest case $I_1 = I_2 = I$, its roots can be written

$$\mu_{1,2} = \pm\frac{i}{2k}\sqrt{q^2(q^2 + 2k\gamma I)}, \quad \mu_{3,4} = \pm\frac{i}{2k}\sqrt{q^2(q^2 - 6k\gamma I)}. \tag{6.7.21}$$

Parametric generation arises if, with no waves entering the system, $u(z=0) = 0$, $v(z=L) = 0$, there exists a nontrivial solution of (6.7.18) corresponding to finite amplitudes for the outgoing disturbances: $u(z=L) \neq 0$, $v(z=0) \neq 0$. Here we assume that the surfaces $z = 0$ and $z = L$ limit the non-linear medium layer. A system of four boundary conditions corresponds to zero input amplitudes:

$$\sum_{i=1}^{4} p_i = 0, \quad \sum_{i=1}^{4} h_i = 0, \quad \sum_{i=1}^{4} m_i \exp(\mu_i L) = 0, \quad \sum_{i=1}^{4} n_i \exp(\mu_i L) = 0. \tag{6.7.22}$$

It is not difficult to express h_i, m_i, n_i in terms of p_i at $i = 1, \ldots, 4$ using (6.7.20). As a result, (6.7.22) for four unknowns $p_1, \ldots, p_4$ will have a non-trivial solution only if its determinant is zero. Thus we have

$$2ST(1 + \cos S \cos T) + (S^2 + T^2)\sin S \sin T = 0, \tag{6.7.23}$$

where $S = -i\mu_1 L$, $T = -i\mu_3 L$. Examination of the solutions of this trans-cendental equation for S and T carried out in [6.12] made it possible to find the threshold value of the intensity $I = I_1 = I_2$ at which absolute instability (parametric generation) arises. At large angles, $\theta^2 \gg (kL)^{-1}$, the condition for parametric FWM generation is $\gamma IL = \pi/2$, see (6.5.6), at $I_1 = I_2$. With decreasing disturbance propagation angle θ, the threshold value diminishes; at $(kL)^{-1} \cdot 0.3 \lesssim \theta^2 \lesssim (kL)^{-1}$ the threshold condition is

$$\gamma IL \approx 0.9, \tag{6.7.24}$$

this being approximately a factor of 1.7 lower than for the threshold of ordinary FWM generation. For still smaller angles the threshold value of γIL increases up to infinity.

If the threshold value is exceeded by about 100%, the time in which the disturbance rises by a factor of e for an instantaneous nonlinearity coincides in order of magnitude with the time $\tau_0 = 2L/c$ the light requires to travel through the medium. If the threshold value is exceeded by a small amount, the instability development time increases, $\tau \sim \tau_0[(I - I_{thr})/I_{thr}]^{-1}$.

6.7.4 Effect of Wave Self-Action on Efficiency and Quality of OPC-FWM

Almost everywhere in the discussion above we considered the reference waves to be plane with constant amplitudes over the entire cross section. In fact, however, a laser beam possesses a limited transverse dimension and, in the absence of special diaphragms, exhibits a Gaussian amplitude distribution. This leads to distortions of the phase-conjugated wave since $P_4(R) \propto E_3^*(R)E_1(R)E_2(R)$, where the factor $E_1(R)E_2(R)$ is not constant any longer. Thus, for example, for a thin medium at $E_{1,2}(R) \propto \exp(-r^2/a_{1,2}^2)$ and $E_3(r) \propto \exp(-r^2/a_3^2)$ we have, in the Born approximation, $E_4 \propto \exp(-r^2/a_4^2)$, where $a_4^{-2} = (a_3^*)^{-2} + a_1^{-2} + a_2^{-2}$. In particular, for three Gaussian beams with planar wavefronts and equal diameters, $a_i = a_i^* = a$ for $i = 1, 2, 3$, we obtain $a_4 \approx 0.58a$, i.e., the phase-conjugated wave is considerably narrower than the incident signal. If $\mathrm{Im}\{a_1^{-2} + a_2^{-2}\} \neq 0$, i.e., the reference waves are not exactly plane and not conjugated, the reflected wave acquires phase distortions. The OPC quality will be high only when the reference waves can be regarded as being homogeneous and conjugated in the localization region of the signal wave; in particular, for Gaussian beams at $|a_3| \ll |a_1|, |a_2|$ we have $a_4 \approx a_3^*$.

The self-focusing of the reference waves distorts their spatial structure and, consequently, the structure of the phase-conjugated wave. Consider a scheme in which the reference wave $E_2(r)$ is obtained by reflection of the wave $E_1(r)$ from a plane mirror placed close to a nonlinear medium. Let the incident wave $E_1(r)$ be of the form

$$E_1(r) = E_1 \exp(-r^2/a_1^2), \qquad \frac{1}{a_1^2} = \frac{1}{w^2} - \mathrm{i}\frac{k}{2F}, \qquad (6.7.25)$$

where F is the wavefront curvature radius and w is the beam radius. If the medium is not too thick, $ka^2 \ll L$, the parameter a can be considered constant along the z axis. Then $E_2(r, z=L) = \sqrt{R}E_1(r, z=L)$, $|E_2(r)|^2 = R|E_1(r)|^2 = R|E_1|^2 \exp(-2r^2/w^2)$ and from (6.7.16) we obtain

$$E_1(r, z)E_2(r, z)$$

$$= \sqrt{R}E_1^2 \exp\left\{-\frac{2r^2}{a_1^2} + \mathrm{i}\frac{\gamma}{2}|E_1|^2[3L(1+R) - z(1-R)]\exp\left(-\frac{2r^2}{w^2}\right)\right\}.$$

$$(6.7.26)$$

It is seen that even at $R \sim 1$ the achievement of an OPC reflectivity of $\gamma L|E_1|^2R \sim 1$ is accompanied by considerable distortion of the phase profile of the phase-conjugated wave. For instance, at $R = 0.5$ and $\gamma|E_1|^2LR = 1$, the phase of the product $E_1(r)E_2(r)$ at $z = 0$ varies by more than π in passing from the beam center $r = 0$ to its periphery $r \gtrsim w$.

In a number of cases the nonlinearity is such that the interference of the counterpropagating waves $E_+^* E_-\exp(2ikz)$ excites the medium noticeably less than the intensity $|E_+|^2$ or $|E_-|^2$ of an unidirectional part of the field. If such is the case, then $|E_4/E_3|^2 \propto |\gamma E_1 E_2 L|^2/4$, and the factor inside the square brackets in (6.7.26) should be replaced by $2L(1-R)$. As a result, at the same value of $|E_4/E_3|^2$ the phase modulation depth increased by a factor of 1.5.

Up to this point, we have considered the phase self-modulation of smooth reference beams. As was shown in Sect. 6.7.3, at $\gamma|E_1|^2 L \sim 0.9$ (for $R=1$), when the steady-state reflectivity $|E_4/E_3|^2 = \tan^2(\gamma|E_1|^2 L) \approx 1.5$, the onset of parametric generation of small-angle disturbances of the reference waves is already observed. Under conditions of developed instability these disturbances irregularly distort the structure of the reference waves and, accordingly, additionally worsen the efficiency and quality of OPC.

6.8 Literature

Following the first publications [6.1 – 3, 13, 14] and more than 60 subsequent publications devoted to the problem of phase-conjugated waves in static holography, ideas were put forward as to how phase-conjugated waves could be generated in real time by dynamic holography, and attempts were made to realize them [6.15 – 20]. An important stimulating role belongs to the work of *Hellwarth* [6.21] in which the principle of OPC-FWM was re-discovered in terms of nonlinear optics. The reports in [6.22, 23] are a continuation of OPC-FWM experiments initiated in [6.21]. In our presentation of the problem we use the results of a number of research groups carried out investigations on OPC-FWM.

The study of selective properties started with [6.24, 25] (angle selection) and [6.26 – 28] (frequency selection). Controllable frequency-angle filters are discussed in [6.29 – 36]. The polarization properties of OPC-FWM are discussed in [6.24, 37 – 39]; complete spatial-polarization OPC-FWM was first realized in [6.8, 9].

The effect of absorption in the working medium on OPC-FWM is considered in [6.10, 40 – 43]. The theory of coupled waves for OPC-FWM was suggested by *Yariv* and *Pepper* in [6.44] where important conclusions were made about the possibility of both signal reflection with amplification and parametric FWM generation. The theory of coupled waves when there are deviations from phase matching was advanced in [6.24]. The model of coupled waves in resonant absorbing media was suggested by *Abrams* and *Lind* [6.10], see also review [6.34]. The depletion of reference waves and related bistability effects were investigated in [6.45 – 48]. Nonstationary effects are considered in [6.49 – 53]. For frequency selection by nonlinearity sluggishness, see [6.54, 55].

The self-focusing of a beam in a nonlinear medium was first discussed theoretically in [6.56, 57] and detected experimentally in [6.58]. The idea that a plane wave in a self-focusing medium may be unstable was suggested by *Khoklov* and by *Bespalov* and *Talanov*, see [6.59]. Instability in the FWM interaction was discovered experimentally in [6.60]. The instability of counter-propagating waves to small-angle disturbances is considered in [6.12]. The problem of OPC compensation of nonlinear distortions and the questions arising about the stability of reference waves in nonlinear media are discussed in [6.61 – 64]. A detailed account of some questions concerning OPC-FWM is given in the reviews [6.65 – 66] and in [6.67].

7. Nonlinear Mechanisms for FWM

A great number of experiments on OPC-FWM have been conducted in recent years. The optical scheme was nearly the same in all of them. The first reference wave was transmitted through a medium and then reflected by a mirror, corner reflector or a triple prism to produce a second reference wave. The signal wave was obtained by branching off a certain amount of the laser beam energy. In some cases, either a phase or an amplitude mask placed in the signal path was used for a quantitative or qualitative determination of the OPC fraction. These experiments differed in the nonlinearity mechanisms used to obtain FWM. This chapter gives an account of the major optical non-linearity mechanisms used therein and reviews, in brief, experimental results.

7.1 Molecular Orientation

Liquid carbon disulfide CS_2 is a classical representative of media with third-order optical nonlinearity. Molecules of CS_2 possess a strongly anisotropic polarizability: $\alpha_\parallel = 15 \times 10^{-24} \text{ cm}^3$, $\alpha_\perp = 5 \times 10^{-24} \text{ cm}^3$. The molecular energy $U = -(1/4)\alpha_{ik}E_i^*E_k$ in a light field is at a minimum when the molecule axis lines up with the field E. In the Boltzmann thermodynamic equilibrium the degree of orientation is determined by a small dimensionless parameter $(\alpha_\parallel - \alpha_\perp)|E|^2/k_B T$, where $k_B T$ is the temperature expressed in energy units. Partial orientation of molecules results in anisotropy of the permittivity, which is proportional to the intensity of the light. The anisotropy is such that the effective permittivity disturbance for the field E is positive. In phenomenological terms we may write

$$\delta\varepsilon_{ik} = \tfrac{3}{8}\varepsilon_2[E_iE_k^* + E_i^*E_k - \tfrac{2}{3}(E \cdot E^*)\delta_{ik}] \, . \tag{7.1.1}$$

This expression is arranged so that the trace of the tensor $\delta\varepsilon_{ik}$ should be zero, corresponding to anisotropy disturbance without changes in the scalar parameters of the medium. The factor $(3/8)$ is chosen such that the expression for the nonlinear dielectric displacement

$$D^{\text{NL}} = \tfrac{3}{8}\varepsilon_2[E^*(E \cdot E) + \tfrac{1}{3}E(E \cdot E^*)]$$

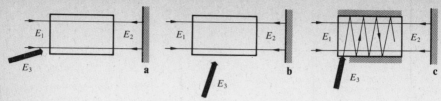

Fig. 7.1a–c. Different geometries for OPC-FWM: (a) longitudinal, (b) transverse, (c) multipass

goes over into $D = (1/2)\varepsilon_2|E|^2E$ for the linear polarization of the wave $E = eE$ when $e = e^*$. The above reasoning enables us to estimate $\varepsilon_2 \sim 4\pi(\alpha_\| - \alpha_\perp)^2N/k_BT$, where N is the number of molecules per unit volume. It is obvious that this mechanism works in any liquid consisting of optically anisotropic molecules. The characteristic time taken to reach the steady-state value of the disturbance (7.1.1) is determined by the relaxation time of molecular orientation in the liquid, which ranges from 10^{-10} to 10^{-12} s for not very large molecules. For CS_2, $\varepsilon_2 \approx 5 \times 10^{-11}$ cm^3/erg and is practically independent of the wavelength of light; $\tau_{rel} \approx 10^{-12}$ s.

The orientational nonlinearity of liquid CS_2 was first used for a purposeful observation of an OPC-FWM signal with a complicated wavefront in [7.1, 2]. An inhomogeneous phase plate was used to control the quality of OPC. Unfortunately, the OPC fraction, i.e., the energy fraction of the exactly conjugated configuration in the reflected signal, was not measured quantitatively. Transverse experimental geometry was employed in [7.1], see Fig. 7.1a, and longitudinal geometry in [7.2], Fig. 7.1b. In [7.1] the effective interaction length determined by the region in which the reference waves were intersected by the signal ray amounted to $l \sim 0.4$ cm, and due to this the reflectivity was low, $\sim 10^{-3}$. In the longitudinal geometry of [7.2], $L \sim 5$ cm and $|E_4/E_3|^2 \approx 0.1$ were realized.

A very long cell, $L \approx 40$ cm, was employed in [7.3], and the reflectivity was far larger than unity. Parametric FWM generation was also realized. In order to increase the interaction length and the power density, a thin light guide filled with CS_2 was used in [7.4].

A multipass transverse scheme, Fig. 7.1c, was employed in [7.5] to increase the length L_3. Its advantages lies in the fact that the length $L_1 = L_2$ of interaction of the powerful reference waves E_1 and E_2 with the medium is small compared to L_3. The harmful effect of the self-focusing and instability of the reference waves is then lessened, while keeping a high efficiency and quality of phase conjugation. We note that the longitudinal scheme in [7.6] provided a reflectivity $R \sim 100$. However, the self-focusing of the reference waves affected the OPC quality. These experiments were followed by a great number of other experiments orientational nonlinearity.

7.2 Saturation Nonlinearities

A high degree of nonlinearity is observed in substances in which the population of atomic and molecular levels, of the electronic levels in the crystal lattice, etc., changes under the action of light. The characteristic feature of this process is the possibility of a large ($\sim 100\%$) change in the population of long-lifetime levels at a relatively low light intensity. Therefore, the description of the system nonlinearity using only the third-order polarizability $\chi^{(3)}$ is insufficient: nonlinearities of higher order corresponding to saturation should be taken into account.

7.2.1 Phenomenological Approach

The simplest phenomenological expression which is commonly used for saturation of absorption is of the form

$$\alpha = \alpha_0 (1 + |E|^2/I_{sat})^{-1}, \tag{7.2.1}$$

where α is the intensity absorptivity [cm^{-1}] and I_{sat} is the saturation intensity. Let us drive this formula by considering a two-level system. On being absorbed by the medium, the light field raises the molecules from the ground (1) to an excited state (2) and changes the population difference $N = N_1 - N_2$ [cm^{-3}], thus varying the absorptivity $\alpha = N\alpha_0/N_0$:

$$\frac{\partial N}{\partial t} + \frac{1}{T}(N - N_0) = -2\frac{N}{N_0}\alpha_0 \mathcal{J}. \tag{7.2.2}$$

Here α_0 is the undisturbed absorptivity, $\mathcal{J} = nc|E|^2/8\pi\hbar\omega$ [$cm^{-2} s^{-1}$] is the density of quantum flux in the light wave, N_0 [cm^{-3}] is the undisturbed population of the ground state, T is the relaxation time of the population, and σ [cm^2] $= \alpha_0/N_0$ is the transition cross section for an individual molecule. In typical conditions the relaxation time T is far shorter than the duration of pulses used; then, considering the steady-state regime, we obtain (7.2.1) from (7.2.2) with I_{sat} given by

$$I_{sat} = \frac{4\pi\hbar\omega N_0}{\alpha_0 ncT}, \quad \mathcal{J}_{sat} = \frac{1}{2\sigma T}. \tag{7.2.3}$$

Thus, saturated absorption occurs at $|E|^2 \sim I_{sat}$. It should be noted that $\mathcal{J} \propto (\sigma T)^{-1}$, i.e., transitions with a large cross section and a long relaxation time are more easily saturable. Equation (7.2.1) corresponds to the dependence of the permittivity on the field intensity:

$$\varepsilon = n_0^2 \left[1 + i\frac{\alpha_0}{k}(1 + |E|^2/I_{\text{sat}})^{-1} \right],\tag{7.2.4}$$

where n_0 is the refractive index and $k = \omega n_0/c$. The nonlinearity constant can be written $\gamma = \omega \varepsilon_2 / 2n_0 c \approx \left(\dfrac{\omega}{n_0 c}\right)\dfrac{\partial \varepsilon}{\partial |E|^2}$, and (7.2.4) yields

$$\gamma = -i\frac{\alpha_0}{I_{\text{sat}}}(1 + |E|^2/I_{\text{sat}})^{-2}.\tag{7.2.5}$$

This is purely imaginary, indicating that the field affects only the absorptivity of the medium.

In the Born approximation the reflectivity $|E_4/E_3|^2$ is equal to $|\gamma|^2 I_1^2 L$, the intensities of the counterpropagating reference waves being supposed approximately equal, $I_1 \sim I_2$. As a result, with increasing reference wave intensity the reflectivity first rises as I_1^2, reaches its maximum at $I_1 \sim I_{\text{sat}}$, and then diminishes. The maximum value of the reflectivity is

$$|E_4/E_3|_{\text{max}}^2 \sim (\alpha_0 L)^2, \quad I_{1,2} \sim I_{\text{sat}}.\tag{7.2.6}$$

This estimate is valid only at $\alpha_0 L \lesssim 1$. At $I_{1,2} \lesssim I_{\text{sat}}$ and $\alpha_0 L \gtrsim 1$, (7.2.6) should be corrected for absorption in the working medium, see Sect. 6.2.

We now proceed to a discussion of system behavior at a high degree of saturation. Equation (7.2.5) would lead to an asymptotic fall in the reflectivity at high saturation to $|E_4/E_3|^2 \sim (\alpha_0 L)^2 (I_{\text{sat}}/I_1)^2$. However, we should recall that (7.2.1, 5) were obtained using a rather rough approximation, and we note some effects which modify these equations.

First, (7.2.2) corresponds to a homogeneously broadened transition line. For an inhomogeneously broadened line, the higher is the value of I/I_{sat}, the larger the number of molecules taking part in the saturation process, i.e., those molecules with a larger value of the detuning $|\omega - \omega_0|$ also contribute. This is a consequence of the spectral dependence of the cross section for an individual molecule, $\sigma(\omega) \propto [1 + (\omega - \omega_0)^2/\Gamma^2]^{-1}$, where Γ is the width of the homogeneously broadened line. As a consequence, we have $\varepsilon(|E|^2) = \varepsilon_0 + i(\alpha_0/k)(1 + |E|^2/I_{\text{sat}})^{-1/2}$, i.e., the nonlinearity does not decrease so strongly at $I \gg I_{\text{sat}}$. Second, the contour of the line for an individual molecule can be varied under the action of a strong field (field broadening). Finally, even in the framework of the local expression (7.2.4), the effective nonlinearity γ determining the FWM coupling coefficient should be estimated more rigorously. The field $E = E_1 \exp(ikz) + E_2 \exp(-ikz)$ consisting of two counterpropagating reference waves of very similar intensities, corresponds to an almost standing wave. Near the nodes of this standing wave, even at $|E_{1,2}|^2 \gg I_{\text{sat}}$, there exist regions (of thickness $\Delta z \ll \lambda$) where the local intensity $|E(z)|^2 \lesssim I_{\text{sat}}$. It is these regions that largely contribute to the FWM constant

γ. As a result, the effective FWM constant γ is asymptotically proportional to $I_1^{-3/2}$ at $I_1 = I_2$. If I_1 and I_2 do not coincide, then $\gamma \propto I_1^{-3}$. The accurate expressions for the effective constant γ in terms of (7.2.1), allowing for the interference of the reference waves, are given in [7.7, 8].

It is important that, in all models and approximations at $I_{1,2} \ll I_{sat}$, the reflectivity is given by $|E_4/E_3|^2 \approx (\alpha L)^2 I_1 I_2/I_{sat}$ and is a maximum at I_1, $I_2 \sim I_{sat}$, see (7.2.6). It is these regimes that are of great interest in OPC and are satisfactorily described by (7.2.5). To calculate a high reflectivity, it is necessary to resort to the theory of coupled waves considered in Sect. 6.5; for details see the review [7.8].

The above results are obtained for a light frequency ω coinciding with the central frequency ω_0 of the transition. If we denote the width of the homogeneously broadened line by Γ and define $\delta = (\omega - \omega_0)/\Gamma \neq 0$, the permittivity at the frequency ω can be written as

$$\varepsilon(\omega) = n_0^2 \left(1 + \frac{\alpha_0}{k} \cdot \frac{1}{(1 + |E|^2/I_{sat})} \cdot \frac{1}{(\delta - i)} \right). \qquad (7.2.7)$$

The linear absorptivity at a shifted frequency is $\alpha = \alpha_0(1 + \delta^2)^{-1}$, and in accordance with (7.2.3), $I_{sat} = I_{sat}(\delta = 0) \cdot (1 + \delta^2)$. From (7.2.7) it follows that $\gamma(\delta) \approx \gamma(0)[(\delta - i)(1 + \delta^2)]^{-1}$, i.e., the effective nonlinearity constant decreases quickly as $|\delta|$ increases. If, however, we increase the intensities of the reference waves as the frequency is defined, so that I_1, $I_2 \sim I \propto (1 + \delta^2)$, and increase the density of particles, so that $\alpha(\delta) = $ const, then the estimate (7.2.3) of the maximum reflectivity in the Born approximation yields

$$|E_4/E_3|^2 \approx |\alpha(\delta)L|^2(1 + \delta^2). \qquad (7.2.8)$$

Such a rise in the reflectivity is connected with the increasing relative role of phase gratings but is achieved at the expense of incrasing the intensity of the reference waves and the number of particles.

We now consider substances in which saturation mechanisms can be realized.

7.2.2 Resonant Gases

We shall consider gases whose particles (molecules or atoms) are capable of a one-photon transition with frequency ω_0 which is close to the frequency ω of the interacting waves. Specific features of resonant gases are the narrow linewidth of transitions and the fact that excited particles (atoms or molecules) are in motion. Since the transition lines are so narrow (small $\Delta\omega$) the contribution of each atom to the permittivity at resonance is noticeable and the nonlinearity is high even for low particle densities. Thus, for example, for vapors of alkali metals an absorptivity of $\alpha \sim 1 \text{ cm}^{-1}$ in the resonance line is achieved at $N \sim 10^{11} - 10^{13} \text{ cm}^3$. Because of the large absorption cross section ($\sigma \propto \Delta \omega^{-1}$) and the rather slow relaxation of the excited states ($T > \Delta \omega^{-1}$), the saturation intensity $\mathscr{I}_{\text{sat}} = (2\sigma T)^{-1}$ is very low; this permits one to operate FWM with the use of continuous wave lasers.

Let us discuss the effect of the motion of the particles on the FWM process. At high pressures the collision linewidth $\Gamma = N v_{\text{T}} \sigma_{\text{col}}$ exceeds the Doppler linewidth $\Delta \omega_{\text{D}} = k v_{\text{T}}$; here σ_{col} is the collision-broadening cross section, which usually coincides in order of magnitude with the gas-kinetic cross section σ_0, and v_{T} is the thermal velocity of the molecules, $v_{\text{T}} \approx \sqrt{k_{\text{B}} T / M}$. Under such conditions the transition is homogeneously broadened. This means that all resonant molecules absorb the incident monochromatic light with an equal cross section in the course of recording a dynamic hologram and scatter it with equal efficiency in the course of reading the hologram.

The number of excited particles is greater in those places where the local field intensity is higher due to interference. Therefore the spatial modulation of the number of excited molecules contains two interference terms – $\delta N_{13} \propto \exp[i(\mathbf{k}_1 - \mathbf{k}_3) \cdot \mathbf{R}]$ and $\delta N_{23} \propto \exp[i(\mathbf{k}_2 - \mathbf{k}_3) \cdot \mathbf{R}]$ with periods $\Lambda_{13} = \lambda/2 \times \sin(\theta_{13}/2)$ and $\Lambda_{23} = \lambda/2 \sin(\theta_{23}/2)$, where θ_{ik} is the angle between the wave vectors i and k. In a typical longitudinal OPC-FWM scheme, $\theta_{13} \ll 1$, $\theta_{23} \approx 180°$, and thus $\Lambda_{13} \approx \lambda/\theta_{13}$, $\Lambda_{23} \approx \lambda/2$. The motion of the excited molecules under these conditions may be described in terms of diffusion since the inequality $\Delta \omega_{\text{D}} < \Gamma$ means that the free path l_{f} of the molecules is less than the light wavelength: $l_{\text{f}} < \lambda$. Allowing for relaxation and diffusion, the equation for the density of excited particles at $N - N_0 \ll N_0$ is of the form

$$\frac{\partial N}{\partial t} + \frac{1}{T} N - D \nabla^2 N = A |E(\mathbf{R})|^2 . \tag{7.2.9}$$

If the field is switched on abruptly, (7.2.9) gives

$$\delta N(\mathbf{R}, t) = \frac{A}{\Gamma_{13}} E_1 E_3^* \exp(i\mathbf{q}_{13} \cdot \mathbf{R}) [1 - \exp(-\Gamma_{13} t)]$$

$$+ \frac{A}{\Gamma_{23}} E_2 E_3^* \exp(i\mathbf{q}_{23} \cdot \mathbf{R}) [1 - \exp(-\Gamma_{23} t)] , \tag{7.2.10}$$

$$\Gamma_{13} = T^{-1} + D(k\theta_{13})^2 , \qquad \Gamma_{23} = T^{-1} + 4k^2 D .$$

Here D [cm²/s] is the diffusion coefficient, $D \sim v_{\text{T}} l_{\text{f}}$. Consider the steady-state condition. If the relaxation time T of the population is shorter than the time within which both gratings become blurred due to diffusion,

$T < (4Dk^2)^{-1}$, the blurring time may be ignored and the particles considered immovable. If T lies between $(4Dk^2)^{-1}$ and $(Dk^2\theta_{13}^2)^{-1}$, diffusion in the steady-state condition blurs the grating $\propto E_2 E_3^*$ of the counterpropagating waves, whereas the grating $E_1 E_3^*$ is recorded in the same way as for immovable molecules. At $T > (Dk^2\theta_{13}^2)^{-1}$ the remaining grating $E_1 E_3^*$ is also blurred by diffusion. The contribution of the grating δN_{13} is a factor $4/\theta_{13}^2$ greater than that of δN_{23}; for example, at $\theta_{13} \sim 10^{-2}$ rad this factor is 4×10^4. If we deal with pulses of a finite duration t, then the times T, $(4Dk^2)^{-1}$, $(Dk^2\theta_{13}^2)^{-1}$ should be compared with t, see (7.2.10).

In the situation described by (7.2.10), the FWM interaction possesses spectral selectivity with respect to the signal frequency ω_3. This selectivity is caused by the sluggishness of the recording mechanism. If the reference waves have the same frequency, $\omega_1 = \omega_2$, the reflectivity maximum is achieved at $\omega_3 = \omega_1$ and the width of the FWM filter is determined by Γ_{13}, which is smaller than Γ_{23}. If the frequencies of the references waves differ, $\omega_1 \neq \omega_2$, the FWM filter has two maxima: at $\omega_3 = \omega_1$ with a width Γ_{13} and reflectivity proportional to Γ_{13}^{-1}, and at $\omega_3 = \omega_1$ with a width Γ_{23} and reflectivity proportional to Γ_{23}^{-2}.

We now estimate the characteristic times of diffusive blur of the gratings δN_{13} and δN_{23} at $P \sim 1$ atm, $v_T \sim 3 \times 10^4$ cm/s and $\theta_{13} = 10^{-2}$ rad. In this case $l_f \sim 3 \times 10^{-6}$ cm, $D \sim 0.1$ cm^2/s, and for $\lambda = 0.6$ μm we have (neglecting the relaxation time T) $\Gamma_{23}^{-1} \sim 2.5 \times 10^{-10}$ s, $\Gamma_{13}^{-1} \sim 10^{-5}$ s. In the infrared region, for $\lambda = 10$ μm, we have $\Gamma_{23}^{-1} \sim 0.6 \times 10^{-7}$ s and $\Gamma_{13}^{-1} = 2.5 \times 10^{-3}$ s.

In the other limiting case, for gases at a relatively low pressure, the collision width becomes less than the Doppler width. This occurs at $kl_f \gtrsim 1$, which corresponds to a buffer gas pressure of $P \leq 0.3$ atm in the visible region and to $P \leq 0.01$ atm in the infrared region ($\lambda \sim 10$ μm). In this case monochromatic radiation of frequency ω_i and wave vector k_i is absorbed only by particles whose velocity satisfies the Doppler condition (to an accuracy of the collision width Γ)

$$|\omega_i - \omega_0 - k_i \cdot v| \leq \Gamma, \tag{7.2.11}$$

where ω_0 is the transition frequency for an immovable particle. For instance, for a reference wave E_1 of frequency ω_1, this condition separates in the velocity space (v_x, v_y, v_z) a layer of thickness $\Delta v \sim \Gamma/k$ oriented perpendicular to the vector k_1 (Fig. 7.2). The copropagating signal wave E_3 of frequency ω_3 separates a layer of its own of the same thickness. Efficient recording of the interference grating of excited molecules by these fields is possible only if both waves interact with the same group of molecules, i.e., if both layers overlap. Such is the case if

$$|\omega_3 - \omega_1| \leq \Gamma \quad \text{and} \tag{7.2.12a}$$

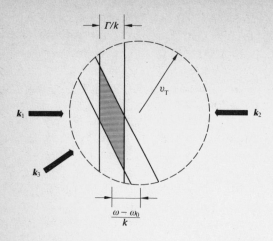

$$\theta_{13} \lesssim \Gamma/k\,v_{\mathrm{T}} = \Gamma/\Delta\,\omega_{\mathrm{D}}. \tag{7.2.13}$$

If, in addition, we want maximum read-out efficiency for this grating, the reference wave $E_2(\omega_2, \mathbf{k}_2)$ should satisfy the condition (7.2.10) for the same group of molecules:

$$|\omega_2 + \omega_3 - 2\omega_0| \lesssim \Gamma; \tag{7.2.12b}$$

the resonance condition of wave E_4 will then be satisfied since $\theta_{24} = \theta_{13}$.

The condition for recording the grating $E_2 E_3^*$ efficiently is similar:

$$|\omega_3 + \omega_2 - 2\omega_0| \lesssim \Gamma. \tag{7.2.14a}$$

The condition (7.2.13) must also be fulfilled. In order that the wave E_1 which reads out this grating, resonantly interacts with the group of molecules forming the grating, it is necessary that

$$|\omega_1 - \omega_3| \lesssim \Gamma. \tag{7.2.14b}$$

Thus, the condition (7.2.12a) for efficient recording of the grating δN_{13} coincides with the condition (7.2.14b) for efficient read-out of the grating δN_{23}, and vice versa.

From a viewpoint of spectroscopical experiments, for a given medium (i.e., for a given value of Γ and at $\Delta\,\omega_{\mathrm{D}} \gg \Gamma$), it is possible to vary four parameters independently: $\omega_1 - \omega_0$, $\omega_2 - \omega_0$, $\omega_3 - \omega_0$, and the angle θ_{13}, where ω_0 is the transition frequency.

Let condition (7.2.13) be satisfied. Simultaneous satisfaction of the conditions (7.2.12a, b) and the equivalent conditions (7.2.14a, b) is possible if

$$\omega_1 + \omega_2 = 2\omega_0 \tag{7.2.15}$$

and $\omega_1 = \omega_3$. These equalities need be satisfied only to an accuracy of the order of the homogeneous width Γ. The non-symmetry of the condition $\omega_1 - \omega_3$ with respect to the reference waves E_1 and E_2 is connected with the fact that it is the wave E_1 which copropagates with the signal E_3.

The dependence of the efficiency of OPC-FWM on the frequency ω_3 of the signal wave is of particular interest for practical purposes. With (7.2.15) satisfied, the qualitative dependence of the square of the signal nonlinearity coefficient on ω_3 has a peak centered at $\omega_1 = \omega_3$ of width $\sim \Gamma$, Fig. 7.3a. The peak value is a maximum not only if $\omega_1 + \omega_2 = 2\omega_0$, but also when $\omega_1 = \omega_0$ or $\omega_2 = \omega_0$. If ω_1 deviates from the transition frequency, the maximum reflectivity decreases following the Doppler contour of width $\Delta\omega_D$.

Tuning the frequency $\omega_1 + \omega_2$ away from $2\omega_0$ makes it impossible to satisfy both conditions (7.2.12a, 14a) simultaneously. If only resonance condition (7.2.12a) is satisfied, the grating δN_{13} is recorded efficiently. However, the read-out of this grating by the wave E_2 is provided not by the resonant maximum of the spectral line of moving particles but by the dispersion wing, with a smaller conjugated wave amplitude, $E_4 \propto [\Gamma + i(\omega_2 + \omega_3 - 2\omega_0)]^{-1}$. Similarly, if only (7.2.14a) is satisfied, the grating δN_{23} is recorded efficiently. Its read-out is provided by the dispersion wing $E_4 \propto [\Gamma + i(\omega_1 - \omega_3)]$. Thus, if there is a certain deviation from the condition (7.2.15), the maximum efficiency diminishes and the profile of the reflectivity spectral line is split into two lines, each with a width approximately equal to the collision width Γ. The qualitative interpolation formula describing these effects is (Fig. 7.3b)

$$|E_4/E_3|^2 \propto |\gamma|^2 \propto \frac{\exp\left[-\left(\dfrac{\omega_1 - \omega_0}{\Delta\omega_D}\right)^2 - \left(\dfrac{\omega_2 - \omega_0}{\Delta\omega_D}\right)^2\right]}{[(\omega_1 - \omega_3)^2 + \Gamma^2][(\omega_2 + \omega_3 - 2\omega_0)^2 + \Gamma^2]}. \qquad (7.2.16)$$

In these considerations we have not accounted for the blurring of the interference gratings $\delta N_{13} \propto \exp(i\mathbf{q}_{13} \cdot \mathbf{R})$ and $\delta N_{23} \propto \exp(i\mathbf{q}_{23} \cdot \mathbf{R})$ due to the motion of the particles. To estimate such blurring, we should compare the accumulation time T of particles on the upper level (population relaxation time) with the time taken by a molecule to travel the interference grating

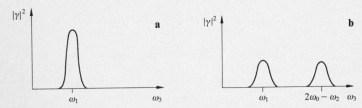

Fig. 7.3a, b. Qualitative dependence of the nonlinearity coefficient $|\gamma|^2$ on the signal frequency ω_3 (a) at $\omega_1 + \omega_2 = 2\omega_0$, (b) at $\omega_1 + \omega_2 \neq 2\omega_0$

period. Suppose that particles move in straight lines with a constant velocity v. For copropagating gratings the spatial period is $\Lambda_{13} = \lambda/\theta_{13}$, and the velocity spread in a group of excited particles projected on $q_{13} = k_1 - k_3$ is $\Delta v_{13} \sim v_T$. Hence, the blurring time of this grating is $\tau_{13} \sim \lambda(\theta_{13} v_T 2\pi)^{-1} \sim (\theta_{13} \Delta \omega_D)^{-1}$. For the counterpropagating grating, $\Lambda_{23} = \lambda/2$ and the velocity spread projected on $q_{23} = k_2 - k_3$ is $\Delta v_{23} \sim v_T \Gamma/\Delta \omega_D$. Therefore, the blurring time of such a grating is $\tau_{23} \sim \lambda(4\pi\Delta v_{23})^{-1} \sim \Gamma^{-1}$. As in the diffusion problem, the gratings of counterpropagating waves are blurred more rapidly since the angle θ_{13} has to satisfy the inequality $|\theta_{13}| \lesssim \Gamma/\Delta \omega_D$.

Typical widths of FWM filters realized experimentally in resonant gases are $\lesssim 50\,\mathrm{MHz}$ [7.9, 10].

Our reasoning has concerned the reflectivity estimated in the Born approximation for weak saturation. There are many publications (see the reviews [7.8, 11]) devoted to strong coupling, deep saturation, Rabi oscillations, transient processes, and to a number of other spectroscopic and polarization effects. There is also a separate series of experiments on OPC-FWM in media with a two-quantum resonance, $\omega_0 \approx 2\omega$. All these questions are rather of interest to nonlinear laser spectroscopy and are somewhat outside the problems discussed in our book.

7.2.3 Absorbing Dyes and Thermal Effects

The use of dye solutions as nonlinear media for quantum electronics was originally based on the effect of saturation of their absorptivity (the bleaching effect) in powerful radiation fields. A dye solution was used in one of the pioneer experiments [7.12] on FWM in a scheme with counterpropagating reference waves. The nonlinearity of dye saturation is often described approximately using the model of a two-level medium. The corresponding formula of the Born approximation is (6.4.4), and for a high reflectivity the theory of coupled waves yields (6.5.10).

However, most experiments on OPC-FWM in dyes showed that pure thermal effects play a major role in the recording of the interference pattern, particularly if the absorbed energy relaxes into heat sufficiently rapidly (on the scale of pulse duration). Spatially inhomogeneous heat release leads to inhomogeneous changes in the permittivity. The nature of such disturbances largely depends on the relationship between the pulse duration and the characteristic times of physical processes in the medium.

Suppose that the absorbed energy relaxes into heat instantaneously. Then a temperature change δT in a condensed medium with specific heat $c_p \approx c_v$ is given by

$$\varrho c_p \left[\frac{\partial(\delta T)}{\partial t} - \chi \nabla^2(\delta T) \right] = \alpha n c |E(R, t)|^2/8\pi . \tag{7.2.17}$$

Here ϱ is the density, $\chi\,[\text{cm}^2/\text{s}]$ is the thermal diffusivity, and $\alpha\,[\text{cm}^{-1}]$ is the absorptivity. The permittivity change is given by

$$\delta\varepsilon = \left(\frac{\partial\varepsilon}{\partial T}\right)_\varrho \delta T + \left(\frac{\partial\varepsilon}{\partial\varrho}\right)_T \delta\varrho. \qquad (7.2.18)$$

In a standard scheme where the signal beam E_3 makes a small angle θ_{13} (in air) with the reference wave E_1 and $\theta_{23} \approx 180°$, the heat release inhomogeneities consist of two gratings with periods $\Lambda_{13} \approx \lambda/\theta_{13}$ and $\Lambda_{23} \approx \lambda/2n$, where n is the refractive index.

For a grating with period Λ thermal expansion does not occur for the time interval $t \ll \Lambda/2v_s$, $\delta\varrho \simeq 0$ and thus $\delta\varepsilon \simeq (\partial\varepsilon/\partial T)_\varrho \delta T$, which is sufficiently small. A noticeable change in the permittivity can be achieved at $t \gtrsim \Lambda/2v_s$, for which the sound wave travels half the period of the grating. Here v_s is the sound velocity. In this case $\delta\varrho \simeq (\partial\varrho/\partial T)_p \delta T$ and then we have

$$\delta\varepsilon(R,t) \simeq \left[\left(\frac{\partial\varepsilon}{\partial T}\right)_p + \left(\frac{\partial\varepsilon}{\partial\varrho}\right)_T \left(\frac{\partial\varrho}{\partial T}\right)_p\right]\delta T(R,t) = \left(\frac{\partial\varepsilon}{\partial T}\right)_p \delta T. \quad (7.2.19)$$

It is the mechanism of thermal expansion which contributes substantially to $\delta\varepsilon$ and gives $(\partial\varepsilon/\partial T)_p < 0$.

It follows from (7.2.17) that if the field is switched on quickly, the temperature grating with a spatial period Λ is built up as

$$\delta T \propto \Gamma^{-1}[1 - \exp(-\Gamma t)], \qquad \Gamma = \chi q^2 \equiv \chi\left(\frac{2\pi}{\Lambda}\right)^2. \qquad (7.2.20)$$

If the intensity is switched on continuously, then for times $t \lesssim \Gamma^{-1}$ the grating amplitude increases proportional to the absorbed energy, and at $t \gtrsim \Gamma^{-1}$ the recording level becomes stabilized. Thus, pulses of duration $\tau_p \lesssim \Gamma^{-1}$ are the most suitable for achieving the highest energy efficiency in recording.

These two requirements, i.e., the necessity for the unloading wave to travel a certain distance and the absence of blurring in the thermal grating due to heat conduction, place the following limitations on the pulse duration:

$$\Lambda/2v_s \lesssim \tau_p \lesssim \Lambda^2/(2\pi)^2\chi. \qquad (7.2.21)$$

If the second condition of (7.2.21) is satisfied, the nonlinearity is accumulated, i.e., thermal FWM proceeds in the nonstationary regime. At $\lambda = 1\,\mu\text{m}$, $n = 1.5$, $\chi = 10^{-3}\,\text{cm}^2/\text{s}$, $v_s = 10^5\,\text{cm/s}$ the gratings of counterpropagating waves make a noticeable contribution to FWM reflection only for nanosecond pulses, since for such gratings (7.2.21) can be written as $1.5\times10^{-10} \lesssim \tau_p \lesssim 3\times10^{-8}$ [s]. For the gratings of copropagating waves, (7.2.21) can be written, for the same parameters, as $5 \times 10^{-10} \theta_{13}^{-1} \lesssim \tau_p \lesssim 3 \times 10^{-7} \theta_{13}^{-2}$. For long laser pulses, $\tau_p \sim 10^{-3}$ s, this inequality yields $\theta_{13} \lesssim 1.7 \times 10^{-2}$ rad. In particular, at $\theta_{23} = 10^{-2}$ rad the gratings of copropagating waves are recorded efficiently at $5 \times 10^{-8} \lesssim \tau_p \lesssim 3 \times 10^{-3}$.

If condition (7.2.21) holds true for both gratings, the reflectivity increases approximately as the square of the energy density of the reference waves:

$$|E_4/E_3|^2 \simeq (W_1 t)(W_2 t) \left(\frac{\omega}{nc}\right)^2 \left[\left(\frac{\partial \varepsilon}{\partial T}\right)_p \frac{1}{\varrho c_p}\right]^2 (\alpha L)^2. \tag{7.2.22}$$

For nanosecond pulses, $\tau_p \sim 10^{-8}$ s, (7.2.22) is valid at $\theta_{13} \gtrsim 2 \times 10^{-2}$ rad. For nanosecond pulses at $\theta_{13} \lesssim 10^{-2}$ rad, only the counterpropagating wave grating works, whereas for long pulses, $\tau_p \gg 3 \times 10^{-8}$ s, only the copropagating wave grating works. In both cases, the right-hand side of (7.2.22) should be reduced by a factor of four.

Equation (7.2.22) is written for $\alpha L \ll 1$; at $\alpha L \sim 1$ we should make, in accord with (6.4.4), the substittion $L \to \alpha^{-1} \exp(-\alpha L)[1 - \exp(-\alpha L)]$, $W_1 W_2 \to W_1^2$. At $\exp(-\alpha L) = 0.5$ [see discussion of (6.4.4), taking this substitution into account] and $(\partial \varepsilon/\partial T)_p = 10^{-4} \text{K}^{-1}$, $\varrho c_p \simeq 1$ J/cm^3 K, we can write (7.2.22) as $|E_4/E_3|^2 \sim 1 \cdot (W_1 t)^2$, where the energy density of the reference wave $W_1 t$ is expressed in J/cm^2. The reflectivity value ~ 1 should be achieved at $W_1 t \sim 1$ J/cm^2, which agrees with experimental results [7.13].

Considerable distortions of the reflected wave which sharply worsen the OPC fraction are usually observed in experiments with an FWM efficiency of ~ 1. One of the sources of such distortions is a thermal lens induced by the reference waves. Its appearance requires a time $\sim a/v_s$, where a is the transverse dimension of the reference beams. Thus, for instance, at $a \sim 1$ cm the thermal lens cannot be excited at $\tau_p \ll 10^{-5}$ s.

7.2.4 Gain Saturation

The major advantage of using amplifying saturable medium as the material for OPC-FWM is that FWM can be realized immediately since the laser active medium uses the radiation generated inside the resonator as the counterpropagating reference waves. Since the process proceeds in the laser active medium, both the incident and reflected signals experience additional amplification. Therefore, it is possible to obtain efficient phase conjugation even for weak saturation [7.14–17].

7.3 Nonlinearities in Semiconductors

Semiconductors represent universally recognized materials with a high optical nonlinearity. There exist many different mechanisms causing semiconductor nonlinearity; some of them will be discussed below; see also [7.18].

7.3.1 Bound Electron Nonlinearity

The contribution of bound electrons to optical nonlinearity can be described qualitatively in terms of anharmonic oscillators. Let us write the potential energy of an electron in the form $U(x) = kx^2/2 - \tilde{v}x^3/3 - \tilde{\mu}x^4/4$. For strongly bound electrons, the natural frequency $\omega_0 = \sqrt{k/m}$ (m is the electronic mass) may be considered much larger than the frequency ω of the light field. Then the electron will follow the field quasi-statically:

$$\frac{\partial U}{\partial x} = kx - \tilde{v}x^2 - \tilde{\mu}x^3 = e\mathcal{E}_{\text{real}}. \tag{7.3.1}$$

To estimate the coefficients k, $\tilde{v}$, and $\tilde{\mu}$, it is convenient to introduce the radius x_0 of the electron orbit and the intra-atomic field strength E_0. Suppose that $k = eE_0/x_0$, so that the linear part of the restoring force kx_0 at a distance x_0 is equal to eE_0. We may also write $\tilde{v} = veE_0/x_0^2$, $\tilde{\mu} = \mu eE_0/x_0^3$, where v and μ are the dimensionless anharmonicity characteristics; in the general case $|v|$ and $|\mu|$ are of the order of 1 and $v \equiv 0$ for centrosymmetric media. Since the intra-atomic field is determined by the Coulomb attraction of the electron to the positive atomic ion, we should take $E_0 = e/x_0^2$. Then from (7.3.1) we can obtain the solution for x expressed as a power series in $\mathcal{E}_{\text{real}} = \mathcal{E}$:

$$x = x_0[\mathcal{E}/E_0 + v(\mathcal{E}/E_0)^2 + (2v^2 + \mu)(\mathcal{E}/E_0)^3 + \ldots]. \tag{7.3.2}$$

The dipole moment per unit volume is $P_{\text{real}} = Nex$, where N is the electron number density. Taking into account that $\mathcal{E} = 0.5[E\exp(-i\omega t) + E^*\exp(i\omega t)]$, we can express the linear $\chi^{(1)}$, second-order $\chi^{(2)}$, and self-focusing third-order $\chi^{(3)}$ polarizabilities as

$$\chi^{(1)} = Nx_0^3, \quad \chi^{(2)} = \frac{v}{2}Nx_0^3/E_0, \quad \chi^{(3)} = \frac{3}{8}(2v^2 + \mu)Nx_0^3/E_0^2. \tag{7.3.3}$$

Let us estimate these quantities numerically. In most dielectrics $Nx_0^3 \sim 0.1$ and thus the refractive index $n = (1 + 4\pi\chi^{(1)}) \approx 1.5$. In dielectrics, $x_0 \sim 10^{-8}$ cm and $E_0 = e/x_0^2 \sim 0.5 \times 10^7$ cgs so that $\chi^{(2)} \approx v \cdot 10^{-8}$ cgs. Experimental values of $\chi^{(2)}$ for non-centrosymmetrical dielectrics are of the order of

10^{-9} cgs, this corresponding to the second-order anharmonicity coefficient $v \sim 0.1$. Finally, according to this model, $\chi^{(3)} = (\mu + 2v^2) \cdot 0.15 \times 10^{-14}$ cm^3/erg. In actual fact, the electronic nonlinearity in dielectrics is characterized by values of $\chi^{(3)}$ between 4×10^{-15} and 4×10^{-16} cm^3/erg, and thus $\varepsilon_2 = 8\pi\chi^{(3)} \sim 10^{-13} - 10^{-14}$ cm^3/erg, corresponding to $\mu \approx 2.5 - 0.25$.

For semiconductors $\varepsilon = 1 + 4\pi\chi^{(1)} \approx 10 - 20$, from which it follows that the electron packing density parameter $Nx_0^3 = \chi^{(1)} \sim 1 - 2$, i.e., approximately an order of magnitude larger than in dielectrics. It is connected, in particular, with a larger orbital size x_0. If we assume $x_0 = 2 \times 10^{-8}$ cm, then $E_0 \approx 10^{-6}$ cgs and $\chi^{(3)} \approx \mu \cdot 10^{-12}$ cm^3/erg. Experimental values of $\chi^{(3)}$ are of the order of $10^{-12} - 10^{-11}$ cm^3/erg; in our rough model this would correspond to $\mu \sim 1 - 10$.

Therefore, the bound-electron third-order nonlinearity in semiconductors proves to be $3 - 4$ orders of magnitude larger than that in solid dielectrics. The most characteristic feature of the bound-electron nonlinearity is its fast response: it is switched on/off instantaneously (on the light period scale) together with the fields acting on the crystal. This nonlinearity may therefore be used for FWM of waves of strongly differing frequencies.

The bound-electron nonlinearity has been used for OPC-FWM in the infrared region at the 10.6 μm CO$_2$-laser wavelength. Germanium was the most popular medium, see [7.19 – 22], for instance. The distinguishing feature of the experimental setup used in [7.19] was a germanium plate (a nonlinear medium) positioned inside the resonator of the laser which generated the reference waves. Owing to this feature the power density of the reference waves in the nonlinear medium was sufficiently high, their exact counterpropagation being achieved automatically.

7.3.2 Nonparabolicity of Conduction Band

In addition to bound electrons (in semiconductors these are in the occupied valence band), free electrons of the conduction band can contribute to the third-order nonlinearity if their density is sufficiently high. The nonlinear character of their motion in the light field is caused by the nonparabolic nature of the energy vs. quasi-momentum dependence. Inasmuch as this nonlinearity does not correspond to the changes in occupation of any resonant levels, it possesses a rather short build-up time, $\tau \sim \hbar/\Delta E \sim 10^{-14} - 10^{-15}$ s, where ΔE is the energy width of the conduction band.

Relaxation and band-to-band transitions being neglected, i.e., at $\hbar\omega \lesssim E_g$ (E_g is the band gap) the change to the quasi-momentum $p(t)$ of an electron residing in a certain energy band caused by the light wave $E(t)$ is given by

$$\frac{dp}{dt} = \frac{e}{2}[Ee^{-i\omega t} + E^*e^{i\omega t}] . \tag{7.3.4}$$

The solution of this equation is trivial: $p(t) = 0.5 \cdot [ieE\exp(-i\omega t)/\omega + \text{c.c.}]$. The nonlinearity of the response, i.e., of the polarization $P = Nex$, is connected with the nonlinear dependence of $v = dx/dt$ on the quasi-momentum. Indeed, according to general theorems of classical and quantum mechanics,

$$\frac{dx}{dt} \equiv v = \frac{\partial W(p)}{\partial p}, \tag{7.3.5}$$

where $W(p)$ is the energy vs. quasi-momentum dependence for a given band (the conduction band in our particular case). In the region near the band bottom, we may take the first two terms in the power series expansion of $W(p)$ in terms of p^2:

$$W(p) = \frac{p^2}{2m^*} + \frac{A}{24}p^4 + \dots, \qquad A = \frac{d^4 W}{dp^4}\bigg|_{p=0}. \tag{7.3.6}$$

Then from (7.3.5) we have $dx/dt = p(t)/m^* + Ap^3(t)/6$. Integrating this relation yields

$$p(t) = \frac{N}{2}\left[-\frac{e^2 E e^{-i\omega t}}{m^*\omega} - \frac{Ae^4 EEE^* e^{-i\omega t}}{16 \cdot \omega^4} + \frac{Ae^4 EEE e^{-3i\omega t}}{16 \cdot 9 \cdot \omega^4} + \text{c.c.} \right]. \tag{7.3.7}$$

The third-order term $\propto \exp(-i\omega t)$ in (7.3.7) corresponds to the self-focusing nonlinearity with the constant $\chi^{(3)} = -Ae^4 N/8\omega^4$. In the Kane model [7.23] it is supposed that $W(p) = 0.5 E_g[(1 + 2p^2/E_g m^*)^{1/2} - 1]$, where m^* is the effective mass near the band bottom. In this case $A = -6/E_g(m^*)^2$ and

$$\chi^{(3)} = \frac{3}{4}\frac{Ne^4}{E_g(m^*)^2\omega^4}. \tag{7.3.8}$$

The electron density N in the conduction band is determined either by temperature (in intrinsic semiconductors), or by the amount of impurities, the intensity of the exciting radiation, and other factors. If we assume $\lambda = 2\pi c/\omega = 10$ μm, $E_g \approx 0.5$ eV $= 0.8 \times 10^{-12}$ erg, $m^* = 0.1\,m_e = 10^{-28}$ g, then using (7.3.8) we have $\chi^{(3)} = (N/2 \times 10^{16}) \cdot 0.75 \times 10^{-10}$ cm^3/erg, where N is expressed in cm^{-3}. An experiment on OPC-FWM making use of the nonlinearity of this type was carried out in [7.24], see also [7.25].

7.3.3 Generation of Free Carriers

If the photon energy $\hbar\omega$ is close to the band gap E_g, light absorption leads to generation of free electrons in the conduction band and an equal number of holes in the valence band, Fig. 7.4. If $N = N_e = N_h$, we may write the equation

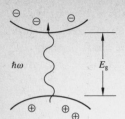

Fig. 7.4. If quantum energy $\hbar\omega$ is close to the band-gap energy E_g of semiconductor, light absorption gives rise to generation of free carriers (electrons) in the conduction band and holes in the valence band

$$\frac{\partial N}{\partial t} - D\nabla^2 N + \frac{1}{\tau_R}N = \eta\,\alpha\,\mathscr{G}(R,t)\,, \tag{7.3.9}$$

where $\mathscr{G}$ [quanta/cm^{2}s] is the light intensity, α is the absorptivity, $\eta \sim 1$ is the quantum efficiency of pair generation, D is the ambipolar diffusion coefficient, and τ_R is the time of electronic relaxation into the valence band. If free electrons and holes are excited, the permittivity of the semiconductor varies by an amount proportional to N. In the simplest model [7.18, 26],

$$\Delta\varepsilon = -(\omega_p^2/\omega^2)f\,. \tag{7.3.10}$$

Here $\omega_p^2 = 4\pi Ne^2(m_e^{*-1} + m_h^{*-1})$ is the plasma frequency for an electron-hole plasma. The correction factor $f \sim 1$ allows for a reduction in the contribution to ε of band-to-band transitions due to electrons being knocked out of the valence band.

Even for angles which are not too large, $\theta_{13} \sim 10^{-2}$ rad, the diffusive relaxation time for the grating $\delta\varepsilon_{13}$, at $D \sim 30$ cm^2/s, $\lambda = 1$ μm, and $q = k\theta_{13} = 3\times10^3$ cm^{-1}, is $\tau_{13} = (Dq^2)^{-1} \sim 3\times10^{-9}$ s, i.e., very short. The intrinsic recombination time is usually $\tau_R \gtrsim 10^{-6}$ s. Therefore, for typical pulse durations $\tau \sim 10^{-8} - 10^{-7}$ s, FWM proceeds in the stationary regime with the accumulation time τ_{13}; on the contrary, the gratings of the counterpropagating waves are blurred almost instantly and are thus of no importance. As a result, we have for N:

$$\delta N_{13} = \eta\,\alpha\tau_{13}ncE_1E_3^*(8\pi\hbar\omega)^{-2} \tag{7.3.11}$$

and the effective constant $\chi^{(3)}$ is given by

$$\chi^{(3)} = -\frac{e^2\eta\alpha nc\tau_{13}}{8\pi\hbar\omega^3}\left(\frac{1}{m_e^*} + \frac{1}{m_h^*}\right)f\,. \tag{7.3.12}$$

For $\theta_{13} = 10^{-2}$ rad, $m^* \sim 3\times10^{-28}$ g, $\lambda = 1$ μm, $\alpha \sim 10$ cm^{-1}, $\eta \sim 1$, $f \sim 10$, we have $\chi^{(3)} \sim 1.5\times10^{-9}$ cm^3/erg. We note the very sharp dependence ($\propto\theta_{13}^{-2}$) of τ_{13} and $\chi^{(3)}$ on the angle θ_{13}.

Experiments on OPC-FWM using this sort of nonlinearity are many; some are reported in [7.27 – 29].

In addition to the mechanisms discussed in Sects. 7.3.1 – 3, the following mechanisms which are possible in semiconductors have been investigated experimentally: saturation of intraband and band-to-band transitions, transitions where impurities or excitons participate processes with multiphoton saturation. A detailed account of these mechanisms is given in the review [7.18].

7.3.4 Photorefractive Crystals

We give here a brief account of FWM in the so-called photorefractive crystals. These are characterized by a strong linear electrooptical effect (the Pockels effect) which manifests itself in refractive index variations $\delta n = n_0^3 r E_0 / 2$, where r is the electrooptical coefficient, E_0 is the static field strenght, and n_0 is the refractive index. Typical representatives of photorefractive materials are ferroelectrics $LiNbO_3$, $LiTaO_3$, $BaTiO_3$, etc., and such semiconductors as CdS, CdTe, $Bi_{12}SiO_{20}$, $Bi_{12}GeO_{20}$, etc. The impurities available in photorefractive crystals can be ionized under the action of light to send electrons into the conduction band, this leading to a refractive index variation. Of many various mechanisms, we shall describe the so-called diffusion mechanism.

At the instant of photoionization no macroscopic charge density is produced since the total charge of the electrons is exactly balanced by the positive charge of ionized impurities. However, owing to their high mobility the electrons quickly start to diffuse from excitation places, whereas the ions remain on their sites. The action of the spatially inhomogeneous excitation in conjunction with the diffusion of electrons towards lower-luminosity regions gives rise to an inhomogeneous spatial charge distribution and, as a consequence, to an inhomogeneous static field which modulates, by the Pockels effect, the refractive index.

To estimate δn in the framework of the diffusion mechanism, we should take account of the fact that the induced static field produces a conduction current of electrons in the direction opposite to the diffusive current. The field strength value in the quasi-static regime is determined by the condition whereby these two flows compensate each other:

$$N(R)\mu E_0(R) = D \nabla N(R) . \qquad (7.3.13)$$

Here μ is the mobility, D is the diffusion coefficient, $D = \mu k_B T / e$ in agreement with the Einstein relation, and e is the electronic charge. As electric current is almost absent, we may write the local equation for the electron density in the grating of the copropagating waves E_1 and E_3 as

$$\frac{\partial N}{\partial t} = \frac{\alpha n c |E_1 \exp(i k_1 \cdot R) + E_3 \exp(i k_3 \cdot R)|^2}{8 \pi \hbar \omega} + \left(\frac{\partial N}{\partial t} \right)_{rel} , \qquad (7.3.14)$$

where $(\partial N/\partial t)_{rel}$ allows for the catching of conduction band electrons by traps (acceptor levels) and for recombination with ionized impurities. If recombination is neglected, we may write

$$N(\boldsymbol{R}, t) = N_0 + A(t)[\,|E_1|^2 + |E_3|^2 + E_1 E_3^* \exp(i\boldsymbol{q} \cdot \boldsymbol{R})$$
$$+ E_1^* E_3 \exp(-i\boldsymbol{q} \cdot \boldsymbol{R})]\,, \tag{7.3.15}$$

where $\boldsymbol{q} = \boldsymbol{k}_1 - \boldsymbol{k}_3$ and N_0 is the equilibrium density of free carriers in the absence of illumination. Substituting the space-averaged electron density $\bar{N} \propto |E_2|^2 + |E_1|^2$ into the left side of (7.3.13) yields (at $|E_2|^2, |E_3|^2 \ll |E_1|^2$)

$$E_0(\boldsymbol{R}) = \frac{k_B T}{e} \cdot \frac{i\boldsymbol{q}\, E_1 E_3^* \exp(i\boldsymbol{q} \cdot \boldsymbol{R}) + \text{c.c.}}{|E_1|^2 + N_0/A(t)}\,. \tag{7.3.16}$$

Finally, the refractive index disturbance is

$$\delta n(\boldsymbol{R}) = i\boldsymbol{q}\, \frac{n_0^3 k_B T}{2e}\, r\, \frac{E_1 E_3^* \exp(i\boldsymbol{q} \cdot \boldsymbol{R})}{|E_1|^2 + N_0/A(t)} + \text{c.c.}\,. \tag{7.3.17}$$

Equation (7.3.17) demonstrates a number of specific features of the photorefractive response in the diffusion mechanism. First, at $|E_1|^2 \gg N_0/A$ the disturbance δn is inversely proportional to the amplitude of the reference wave E_1, this being connected with the short-circuiting effect of the uniform background of the electron density in the conduction band. Second, the grating δn appears to be phase-shifted by $\pi/2$ [the factor i in (7.3.17)] with respect to the interference pattern $E_1 E_3^* \exp(i\boldsymbol{q} \cdot \boldsymbol{R})$ of the interacting fields. As a result, energy interchange appears possible [7.30] between the waves E_1 and E_3 by a mechanism similar to that of stimulated light scattering, cf. Sect. 2.1. Finally, δn appears to be proportional to the wave vector $\boldsymbol{q}$ of the grating, i.e., it depends sharply on the angle between the interfering waves.

The phase-conjugated wave arises as a result of the scattering of the second reference wave $E_2 \exp(i\boldsymbol{k}_2 \cdot \boldsymbol{R})$ by the grating (7.3.17), and in the Born approximation we have

$$|E_4/E_3|^2 = B^2 q^2 \frac{I_1 I_2}{[I_1 + I_2 + N_0/A(t)]^2}\,, \tag{7.3.18}$$

where $B = \omega n_0^3 k_B T r L/2ec$; the contribution of the second reference wave to the background illumination is taken into account. Let us estimate numerically the reflectivity. At $I_1 = I_2 \gg N_0/A$, $q = (\omega/c) 2 n_0 \sin(\theta_{13}/2)$, $\theta_{13} \approx 20°$, $\omega/c = 10^5$ cm^{-1}, $n_0 = 1.5$, $L = 1$ cm, $r = r_{33} \approx 10^{-6}$ cgs, and at room temperature, we have $|E_4/E_3|^2 \sim 4\%$. It is important to note that even if relaxation is neglected, a sufficiently high energy density $\int I_{1,2} dt$ of the reference waves is required in order that the photoelectron density in the conduction band should be noticeably higher than the equilibrium value N_0. Thus, for instance, for nominally pure $LiNbO_3$, $LiTaO_3$ [7.31, 32] the incident energy density $\int I\,dt \sim 1 - 10$ J/cm^2 appeared quite sufficient. If the incident energy is increased further (by increasing either the intensity or the exposure), the reflectivity $|E_4/E_3|^2$ reaches its steady-state value.

In continuous operation at the He $-$ Cd-laser wavelength $\lambda = 0.44$ μm at a reference wave power density of ~ 1 W/cm^2 [7.31], the reflectivity reached its steady-state value for a time of the order of 20 s.

A slightly different photorefractive response mechanism can be observed if a sufficiently strong external electrostatic field E_0 is applied to the crystal [7.33]. This field carries electrons away from positive ions, resulting in the spatial inhomogeneity of the charge, field, and refractive index.

An interesting feature of photorefractive media is that in some cases dynamic holograms recorded in them are stored for a long time after exposure to light has ceased. The recorded information can be erased by, for example, thermal exposure or uniform illumination of the crystal.

One more specific feature of photorefractive nonlinearity is the very strong anisotropy of the coefficient r. Thus, for example, if q is directed along an optical axis in $LiNbO_3$, the coefficient r_{33} is approximately 5 times larger than r_{22} with q oriented transverse to the axis.

There are many experiments in which OPC-FWM by photorefractive media has been realized [7.30 $-$ 37]. In [7.36, 37] it was possible, due to the low intensity level of the reference waves, to realize a resonator with PCM in the continuous operation of a lower-power laser.

7.4 SBS Nonlinearity and Parametric SBS Generation

Particularly interesting results have been obtained in investigations of OPC-FWM using SBS nonlinearity, in which counterpropagating waves of slightly different frequencies resonantly build up hypersonic waves [7.38 $-$ 41]. The SBS nonlinearity can produce FWM phase conjugation only due to the grating $\delta\varepsilon_{23} \propto E_2 E_3^* \exp[-i(k_2 + k_3)z - i(\omega_2 - \omega_3)t]$. For this interference term to efficiently excite hypersound, the condition $|\omega_2 - \omega_3| = (k_2 + k_3)v_s \approx \Omega_B$ should be satisfied. Here $\Omega_B = 2kv_s$ is the frequency shift in backward SBS, v_s is the sound velocity, and the above condition should be satisfied to an accuracy of the scattering linewidth Γ. The signal E_3 may be either Stokes-shifted, $\omega_3 \approx \omega_2 - \Omega_B$, or anti-Stokes-shifted, $\omega_3 \approx \omega_2 + \Omega_B$, with respect to the counterpropagating reference wave E_2. Then we have at our disposal a free parameter $-$ the frequency ω_1 of the copropagating reference wave E_1. The phase matching condition specifies this frequency to be in the vicinity of

ω_3 to an accuracy of $\pi c/L$, where L is the effective interaction length: $|\omega_1 - \omega_3| \lesssim \pi c/Ln$. At $L \sim 3$ cm, $n = 1.5$ we have $|\omega_1 - \omega_3|/2\pi c \lesssim 10^{-1}$ cm^{-1}, this exceeding, as a rule, the shift $\Omega_B/2\pi c \sim 0.05$ cm^{-1} in SBS. The phase-conjugated wave E_4 has the frequency $\omega_4 = \omega_1 + \omega_2 - \omega_3$, i.e., either $\omega_4 = \omega_1 + \Omega_B$ or $\omega_4 = \omega_1 - \Omega_B$. In both cases, the interference of the fields E_1 and E_4^* resonantly builds up the same hypersonic wave, the hypersound amplitude being

$$Q + Q^* = \text{const} \cdot \left[\frac{(E_3^* E_2 + E_4 E_1^*) \exp(-2ikz + i\Omega t)}{\Omega_B^2 - 2i\Omega\Gamma - \Omega^2} + \text{c.c.} \right], \qquad (7.4.1)$$

where $\Omega = \omega_3 - \omega_2$. The change in the amplitude of the wave E_4 propagating in the $(-z)$ direction is determined by the scattering of the wave $E_1 \exp(ikz - i\omega_1 t)$ by the grating $\delta\varepsilon \propto Q$:

$$\frac{dE_4}{dz} = \frac{G}{2}(|E_1|^2 E_4 + E_1 E_2 E_3^*) m^*, \qquad (7.4.2)$$

where $m = 2\Omega_B \Gamma [i(\Omega_B^2 - \Omega^2) - 2\Omega\Gamma]^{-1}$. In fact, $\Omega = \Omega_B + \Delta$ or $\Omega = -\Omega_B + \Delta$, where $|\Delta| \lesssim \Gamma \ll \Omega_B$, is of interest only. Thus the expression for m can be simplified:

$$m = -\frac{\Omega}{|\Omega|} \cdot \frac{1}{1 + i\Delta/\Gamma}. \qquad (7.4.3)$$

The coefficient G characterizes the SBS nonlinearity constant (Chap. 2). The change in the amplitude of the wave E_3 is determined by the scattering of the reference wave $E_2 \exp(-ikz - i\omega_2 t)$ by the grating $\delta\varepsilon \propto Q^*$, see (7.4.1). Hence, the system of coupled equations for E_3 and E_4^* is

$$\frac{dE_3}{dz} = \frac{G}{2} m(|E_2|^2 E_3 + E_1 E_2 E_4^*), $$

$$\qquad (7.4.4)$$

$$\frac{dE_4^*}{dz} = \frac{G}{2} m(|E_1|^2 E_4^* + E_1^* E_2^* E_3). $$

The particular case of this system with $\omega_1 = \omega_2$, $\Omega = +\Omega_B$ is considered in Sect. 2.7, whereas (7.4.4) is valid at an arbitrary value of ω_1 within the limits of the phase matching condition.

Different values of ω_1 correspond to different physical situations. If $\omega_1 \approx \omega_3$, we have deal OPC-SBS of a weak signal E_3 on the background of OPC-SBS of a powerful pump E_1; this is the so-called thresholdless OPC-SBS which was discussed in Sect. 5.7.

The most interesting results were obtained in schemes where the reference waves were of equal frequencies, $\omega_1 = \omega_2$. The wave E_2 was produced by reflection of the wave E_1 from a conventional mirror. Since in this case the SBS nonlinearity does not lead either to the self-action or to the interaction of the waves E_1 and E_2, the coefficients $|E_1|^2$, $|E_2|^2$, and $E_1 E_2$ in (7.4.4) can be considered constant, and it is thus possible to obtain an explicit analytical solution. For the boundary conditions $E_4(z=L) = 0$, $E_3(z=0) = E_3$, we have

$$\left| \frac{E_4(z=0)}{E_3(z=0)} \right|^2 = \frac{I_1 I_2 |1 - e^{\mu L}|^2}{|I_2 + I_1 e^{\mu L}|^2}, \qquad \mu = \frac{1}{2} m G(I_1 + I_2) . \tag{7.4.5}$$

Let us discuss the result obtained. Suppose that $\Omega = \Omega_B$, and the signal ω_3 is shifted, with respect to the reference waves $\omega_1 = \omega_2$, into the anti-Stokes region, $\omega_3 = \omega_1 + \Omega_B$. In this case $m = -1$ and at $|\mu L| = G(I_1 + I_2)L/2 \gtrsim 1$ we have $0 < \exp(\mu L) \ll 1$. If $I_1 = I_2$, then it follows from (7.4.5) that $|E_4/E_3|^2 \approx I_1/I_2$. To achieve a high reflectivity in this case, it is reasonable to decrease the intensity of the counterpropagating reference wave to the optimal value $I_2 \approx I_1 \exp(-GI_1 L/2)$ rather than increase it, and then we have $|E_4/E_3|^2 \approx (1/4)I_1/I_2 \approx (1/4) \exp(GI_1 L/2)$.

If ω_3 is shifted into the Stokes region, i.e., $\omega_1 = \omega_2 = \omega_3 + \Omega_B$, then $m = +1$ and the above considerations remain in force if the substitution $I_1 \rightleftarrows I_2$ is made. In particular, the optimum reflectivity is observed at a low intensity of the copropagating reference wave, $I_1 \approx I_2 \exp(-GI_2 L/2)$. In this case $|E_4/E_3|^2 \approx 0.25 \exp(GI_2 L/2)$.

We have so far dealt with the cases of precise resonance, $\Omega = \pm \Omega_B$. Though the reflectivity given by (7.4.5) may be high in such cases, it is always finite since the denominator does not vanish. If, however, there is detuning, $\Delta \neq 0$, a reactive component appears in the nonlinear coefficient $\propto m$. As a result, at a sufficient reference wave intensity an infinitely strong FWM reflection is possible, this being equivalent to parametric FWM generation. The condition for parametric generation is (2.7.12).

Parametric FWM generation by SBS nonlinearity has much in common with FWM generation by reactive fast-responding nonlinearities, which was considered in Chap. 6. Indeed, if the reference waves' intensity exceeds the threshold, the reflectivity for incoming weak signals quickly rises with time. This rise is limited either by the pump pulse duration or by depletion of a reference wave.

In addition to the actual input signal E_3, various kinds of parasitic scattering of the reference waves by medium inhomogeneities may serve as primings for FWM generation.

FWM-SBS differs, however, from reactive FWM in parametric SBS generation in two essential aspects, which enable one to achieve a highly efficient

OPC-FWM without a loss of quality. First, strong self-focusing distortions of the transverse structure of the reference waves are observed in reactive fast-responding nonlinear media. For small-angle disturbances of the counter-propagating reference waves, the rise of distortions corresponds to an absolute instability and, as shown in [7.42] (see Sect. 6.7), the threshold of this instability may be lower than in ordinary FWM generation. These distortions of the reference waves are likely to be the reason for the low quality of OPC under high-reflectivity conditions in reactive media [7.43]. Second, in the case of degenerate FWM ($\omega_1 = \omega_2 = \omega_3 = \omega_4$) parasitic signals, which also initiate generation, are plentiful due to the scattering of reference waves by static inhomogeneities of the optical setup.

The resonance SBS nonlinearity constant is usually noticeably larger than the Kerr (reactive) constant for the same medium. Therefore, at intensities corresponding to FWM-SBS generation, the self-focusing instability is absent and there are no distortions of the reference waves. Owing to a considerable frequency shift ($\Omega_B/2\pi \gtrsim 10^{10}$ Hz), the parasitic signal in the region $\omega = \omega_3$ may be produced from the reference waves only if they are scattered from thermal excitation of hypersound in the medium. The corresponding intensity is very small, a rough estimate being $I_3' \sim I_1 R L \Delta\Theta/4\pi$. Here, $R \sim 10^{-5}$ cm^{-1} is the spontaneous extinction coefficient in BS and $\Delta\Theta$ is the solid angle of signals for which FWM generation is allowed. At $L \sim 3$ cm, $\Delta\Theta \sim (D/L)^2$ $\sim 3 \times 10^{-3}$ sterad, we have $I_3' \sim 10^{-8} I_1$. In other words, if the intensity I_3 of the signal to be conjugated exceeds this figure (by a reasonable amount, in view of the rough estimate), then it is the signal itself which initiates FWM generation, and high-quality OPC with a high reflectivity can be realized.

OPC with a high reflectivity up to 7×10^5 was observed in FWM generation with over-threshold conditions, and the signal was conjugated without distortions [7.39]. The anti-Stokes signal, $\omega_1 = \omega_2 = \omega_3 - \Omega_B$, was used in this work. The intensity $|E_4|^2$ of the reflected wave was limited by saturation effects, so that the percentage of energy transferred from the reference wave E_1 amounted to $|E_4/E_1|^2 \sim 10\%$. If the intensity I_3 of the input signal was less than a certain level ($\lesssim 1$ W, $\tau_p \sim 10^{-8}$ s), the phase-conjugated wave was overwhelmed with noise initiated by spontaneous primings.

7.5 Literature

Experiments on OPC-FWM by orientational nonlinearity of liquid CS_2 were reported in [7.1 – 5] and by others. The coupled wave model for absorbing saturable media was suggested by *Abrams* and *Lind* [7.7]. OPC-FWM in resonant gases is described in [7.6, 8 – 11] and experiments on OPC in dyes are reported in [7.12, 13, 44 – 46]. Gain saturation nonlinearity in active media was investigated in application to OPC-FWM in [7.14 – 17]. Bound electron nonlinearity in germanium at $\lambda = 10.6\,\mu m$ is considered in [7.19 – 22] and free-electron nonlinearity in the nonparabolic conduction band in [7.24, 25]. For nonlinearity due to carrier generation in semiconductors see [7.26 – 29]. An excellent review of different nonlinearities in semiconductors which concern OPC-FWM is given in [7.18] and dynamic holography as a whole and photorefractive media are excellently reviewed in [7.30], see also [7.47]. The diffusion mechanism for recording photorefractive dynamic holograms was suggested in [7.48]. OPC-FWM in photorefractive crystals were first observed in [7.31, 43]; refer also to [7.30, 32 – 37]. References [7.38 – 41, 49, 50] are devoted to SBS nonlinearity in a FWM scheme.

8. Other Methods of OPC

In the previous chapters we have considered in detail the physics of two wide-spread OPC methods, namely, OPC by backward stimulated scattering and OPC by four-wave mixing. A number of other methods of OPC have also been suggested and investigated. It is to these methods that this chapter is devoted.

8.1 OPC by a Reflecting Surface (OPC-S)

8.1.1 Principle of OPC-S

Suppose that a plane reference wave $E_0 \exp(-ikz)$ is incident strictly normally on a mirror surface, and that a weak signal wave $E_3(R) \exp(ik_3 \cdot R)$, coherent with the reference wave, is incident so as to make a small angle $\theta_3 = (\theta_{3x}, \theta_{3y})$ with the normal e_z (Fig. 8.1). Let us also suppose that the surface possesses nonlinear properties, that is to say, the complex amplitude reflectivity ϱ is dependent on variations of the incident radiation intensity $I(r) = |E(r)|^2$:

$$E_{\text{ref}}(x, y) = \varrho E_{\text{inc}}(x, y), \quad \varrho(I) = \varrho(I_0) + \frac{\partial \varrho}{\partial I}[I(r) - I_0] + \dots . \quad (8.1.1)$$

As a result, the interference pattern of the waves E_0 and E_3 represents the reflectivity modulation

$$\delta\varrho(r) = \beta[E_0 E_3^*(r) \exp(-ik_3 \cdot r) + E_0^* E_3(r) \exp(ik_3 \cdot r) + |E_3|^2] . \quad (8.1.2)$$

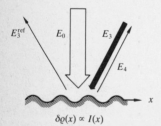

E_3^{ref} E_0 E_3

E_4

x

$\delta\varrho(x) \propto I(x)$

Fig. 8.1. In OPC by a surface the interference pattern of the signal E_3 and reference E_0 waves is reproduced in the reflectivity hologram $\delta\varrho(r) \propto \delta I(r)$; the reference wave itself reads out this hologram in real time while reflecting

Here $\beta = (\partial\varrho/\partial I)$ and r is the radius vector in the plane (x, y). In the plane $z = 0$ the reflected field is of the form

$$E_{\text{ref}}(r) = \varrho(r)[E_0 + E_3(r)\exp(\mathrm{i}k_3 \cdot r)]$$

$$= \varrho_0 E_0 + (\varrho_0 + \beta|E_0|^2)E_3(r)\exp(\mathrm{i}k_3 \cdot r) + \beta E_0^2 E_3^* \exp(-\mathrm{i}k_3 \cdot r) , \quad (8.1.3)$$

terms $\propto E_3^2$ being neglected. Thus, the reflected field incorporates: 1) the mirror-reflected reference wave $\varrho_0 E_0 \exp(\mathrm{i}kz)$, 2) the mirror-reflected signal $(\varrho_0 + \beta|E_0|^2)E_3(r, -z)\exp(\mathrm{i}k_3 \cdot r - \mathrm{i}k_{3z}z)$, and 3) the phase conjugate of the signal wave

$$E_4(R) = \exp(-\mathrm{i}k_3 \cdot R)E_3^*(r)\beta|E_0|^2 . \tag{8.1.4}$$

In the model where $\varrho = \varrho_0 + \beta(I - I_0)$, the OPC reflectivity is expressed by

$$\left|\frac{E_4}{E_3}\right|^2 = I_0^2|\beta|^2 , \tag{8.1.5}$$

where $I_0 = |E_0|^2$ is the reference wave intensity.

Let us discuss the properties of such surface dynamic holograms. First, at a sufficiently high reference wave intensity I_0 or for a sufficiently sharp dependence of the reflectivity on the intensity, the power $|E_4|^2$ of the phase-conjugated wave may considerably exceed the power $|E_3|^2$ of the signal. In this case the phase-conjugated wave experiences amplification due to the energy transferred from the incident reference wave E_0. Further, under these conditions energy is also transferred from E_0 to the mirror-reflected signal, leading to its amplification by a factor of $|\varrho_0 + \beta I_0|^2$. We should note, however, that for almost all nonlinearity mechanisms the value of $\beta = d\varrho/dI$ is strongly dependent on the intensity I_0 or energy $\int I_0 dt$ corresponding to the illumination by the reference wave. Therefore, in specific cases optimization of (8.1.5) does not lead to a trivial increase of I_0.

For precise OPC to be achieved with this method, the reference wave should be strictly normal to the surface, i.e., its wavefront should coincide with the reflecting surface. Curved surfaces may be used in meeting this requirement. If, however, the reference wave deviates from the normal by a small angle θ_0, then $E_0(z=0, r) = E_0 \exp(\mathrm{i}k\theta_0 \cdot r)$ and the conjugated signal acquires the central direction characterized by the angle

$$\theta_4 = -\theta_3 + 2\theta_0 . \tag{8.1.6}$$

In other words, in this case we are dealing with OPC with tilting on an angle $\approx 2\theta_0$, cf. Sect. 6.2.

8.1.2 Mechanisms of Surface Nonlinearity

a) The mechanism underlying the pioneer realization of OPC-S was the actual destruction of a mirror under the action of intense radiation. In [8.1] a reflecting aluminium film, $\sim 1\,\mu m$ thick, deposited on a glass substrate was employed. We may accept the following mathematical model as suitable for intense laser pulses of duration $\sim 10^{-8}$ s. The destruction, leading to an abrupt decrease in the mirror reflectivity from unity to zero, begins at an instant when the energy density $\int^t I(r, t')dt'$ at a given point reaches a critical value, $W_c \approx 0.1$ J/cm. If the profile $I(r)$ of the incident intensity has inhomogeneities of an interference nature, like (8.1.2), the destruction begins at the maxima of the interference pattern at an instant t_1 such that $I_{max}t_1 \approx (I_0 + 2\sqrt{I_0 I_3})t_1 \approx W_c$. With the passage of time the width of the dark lines of the induced holographic grating rises, and the instant t_3 when the mirror completely burns out is determined by an approximate relation $I_{min}t_3 \approx (I_0 - 2\sqrt{I_0 I_3})t_3 \approx W_c$. The maximum efficiency of the diffraction of the reference wave E_0 into the conjugated field E_4 is observed at an instant $t_2 \approx W_c/I_0$ when the widths of the dark and light lines are equal:

$$\left|\frac{E_4}{E_3}\right|^2_{t=t_2} = \left|\frac{E_0}{E_3}\right|^2 \cdot \left|\frac{E_4}{E_0}\right|^2 \approx \frac{I_0}{I_3} \cdot 0.1 . \tag{8.1.7}$$

Here the factor 0.1 stands for π^{-2}, which characterizes the efficiency of first-order diffraction by a rectangular amplitude grating with symmetric lines.

Equation (8.1.7) shows that high-amplification OPC is possible if $I_3 \ll I_0$. Indeed, the experiments described in [8.1, 2] demonstrated OPC with a 1.3-fold and 7-fold gain, respectively, in the instantaneous intensity. The specific feature of this OPC-S mechanism is the shorter duration and sharper leading edge of the conjugated signal compared with the incident signal.

b) Strong absorption of light with a transversely inhomogeneous intensity profile $I(r)$ causes non-uniform heating of the surface layers of the mirror. Due to thermal expansion in the intensity maxima the mirror bulges towards the beam, i.e., the relief grating $\delta h(r) = h_1 \cos(k\theta_3 \cdot r + \phi) = A[E_0 E_3^* \times \exp(-ik\theta_3 \cdot r) + c.c.]$ is formed. The diffraction efficiency of the phase-relief hologram produced is (at $\delta h \ll \lambda$)

$$\left|\frac{E_4}{E_0}\right|^2 \approx \left|\varrho_0 \frac{\omega}{c} A E_0 E_3\right|^2 = |\varrho_0 k h_1/2|^2 . \tag{8.1.8}$$

In the steady-state regime the relief modulation can be estimated by $\delta h \approx \delta T \alpha l_{heat}$, where δT is the surface layer temperature variation, α is the coefficient of thermal expansion, and l_{heat} [cm] is the thickness of the nonuniformly heated layer coinciding with the transverse inhomogeneity dimension, $l_{\text{heat}} \approx (k\theta_3)^{-1}$. The steady-state value $\delta T \approx 2\sqrt{I_0 I_3}(1 - |\varrho_0|^2) t_{\text{st}}/ \varrho c_p l_{\text{heat}}$. Here $(1 - |\varrho_0|^2)$ is the fraction of energy absorbed, ϱc_p [J/cm^{3}K] is the heat capacity per unit volume, $t_{\text{st}} = (\chi k^2 \theta_3^2)^{-1}$, and $\chi = \Lambda/\varrho c_p$ [cm^2/s] is the thermal diffusivity. Then the OPC reflectivity, in the steady-state regime, is

$$\left| \frac{E_4}{E_3} \right|^2 \approx \frac{\alpha^2 |\varrho_0|^2 (1 - |\varrho_0|^2)^2 I_0^2}{\Lambda^2 k^2 \theta_3^4}, \qquad (8.1.9)$$

and the process of reaching the steady-state regime may be described by the approximate formula $|E_4|^2 \propto [1 - \exp(-t/t_{\text{st}})]^2$. In [8.3], at a reference wave power density of ~ 1 W/cm^2 (He–Ne laser radiation) and at $\theta_3 \sim 3 \times 10^{-3}$ rad, a reflectivity $|E_4/E_3|^2 \sim 2\%$ was obtained in a build-up time of $\sim 10^{-3}$ s.

c) OPC-S is possible through nonlinear reflection from narrow-band semiconductor surfaces. The reflection of light is determined by the thin near-surface layers of thickness $\sim \lambda/2\pi$, where λ is the light wavelength. In order to obtain a noticeable modulation of the reflectivity, it is necessary to vary the optical properties of a semiconductor strongly in the vicinity of its surface. Such variations can be achieved when the absorption of incident light results in the transition of electrons from the valence band to the conduction band. Therefore, the band-gap energy E_g should be small, $E_g < \hbar\omega$, where ω is the incident light frequency.

The expression for the Fresnel amplitude reflectivity, $\varrho = (\sqrt{\varepsilon} - 1)/ (\sqrt{\varepsilon} + 1)$ is also valid for complex ε. For semiconductors, we may accept $\varepsilon = \varepsilon_0 + i \text{Im}\{\varepsilon\} - a$, $a = 4\pi N e^2/m^* \omega^2 = \text{const} \cdot N$. The dependence $\varrho(N)$ is strongest when the negative contribution $(-a)$ of the electron-hole plasma approximately compensates for the positive contribution ε_0 of the bound electrons, $\varepsilon_0 = n_0^2$, where $n_0 = 3 - 5$ is the initial refractive index. It is remarkable that the density N of electron-hole pairs required for this condition proves to be moderate due to the smallness of m^*, viz., $N \sim 10^{21}$ cm^{-3} at $m^* = 0.03 \, m_e$ and $\lambda \sim 1$ μm. Such density values are really obtained when the narrow-band semiconductor surface is exposed to radiation from a high-power visible-region pulsed laser. At such values of N, the pair recombination rate depends on N in a complicated manner. We shall use the approximate relations $N = \text{const} \cdot I^{1/p}$, $da/dI = p^{-1}/I$, where $p \sim 1 - 3$, when the processes of excitation and recombination of carriers in the near-surface layer are in equilibrium. Therefore the phase-conjugation efficiency is

$$\left| \frac{E_4}{E_3} \right|^2 = \left| \frac{\partial \varrho}{\partial I} I_0 \right|^2 = \frac{1}{p^2} \left| \frac{d\varrho}{da} a_0 \right|^2, \qquad (8.1.10)$$

where $a_0 = 4\pi N_0 e^2 / m^* \omega^2$ corresponds to the density N_0 produced by illumination with the reference wave I_0 (without the signal E_3).

For $\varepsilon = \varepsilon_0 - a + i\,\mathrm{Im}\{\varepsilon\}$ we have

$$\frac{d\varrho}{da} = \frac{1}{\sqrt{\varepsilon(a)}\,(\sqrt{\varepsilon(a)}+1)^2}. \tag{8.1.11}$$

The maximum value of (8.1.10) achieved at $a \approx \varepsilon_0$ is roughly estimated as

$$\left|\frac{E_4}{E_3}\right|^2 \approx \frac{1}{8p^2}\frac{n_0}{\varkappa^3}, \tag{8.1.12}$$

where $\varkappa \approx 0.5\,c\alpha/\omega$ and α is the intensity absorptivity $[\mathrm{cm}^{-1}]$ of the near-surface layer of the excited semiconductor.

An experimental study of OPC-S for Ge, GaAs, InSb, and Si was conducted in [8.4, 5]. Below a certain threshold, the density of excited pairs is small, the parameter $a \ll \varepsilon_0$, and the slope of the function $\varrho(I)$ is low. At $W \lesssim 0.1$ J/cm^2 no reflection was observed. At higher energies, the reflectivity $|E_4/E_3|^2$ was dependent on the reference wave energy (on the parameter a_0 in terms of the theory), a pronounced maximum of $\sim 5\%$ corresponding to the plasma resonance being observed.

d) A controllable liquid-crystal optical valve incorporates two layers: a photosensitive semiconductor and an oriented liquid crystal (LC), both layers being subject to a voltage from a power supply. The illumination of a semiconductor varies its electrical impedance. As a result, the local electric field applied to the LC varies. This leads to reorientation of the unit vector along the optical axis of the LC and, due to optical anisotropy, to a phase change of the light passing through the LC. Thus, it is possible to create dynamic holograms using the valve and even to make them reflection holograms if the valve is provided with a mirror.

The idea of using a controllable LC valve for OPC-S was realized by *Garibyan* et al. in [8.6]. A remarkable feature of the LC valve is its extremely high sensitivity: a 1-radian phase change of the reflected light was achieved at a power density of 3×10^{-5} W/cm^2. The experiment in [8.6] demonstrated OPC-S of a signal of an extremely low power, $\sim 3 \times 10^{-6}$ W, with an efficiency of $\sim 1\%$ and a build-up time of ~ 10 s.

Among other mechanisms of reflection nonlinearity we should note the change of shape of a liquid surface due to the pressure exerted by radiation. The pressure may be produced either by light or sound, see Sect. 8.4, or be due to the reaction of an evaporating substance. An OPC-S with a mercury-glass mirror interface was reported in [8.7]. The distinguishing feature of this work was the use of a free-running pulsed laser of relatively low power. The recording mechanism for the hologram was possibly the vaporization of the mercury and its separation from the glass.

8.2 OPC in Three-Wave Mixing (TWM)

For non-centrosymmetric crystals the series expansion of the nonlinear polarization P^{NL} in terms of powers of E begins with a second-order term [8.8 – 11]

$$P_{real}^{NL} = \chi^{(2)} E_{real}^2 . \tag{8.2.1}$$

If the field E_{real} contains several frequency components, $E_1 \cos(\omega_1 t + \phi_1), \ldots,$ $E_k \cos(\omega_k t + \phi_k)$, then the polarization, see (8.2.1), corresponds to all possible frequency combinations of the form

$$\omega = |\omega_i \pm \omega_k| . \tag{8.2.2}$$

When ω_i and ω_k coincide, we have second harmonic generation, at $\omega = \omega_1 + \omega_2$ we have sum frequeny generation, and at $\omega = \omega_1 - \omega_2$ difference frequency generation. The latter process is described by the term

$$P_j^{NL}(\omega_1 - \omega_2) \exp[-i(\omega_1 - \omega_2)t] = \chi_{jkl}^{(2)} E_k(\omega_1) \exp(-i\omega_1 t) E_l^*(\omega_2)$$
$$\times \exp(+i\omega_2 t) . \tag{8.2.3}$$

The appearance of the tensor $\chi_{jkl}^{(2)}$ is mainly due to the bound electron nonlinearity, its build-up time being of the order of the light period. The structure of this tensor is determined by the symmetry properties of the crystal. The value of $\chi^{(2)}$ was estimated in Sect. 7.3; for typical crystals $\chi^{(2)} \sim 10^{-8} - 10^{-9}$ cgs.

In OPC by TWM a complex-conjugated wave is obtained if the signal $E_3(R) \exp(ik_3 \cdot R - i\omega t)$ and a plane reference wave of twice the frequency, $E_0 \exp(ik_0 \cdot R - 2i\omega t)$, are directed onto a nonlinear crystal (Fig. 8.2). Then the difference frequency $\omega_4 = 2\omega - \omega$ will coincide with the signal frequency, and the space dependence of the polarization P_4' will be

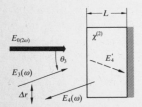

Fig. 8.2. In OPC-TWM the plane reference wave $E_0(2\omega)$ and signal $E_3(\omega, r)$ are directed onto a nonlinear crystal. The generated complex-conjugated wave E_4' is copropagating; its reflection by a mirror produces the phase-conjugated wave $E_4(\omega, r)$ with a slight transverse shift $\Delta r \sim L\theta_3$

$$P_4'(R) \exp[i(k_0-k_3) \cdot R] = \chi^{(2)} E_0 E_3^*(R) \exp[i(k_0-k_3) \cdot R] . \qquad (8.2.4)$$

For the polarization to excite an electromagnetic wave efficiently, the phase matching condition should be satisfied:

$$\frac{\omega}{c}\sqrt{\varepsilon} = |k_0 - k_3| . \qquad (8.2.5)$$

Since $k_0 = 2\omega\sqrt{\varepsilon}/c$, $k_3 = \omega\sqrt{\varepsilon}/c$, the condition (8.2.5) is satisfied only when E_0 and E_3 are approximately copropagating. A more rigorous consideration of (8.2.5) requires one to take into account the dependence of the effective refractive index $\sqrt{\varepsilon}$ on ω due to the medium dispersion and on the propagation direction k/k and the type of polarization due to the anisotropy of the linear optical properties of the crystal, [8.8 – 10].

To determine the width of the phase matching curve, let us suppose that the z axis (the normal to the crystal faces) is directed at the exact phase matching angle for the interaction $1° + 1° \rightarrow 2^e$ in a uniaxial crystal with $n^e < n^0$. Let the reference wave E_0 propagate in the direction $n_0 \approx e_z(1 - \theta_0^2/2) + \theta_0$ characterized by a transverse angle $\theta_0 \equiv (\theta_{0x}, \theta_{0y})$ of deviation from the z axis, and the signal E_3 propagate at an angle θ_3. Then the wave P_4' is directed at an angle $\theta_4 \approx 2\theta_0 - \theta_3$, and the wave detuning Δk_z is given by

$$\Delta k_z = (k_0 - k_3 - k_4)_z \approx k_3[(\theta_0 - \theta_3)^2 - 2\eta \cdot \theta_0] . \qquad (8.2.6)$$

The energy efficiency of TWM is proportional to $\sin^2 x/x^2$, where $x = \Delta k_z L/2$, and L is the layer thickness. Here $\eta = \eta e_x$, the x axis lying in the plane of the diagram (Fig. 8.3). The quantity η is the angle of intersection of the ellipsoid $k(2\omega,n)$ of the e-wave of frequency 2ω with the sphere $2k(\omega,n)$ of the o-wave of frequency ω.

At $\theta_0 = 0$ we have $\theta_4 = -\theta_3$, and the phase matching condition $\Delta k_z L \lesssim 2\pi$ takes the form $\theta_3^2 L/\lambda \lesssim 1$, where λ is the wavelength in the medium. Allowing for $\theta_3 = \theta_3^{air}/n$, $\lambda = \lambda_0/n$, we have $\theta_3^{air} \lesssim 10^{-2}$ rad at $L = 1$ cm and $\lambda_0 = 0.7$ μm. Thus, TWM makes it possible to conjugate signals only within limited angle of vision $\Delta\Theta \sim \theta_3^2 \sim \lambda/L$, i.e., $\Delta\theta_3 \sim \sqrt{\lambda/L}$[1]. At $\theta_0 \neq 0$ there is a cone of optimal angles θ_3^{opt} for which $\Delta k_z = 0$, $|\theta_3^{opt}| \approx \sqrt{2\eta \cdot \theta_0} \gg |\theta_0|$. Under such conditions $\Delta k_z \approx 2k\theta_3^{opt} \cdot \Delta\theta_3$ and the limitation placed on the deviation $\Delta\theta_3$ from the cone of optimal directions becomes more severe: $\theta_3^{opt} \cdot \Delta\theta_3 \lesssim \lambda/2L$.

---

1 Note that the phase matching condition for OPC-TWM differs from that for second-harmonic generation by a single plane wave of frequency ω. In the latter case $\Delta\theta_x \lesssim \lambda/L\eta$, and thus the limitation placed on $\Delta\theta_x$ is more severe, since $\eta \sim 10^{-1} - 10^{-2}$. In contrast, $\Delta\theta_y$ is not limited by the phase matching condition for second-harmonic generation

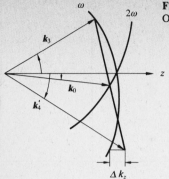

Fig. 8.3. Calculation of detuning from phase matching in OPC-TWM

So far we have discussed the excitation process of the complex-conjugated wave E_4' propagating in the signal direction at an angle of $2(\theta_0 - \theta_3)$ to the signal. To generate the backward-going phase-conjugated wave, a mirror is usually employed, placed exactly normal to the propagation direction of the reference wave E_0 (Fig. 8.2).

If the angle of incidence of the signal wave on this mirror is other than zero, i.e., $|\theta_0 - \theta_3| \neq 0$, its propagation through the crystal gives rise to a transverse shift $\Delta r \sim L(\theta_3 - \theta_0)$ with respect to the exactly conjugated signal wave: $E_4(r) \propto E_3^*(r + \Delta r)$. High-quality OPC requires that the shift $|\Delta r|$ be less than the characteristic dimension of the transverse inhomogeneity of the signal E_3, i.e., $\lambda/\Delta\theta_3 \lesssim |\Delta r| \sim L|\theta_3 - \theta_0|$. This condition is easily seen to coincide with the limitations resulting from the phase matching condition.

In the Born approximation, under precise phase matching conditions the OPC reflectivity is given by

$$\left|\frac{E_4}{E_3}\right|^2 = \left|\frac{2\pi\omega L \chi^{(2)} E_0}{nc}\right|^2. \tag{8.2.7}$$

At $\chi^{(2)} = 10^{-9}$ cgs (KDP crystal), $n = 1.5$, $\lambda_0 = 1\,\mu\text{m}$, $L = 1\,\text{cm}$, $nc|E_0|^2/8\pi$ $= 10^{14}\,\text{erg/cm}^2\text{s} = 10\,\text{MW/cm}^2$, we obtain $|E_4/E_3|^2 \approx 5 \times 10^{-3}$.

For experimental investigations on OPC-TWM the reader may refer to [8.12 – 15].

8.3 Medium Parameter Modulation at Doubled Frequency

Consider a medium whose permittivity is synchronously time-dependent through the whole volume:

$$\varepsilon(t) = \varepsilon_0 + 2|a|\cos(2\omega t - \phi) \equiv \varepsilon_0 + a\,\text{e}^{-2i\omega t} + a^*\text{e}^{2i\omega t}, \tag{8.3.1}$$

where $a = |a|\exp(\mathrm{i}\phi)$ is a complex constant independent of space coordinates, $|a| \ll \varepsilon_0$. If a wave $E_{3\,\text{real}}(R, t) = 0.5\,[E_3(R)\exp(-\mathrm{i}\omega t) + E_3^*(R)\exp(\mathrm{i}\omega t)]$ propagates in such a medium, then an additional term is present in the dielectric displacement:

$$\delta D = 2\,|a|\cos(2\omega t + \phi)E_{3\,\text{real}}(R, t)$$

$$= [E_3(R)a\mathrm{e}^{-3\mathrm{i}\omega t} + E_3^*(R)a\mathrm{e}^{-\mathrm{i}\omega t}] + \text{c.c.} \tag{8.3.2}$$

The term containing the threefold frequency does not give rise to an electromagnetic wave because $E_3(R) \propto \exp(\mathrm{i}k_3 \cdot R)$ does not satisfy the wave equation at 3ω: $|k_3| \neq 3\omega\sqrt{\varepsilon_0}/c$. However, the term $\propto E_3^*(R)\exp(-\mathrm{i}\omega t)$ synchronously excites the wave E_4 which is phase conjugated with respect to the incident signal. The excited wave $[E_4(R)\exp(-\mathrm{i}\omega t) + \text{c.c.}]$ is partially converted by the same mechanism into the original wave $E_3(R) \propto E_4^*(R)$.

The system of shortened equations for the coupled waves $E_3\exp(\mathrm{i}k_3 z)$ and $E_4^*\exp(\mathrm{i}k_3 z)$ is

$$\frac{dE_3}{dz} = -\mathrm{i}\,\frac{\omega a}{2 n_0 c}\,E_4^*, \qquad \frac{dE_4^*}{dz} = -\mathrm{i}\,\frac{\omega a^*}{2 n_0 c}\,E_3. \tag{8.3.3}$$

From (8.3.3) we find the conjugation efficiency

$$\left|\frac{E_4}{E_3}\right|^2_{z=0} = \tan^2\left(\frac{\omega\,|a|}{2 n_0 c}\,L\right), \tag{8.3.4a}$$

where $n_0 = \sqrt{\varepsilon_0}$ and L is the modulation layer thickness along the signal ray. In the Born limit, we may substitute tan by its argument, and (8.3.4a) is simplified to the form

$$|E_4/E_3|^2 = |a/2\varepsilon_0|^2 (kL)^2, \tag{8.3.4b}$$

where $k = \omega n_0/c$.

At first sight, the creation of a modulation $\delta\varepsilon(t)$ homogeneous within in the whole volume with the light frequency seems impossible. Indeed, for a time equal to the light period $T = 2\pi/2\omega$, the fastest signal, i.e., an electromagnetic wave, propagates only through a thickness of the order of the light wavelength λ and synchronizes the modulation only within this thickness. However, homogeneous modulation is really observable in a four-wave mixing scheme with the use of a fast-responding electron nonlinearity medium. In this case we may write

$$\delta\varepsilon(R, t) = 4\pi\chi^{(3)}E_{\text{real}}^2(R, t). \tag{8.3.5}$$

If the field $E_{real}(R, t)$ comprises two counterpropagating reference waves, $E_{real} = \{0.5 \cdot [E_1 \exp(ikz - i\omega t) + E_2 \exp(-ikz - i\omega t)] + \text{c.c.}\}$, then (8.3.5) contains the term

$$2\pi\chi^{(3)}E_1E_2 e^{-2i\omega t} + 2\pi\chi^{(3)}E_1^* E_2^* e^{2i\omega t} . \tag{8.3.6}$$

It is important that owing to the exact counterpropagation of the plane waves E_1 and E_2 the product $E_1(R)E_2(R)$ proves to be independent of coordinates.

We note that TWM caused by the fast (electron) second-order nonlinearity can be described in terms of modulation of the permittivity at the doubled frequency:

$$\delta\varepsilon = 4\pi\chi^{(2)}E_{real} = 2\pi\chi^{(2)}[E_0 \exp(2ikz - 2i\omega t)$$
$$+ E_0^* \exp(-2ikz + 2i\omega t)] . \tag{8.3.7}$$

In this case, however, the modulation represents a traveling wave $\propto \exp(2ikz)$. As a result, the conjugated wave is generated not in the backward but in the forward direction and only if the rigorous phase matching condition is satisfied, see Sect. 8.2.

OPC can also be achieved through a complex-reflectivity modulation synchronous within some area of the reflecting surface at the doubled frequency:

$$\varrho = \varrho_0 + 2|b|\cos(2\omega t - \phi) . \tag{8.3.8}$$

In this case the OPC efficiency is

$$|E_4/E_3|^2 = |b|^2 . \tag{8.3.9}$$

Such procedure is called OPC by a twinkling surface.

8.4 OPC of Sound Waves

So far we have considered phase conjugation of electromagnetic waves. The problem of phase conjugation may be raised for radiation of another physical nature, say, for sound waves. The first experiments aimed at generating a phase-conjugated sound wave employed the reflecting surface technique [8.16]. The nonlinearity associated with the change of the surface profile of a liquid due to radiation pressure (originally suggested for OPC-S of light waves [8.17]) appeared to be much more efficient in the sound region. Indeed, the radiation pressure on a completely reflecting surface is $p = 2I/v$, where $I = |A|^2$ is the power density and v is the velocity of radiation. Since the sound velocity is a factor of $10^5 - 10^6$ lower than the light velocity, then, at equal power values, sound effects of the surface profile are stronger than light effects by the same factor.

In the PC-S problem we are interested in the surface height disturbance $\delta h \propto A_0 A_3^* \times \exp(iqx)$, where $q = k \sin\theta_3$ and θ_3 is the angle between the vector k_3 and the normal to the surface. The equation for $\delta h(x, t) = [h \exp(iqx) + \text{c.c.}]$ can be written as

$$\ddot{h} + 2\Gamma\dot{h} + \Omega_0^2 h = \frac{2qA_0 A_3^*}{\varrho v}, \tag{8.4.1}$$

where $\Gamma = 2vq^2/\varrho$, $\Omega_0^2 = \sigma q^3/\varrho + qg$, v is the viscosity of the liquid, σ and ϱ are its surface tension and density, respectively, $g = 10^3$ cm/s^2 is the acceleration due to gravity; usually $qg \lesssim \sigma q^3/\varrho$. Let us clarify the sense of (8.4.1) by the following arguments. In surface wave, the wave vector $q_x = q$ arises mainly due to the motion of particles present in the near-surface layer of thickness $\sim q^{-1}$. If we denote the layer width in the y direction by Δy, then the mass involved in the motion, over a spatial period, is $m \sim \varrho q^{-2} \Delta y$. The external force F acting on this mass is equal to the interference term in the pressure, $2A_0 A_3^*/v$, multiplied by the area $q^{-1}\Delta y$. Using Newton's second law we have $\ddot{h} = 2qA_0 A^*/\varrho v$. Reasoning similarly, we can also obtain the term $\propto \Omega_0^2$ describing the restoring force and the viscosity term $\propto \Gamma$. Here Ω_0 is the natural frequency of the surface wave with the wave vector q and Γ is the damping coefficient. In typical conditions $q \sim 10$ cm^{-1}; in this case the restoring force $\propto \sigma q^2$ mostly results from surface tension.

Let us make numerical estimates for water: $v = 1.5 \times 10^5$ cm/s, $\varrho = 1$ g/cm^3, $\sigma = 75$ dyne/cm, and $v \approx 1.3 \times 10^{-2}$ poise. For the sound frequency $f = \omega/2\pi = 10^6$ Hz, $\lambda = 0.15$ cm and at $\theta_3 = 0.3$ rad we have $q = 2\pi \sin\theta_3/\lambda \approx 12$ cm^{-1}, $\Omega_0/2\pi \approx 60$ Hz, $\Gamma \approx 3.7$ s^{-1}. In experimental situations the surface wave may propagate through the interaction area for a time shorter than Γ^{-1}, and, therefore, the effective damping may be even stronger, $\Gamma_{\text{eff}} \sim \Omega_0(ql)^{-1}$, where l is the excitation region dimension.

For a pulse duration $t \lesssim \Omega_0^{-1}$, the solution of (8.4.1) is $h = qt^2 A_0 A_3^*/\varrho v$, so the PC reflectivity is

$$|A_4/A_3|^2 = (kh)^2 I_0/I_3 = t^4 k^4 \sin^2\theta_3 I_0^2/\varrho^2 v^2. \tag{8.4.2}$$

At $t > \Omega_0$ the reflection efficiency begins to oscillate with a frequency Ω_0 about the steady-state value $h = 2qA_0 A_3^*/\varrho v \Omega_0^2$:

$$\left|\frac{A_4}{A_3}\right|^2 = \frac{4I_0^2}{k^2 v^2 \sigma^2 \sin^4\theta_3}. \tag{8.4.3}$$

For a time $\sim \Gamma^{-1}$ these oscillations will be damped. If the reference wave intensity $I_0 \sim 3$ W/cm^2, after $t \approx 10^{-3}$ s we have (for the above parameters) $|A_4/A_3|^2 \sim 10^{-2}$, the steady-state value being $|A_4/A_3|^2 \sim 1$.

In the experiment in [8.16], the coherent sound wave and signal had a frequency $f = 3\,\text{MHz}$, the pulse duration being $\sim 4 \times 10^{-4}\,\text{s}$. A signal corresponding to the phase-conjugated wave was registered. The conjugated signal was also observed (at a lower reference wave intensity) in the steady-state regime, the efficiency of diffraction of the reference wave into the conjugated one being $\sim 5\%$.

Brysev et al. [8.18] achieved PC of sound waves by modulating medium parameters at the doubled frequency. The velocity of a sound wave in a LiNbO$_3$ crystal was modulated by a uniform electric field taken from an oscillator with a frequency $f = 16\,\text{MHz}$, i.e., twice the frequency of the signal wave. The external field strength $E \sim 30\,\text{kV/cm}$ made it possible to achieve a PC reflection efficiency of $\sim 4 \times 10^6$ using an interaction length of $\sim 2\,\text{cm}$.

PC of sound waves by a twinkling surface was achieved in [8.19]. The complex reflectivity $\varrho(t)$ was modulated as follows. For reflection of the wave incident approximately normally to the surface, $\varrho = \exp(2ikz_0)$, where z_0 is the surface coordinate. If z_0 varies according to $z_0 = \text{const} + \varDelta z \cos(2\omega t - \phi)$, the coefficient b in (8.3.8) is equal to $k\varDelta z$. An experimentally obtained PC-reflection efficiency of $|A_4/A_3|^2 \sim 10^{-4}$ is reported in [8.19]; this was limited by the fact that at larger vibration amplitudes $\varDelta z$ the sound radiated by the surface at 2ω became too intense.

8.5 Quasi-OPC: Retroreflectors

A conventional plane mirror reverses the direction of the z component of the wave vector k of an incident plane wave $\exp(i\boldsymbol{k} \cdot \boldsymbol{R})$, namely, $(k_x, k_y, k_z) \to (k_x, k_y, -k_z)$. Here z is the direction of the normal to the mirror plane. If we neglect the diffractive effects on the edge of the mirror diaphragm, then because of the mirror image principle we can obtain the expression for the general field:

$$E_{\text{ref}}(x, y, z) = \varrho E_{\text{inc}}(x, y, -z + 2z_0) . \qquad (8.5.1)$$

Here z_0 is the mirror coordinate and ϱ is the complex reflectivity.

If we have two mirrors forming a dihedral angle of 90° (for example, mirrors $z = z_0$ and $y = y_0$), the reflection by such a pair of mirrors will transform the field as

$$E_{\text{ref}}(x, y, z) = \varrho^2 E_{\text{inc}}(x, -y + 2y_0, -z + 2z_0) . \qquad (8.5.2)$$

For a plane wave, such a reflection is equivalent to the transformation (Fig. 8.4) $(k_x, k_y, k_z) \to (k_x, -k_y, -k_z)$.

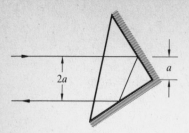

Fig. 8.4. Backward reflection by a retroreflector leads to a shift of the ray by an amount $2a$, where a is the impact parameter of the ray with respect to the vertex

If we place one more mirror at right angles to the first two to form a cube corner, the incident ray experiences reflection by all three mirrors: (k_x, k_y, k_z) $\rightarrow (-k_x, -k_y, -k_z)$. The law of field transformation by such a corner reflector is of the form of inversion,

$$E_{\text{ref}}(R) = \varrho^3 E_{\text{inc}}(2R_0 - R) \,, \tag{8.5.3}$$

with respect to the vertex R_0 of the corner reflector. Such a corner reflector is also called a triple prism.

A striking feature of the corner reflector is the reversal of the propagation direction $k \rightarrow -k$ of any plane wave. It is of no less importance that this feature is independent of the orientation of the reflector: only the vertex coordinate R_0 appears in (8.5.3). Such corner reflectors are widely used on automobiles, motorcycles, bicycles as cat's eyes; sometimes they are called retroreflectors.

It may seem that the operation $k \rightarrow -k$ resembles phase conjugation. However, (8.5.3) shows that no PC is observed in this case. Really, if the incident field represents two plane waves, $E_{\text{inc}}(R) = c_1 \exp(ik_1 \cdot R)$ $+ c_2 \exp(ik_2 \cdot R)$, the PC operation would give

$$E_{\text{ref}}(R) = \varrho' [E_{\text{inc}}(R)]^* = \varrho' [c_1^* \exp(-ik_1 \cdot R) + c_2^* \exp(-ik_2 \cdot R)] \,,$$
$$\tag{8.5.4}$$

whereas the corner reflector gives, according to (8.5.3),

$$E_{\text{ref}}(R) = \text{const} \cdot [c_1 \exp(2ik_1 \cdot R_0) \exp(-ik_1 \cdot R)$$
$$+ c_2 \exp(2ik_2 \cdot R_0) \exp(-ik_2 \cdot R)] \,. \tag{8.5.5}$$

In other words, in the case of true phase conjugation we deal not only with the reversal of the direction of each of the plane waves but also with the sign reversal of their phase difference. Unlike this, the linear (without complex conjugation) operation (8.5.3) reverses only the propagation direction. In terms of geometrical optics, a transverse shift of an individual ray by an amount $2a$ (Fig. 8.4) corresponds to an unchanged at $R_0 = 0$ phase relationship.

A spherical wave diverging from a source with coordinate R_1 is transformed by the corner reflector into the wave diverging from a new source

with the coordinate $R_2 = 2R_0 - R_1$, unlike the OPC case in which the transformed wave would converge towards R_1. Thus, the conjugation of the
central direction but not of the curvature of the spherical wave beam occurs
within the limits of the corner reflector aperature. Let the cross-sectional
dimension D of the corner reflector be limited. The radius of curvature of the
incident spherical wavefront is denoted by r_0, and $r_0 \gg D$. Then the deviation
of the wavefront with respect to the plane normal to the central propagation
direction is $\delta h \sim D^2/r_0$. If $\delta h \ll \lambda$, then the operation (8.5.3) within the
aperature D corresponds, to within a correction factor $(\delta h/\lambda)^2$, to the operation of optical phase conjugation.

For the system efficiency to be increased, a single corner reflector is often
replaced by a set of smaller reflectors characterized by the set of coordinates
$R_1, \ldots, R_N$ of their vertices. In this case, the field entering the aperture of the
ith reflector is transformed according to (8.5.3) with a corresponding $R_0 = R_i$.
Each of the reflectors therefore reverses the central direction (i.e., the average
inclination of the wavefront), whereas the maximum transverse shift is limited
by the dimensions of one reflector.

However, there is a reasonable limit to the smallness of individual reflectors. Even if we realize OPC with each of the reflectors, the fields reflected by
different reflectors are phase shifted with respect to one another by an amount
$2k \cdot (R_i - R_j)$. In the general case, the wavefront of the reflected radiation
experiences jumps inbetween the elements. Due to these jumps the reflected
radiation acquires an additional divergence $\delta\theta \sim \lambda/D$ corresponding to
diffraction by the aperture of one element.

Let us consider a double-pass scheme using a set of corner reflectors, each
of dimension $2D$, to compensate for distortions introduced by an amplifier
(Fig. 8.5). Let a plane wave be distorted in the first pass so that a certain
irregular distribution $\theta(x)$ of the ray inclination angle θ as a function of the
transverse coordinate x is observed. In the geometrical optics approximation
each element reverses the direction of the incident ray with a transverse shift:
$\theta_{\mathrm{ref}}(x) \approx -\theta_{\mathrm{inc}}(x+a)$, where $a \lesssim D$. As a result, an angular compensation

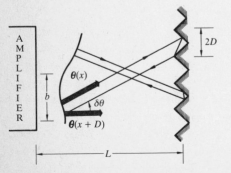

Fig. 8.5. Amplifier distortion compensation
double-pass scheme with a retroreflector

inaccuracy $\delta\theta_1 \sim D \cdot d\theta/dx$ arises on the return pass of the wave through the amplifier. The diffraction by the aperature of each element also inserts a divergence $\delta\theta_2 \sim \lambda/D$. This divergence leads to an uncertainty in the transverse coordinate of the ray incident on the amplifier during the return pass, $\delta x \sim L\lambda/D$, where L is the distance of the retroreflector from the amplifier. An angular compensation inaccuracy $\delta\theta_3 \sim (d\theta/dx)L\lambda/D$ arises for the same reason. Furthermore, the wavefront of the corrected radiation consists of independent pieces, each of dimension $\sim D$, this resulting in additional angular distortions, $\delta\theta \sim \delta\theta_2$. Thus, an estimate of the divergence after two passes with reflection by a retroreflector in between is:

$$\delta\theta = \delta\theta_1 + \delta\theta_2 + \delta\theta_3 = \frac{d\theta}{dx}D + \frac{\lambda}{D}\left(1 + L\frac{d\theta}{dx}\right). \tag{8.5.6}$$

To estimate the compensation efficiency, let us compare $\delta\theta$ from (8.5.6) with θ_0 observed after a one-pass amplification. In the latter case, $d\theta/dx \sim \theta_0/b \sim R^{-1}$, where b is the transverse dimension of the inhomogeneity of the function $\theta(x)$ and R is the characteristic local radius of curvature of the distorted wavefront. The ratio of these two divergences is

$$\frac{\theta_0}{\delta\theta} \sim \left[\frac{D}{b} + \frac{\lambda}{D\theta_0}\left(1 + \frac{L}{R}\right)\right]^{-1}. \tag{8.5.7}$$

Consider first the case when the retroreflector is installed close to the amplifier, $L \to 0$. Then (8.5.7) turns out to be greater than unity (i.e., there is a certain advantage in the double-pass scheme) at

$$\frac{\lambda}{\theta_0} \lesssim D \lesssim b. \tag{8.5.8}$$

It can be easily seen that if the retroreflector is employed in this scheme, the advantage is considerable only at $\lambda/\theta_0 \ll b$, i.e., if distortions in the amplifier are high and the local phase disturbance $\delta\phi \sim k\theta_0 b \gg 1$. The optimal advantage, of the order of $\theta_0/\delta\theta \sim (\theta_0 b/\lambda)^{1/2}$, is achieved at $D \sim (b\lambda/\theta_0)^{1/2}$. Consider a numerical example. At $\theta_0 \sim 10^{-2}$ rad, $b \sim 1$ cm, $\lambda \sim 10^{-4}$ cm, the maximum advantage $\theta_0/\delta\theta \sim 10$ at $D \sim 0.1$ cm. The divergence obtained is 10 times the diffraction limit λ/b at $b = 1$ cm but, nevertheless, 10 times better than the non-corrected divergence θ_0. If the dimension D of the elementary reflectors exceeds the optimal value, the advantage reduces to $\sim b/D$.

In the other limiting case (less advantageous), when $L \gg R \approx b/\theta_0$, we have $D_{opt} \approx (\lambda L)^{1/2}$, and the advantage $\theta_0/\delta\theta \approx b/D_{opt} \approx (b^2/\lambda L)^{1/2}$ proves to be larger than unity only at $L \lesssim b^2/\lambda$.

Up to this point, we have assumed that disturbances of the wavefront are large compared to the wavelength, and resorted to geometrical optics. If phase distortions are less than or of the order of 1 rad, other methods should be considered. Referring the reader for details to the reviews [8.20, 21], we wish to point out that in this case retroreflectors are capable of compensating for part of the phase distortion (namely, the odd, with respect to the reflector middle, component of a phase disturbance).

The compensating abilities of retroreflectors were demonstrated experimentally in [8.22] where the divergence introduced by a neodymium-glass amplifier was reduced by a factor of four in the double-pass scheme. In [8.23] a set of retroreflectors was employed as one of the mirrors of the laser resonator, which permitted a 10-fold decrease in the beam divergence for an inhomogeneous active medium laser.

8.6 Exotic Methods of OPC

8.6.1 Photon Echo

The phenomenon of photon echo [8.24, 25] is an analogue of spin echo in magnetic resonance and can be described as follows. Two short resonance radiation pulses are delivered at a two-level atomic system at instants t_1 and $t_2 = t_1 + T$, respectively; then the system emits a radiation pulse at a time $t = t_1 + 2T$ at the same resonant frequency. It is this pulse that is called the echo signal. Let us consider the simplest model of photon echo.

Let a two-level system with the inhomogeneous broadening profile $g(\delta\omega)$ be subject to a short pulse of resonance radiation of amplitude $\propto E_1$ at an instant $t = t_1$. The slowly varying complex dipole moment $d_{12}\varrho_{12}$ of an excited individual atom will contain the term

$$d(t) = d_{12}\varrho_{12}(\delta\omega, t) \propto E_1 \exp[-i\delta\omega(t - t_1)] . \qquad (8.6.1)$$

Here ϱ_{12} is a non-diagonal element of the system density matrix. At a time such that $\Delta\omega_{in}(t - t_1) \gtrsim 1$ the interference of the contributions of different atoms cancels, due to phase mismatch, the total dipole moment per unit volume; here $\Delta\omega_{in}$ is the inhomogeneous linewidth. In the absence of relaxation, however, the dipole moment of individual atoms is not cancelled.

Now suppose that at an instant $t = t_2$ a short resonance pulse with a complex amplitude E_2 is delivered. The change δn in the population of the upper level is proportional to the power lost by the field,

$$\delta n \propto \int E \frac{dd}{dt} dt \propto i\omega(E_2\varrho_{21}^* - E_2^*\varrho_{21}) . \qquad (8.6.2)$$

There is a population change in those atoms with a given detuning $\delta\omega$ from resonance:

$$\delta n(\delta\omega) \propto iE_2E_1^* \exp(i\delta\omega(t_2-t_1)) - iE_1E_2^*[-i\delta\omega(t_2-t_1)] . \qquad (8.6.3)$$

In the absence of relaxation the value of δn given by (8.6.3) will not vary with time.

Finally, let a third short resonance pulse with a complex amplitude $\propto E_3$ be fed to the system at $t = t_3$. This pulse gives rise to a dipole moment component which, for a given atom, is

$$d(t) = d_{12}\varrho_{21}(\delta\omega, t) \propto E_3 \exp[-i\delta\omega(t-t_3)](n_0+\delta n) . \qquad (8.6.4)$$

The term proportional to $E_1^* E_2 E_3$ in the dipole moment of unit volume is of interest to us. According to (8.6.3, 4) it is expressed as

$$\delta P(t) \propto E_1^* E_2 E_3 \int g(\delta\omega) \exp[-i\delta\omega(t-t_3)] \exp[i\delta\omega(t_2-t_1)] d\delta\omega . \qquad (8.6.5)$$

It is clear that at an instant $t_4 = t_3 + t_2 - t_1$ the vibrations of excited atoms are in phase whatever the detuning $\delta\omega$ may be, and an amplitude burst of the dipole moment $P(t)$ arises, its duration being of the order of $\sim \Delta\omega_{in}^{-1}$. It is this burst that is called "spin echo" in the case of magnetic resonance. In experimental situations, a single pulse E_2 of somewhat larger amplitude is often employed instead of two separate pulses E_2 and E_3. In this case

$$\delta P(t) \propto E_1^* E_2^2 \int g(\delta\omega) \exp[-i\delta\omega(t-t_2)] \exp[i\delta\omega(t_2-t_1)] d\delta\omega \qquad (8.6.6)$$

and the echo appears at $t_4 = 2t_2 - t_1$.

In the case of photon echo we should take into account the fact that the sample dimensions are usually far larger than the wavelength of light. Therefore, the phases of the fields E_1, E_2, E_3 are coordinate dependent: $E_i \propto \exp(i k_i \cdot R)$, where k_i is the wave vector. As a result, the polarization $\delta P(R)$ responsible for photon echo possesses a spatial structure

$$\delta P_4(R) \propto E_1^* E_2 E_3 \exp[i(k_2+k_3-k_1) \cdot R] . \qquad (8.6.7)$$

For echo waves to be observed, the phase matching condition

$$\left| |k_2+k_3-k_1| - \frac{\omega}{c}\sqrt{\varepsilon_0} \right| L \leq 1 \qquad (8.6.8)$$

must be satisfied, where L is the interaction length.

The first photon echo observation [8.24] clearly demonstrated that (8.6.7) was satisfied. The reading-out pulse E_3, which identically coincided with E_2, propagated at a small angle β to the propagation direction of the pulse E_1. In this case,

$$k_1 \approx k(e_z + \beta e_x), \quad k_2 \equiv k_3 = k e_z,$$
$$k_4 = k_2 + k_3 - k_1 \approx k(e_z - \beta e_x),$$

$(8.6.9)$

and the echo pulse was excited at an angle of 2β to the signal E_1 direction.

The idea of OPC by photon echo is based on the fact that $(8.6.5-7)$ contain the complex conjugate of the amplitude of the signal $E_1(R)$. In the experiment described in [8.24], this resulted in the copropagating conjugated wave, $E_4 \propto E_1^*$. In subsequent experiments [8.26, 27] the reference waves E_2 and E_3 were counterpropagating; this fact enabled one to realize OPC in its proper sense. To avoid misunderstanding, we note that in this section the signal to be conjugated is denoted by $E_1(R)$, while the reference waves are denoted by E_2 and E_3. Such indexing (which does not coincide with that adopted elsewhere in this book) is connected with the order in which the waves follow in time.

To estimate the reflection efficiency, it is convenient to introduce the dimensionless pulse area $\theta = h^{-1} d_{12} \int E \, dt$, see [8.25]. The duration of the conjugated echo pulse is $\tau \sim \Delta \omega_{in}^{-1}$, while its area can be estimated by

$$(\theta_4 / \theta_1)^2 \approx \theta_2^2 \theta_3^2 (\alpha L)^2 . \tag{8.6.10}$$

Here $\alpha [\text{cm}^{-1}]$ is the absorption coefficient at the center of the inhomogeneously broadened line, and we assume that $\alpha L \lesssim 1$.

We have considered photon echo by the perturbation procedure with respect to all three fields E_1, E_2, E_3. The true small perturbation parameters which validate such a procedure are the pulse areas θ_1, θ_2, θ_3. At $\theta_i \sim 1$ extremely interesting effects, such as self-induced transparency [8.25, 28], which significantly vary quantitative characteristics of OPC [8.29], are observable.

8.6.2 Bragg's Three-Wave Mixing

As mentioned in Sect. 8.2, three-wave mixing (TWM) gives rise to a phase-conjugated wave copropagating with the signal. This fact may be useful for correcting distortions of an image transmitted through a light guide, see Sect. 1.3. However, in order to succeed in OPC by TWM, it is necessary to reflect the conjugated wave backwards by a mirror, which introduces additional distortions. The phase matching condition also places severe limitations on permissible values of the input angle of the signal (Sect. 8.2).

These difficulties, typical of TWM, can be eliminated if we make use of a special nonlinear medium. Let us assume that the medium possesses a second-order nonlinearity $\chi^{(2)}(R)$ spatially modulated with a period exactly equal to the wavelength $\lambda_2 = 2\pi/k_2$ at the second-harmonic frequency in the medium:

$$\chi^{(2)}(R) = a \exp(ik_2 z) + a^* \exp(-ik_2 z) + \dots . \tag{8.6.11}$$

Here $k_2 = 2\omega\sqrt{\varepsilon(2\omega)}/c$ and the remaining Fourier spatial components have not been written. If the reference wave of frequency 2ω in the z direction, $E_0 \exp(ik_2 z - 2i\omega t)$, and the signal, $E_3 \exp(ik_3 R - i\omega t)$, are sent into such a medium, then the nonlinear polarization will contain the term

$$P_4^{NL}(R, t) = a^* E_0 E_3^* \exp(-i\omega t - ik_3 \cdot R) . \tag{8.6.12}$$

This polarization exactly satisfies the wave equation at ω for any k_3 and, hence, efficiently radiates the exactly conjugated wave with respect to the signal. In the Born approximation the phase-conjugation efficiency is

$$\left|\frac{E_4}{E_3}\right|^2 = \left(\frac{2\pi\omega L}{nc}\right)^2 |aE_0|^2 . \tag{8.6.13}$$

This method is called Bragg three-wave mixing (BTWM).

If the amplitude a of the required spatial Fourier component is of the order of $\chi^{(2)}$ for a homogeneous crystal, the OPC-BTWM efficiency will be the same as in ordinary TWM at the same values of L and $|E_0|^2$.

To make it easier to understand the BTWM process, we recall that the second-order nonlinearity $\chi^{(2)}$ can be interpreted in terms of an addition $\delta\varepsilon$ to the permittivity proportional to the field strength: $\delta\varepsilon = 4\pi\chi^{(2)} E_{real}(R, t)$. If the field is of the form

$$E_0 \exp(-2i\omega t + ik_2 z) + E_0^* \exp(2i\omega t - ik_2 z) ,$$

and the constant $\chi^{(2)}$ is spatially modulated in accord with (8.6.11), then the contribution to the permittivity modulation is

$$\delta\varepsilon(t) \propto E_0 a^* e^{-2i\omega t} + E_0^* a e^{2i\omega t} . \tag{8.6.14}$$

Since it is at twice the light frequency, this contribution is uniform everywhere in space, due to the mutual compensation of the spatial inhomogeneity of the nonlinearity and the fast-varying phase factor of the traveling reference wave (Sect. 8.3). This method has not been realized yet. As feasible media we can suggest ferroelectric crystals in the vicinity of their transition to the non-commensurate phase.

8.6.3 OPC by Superluminescence

The discrimination mechanism in OPC-SS is based on the coincidence of local maxima of the exciting field with those of the field to be amplified. It is clear that this mechanism must work, in principle, for any kind of amplification if the local gain increases as the local intensity of the speckle-inhomogeneous pump rises.

Consider a medium in which the population inversion, and thus the gain, at a certain luminescence frequency ω is caused by an optical pump of frequency ω_L (usually $\omega_L > \omega$). Suppose that the pump field E_L is monochromatic and coherent and, accordingly, can produce an interference pattern of the speckle-structure type. The density of excited particles is higher in those places where the intensity of the exciting radiation $I_L(R) = |E_L(R)|^2$ is higher. As a result, the gain of the wave profile $E(\omega, R)$, spatially correlated with the pump $[E(\omega, R) \propto E_L^*(\omega_L, R)$ in the case of counterpropagation], will be larger than that of uncorrelated profiles, see Sect. 4.1. Therefore, under superluminescence conditions, i.e., in single-pass amplification with a large exponential factor, the radiation excited by spontaneous noise can acquire, on its backward pass, the structure of the phase-conjugated wave.

To estimate the degree of discrimination, we should take into consideration the fact that gain saturation is observed at a sufficiently high local pump intensity due to, say, the depletion of non-excited particles. Let us accept the following model for the dependence of the gain $g(\omega, I_L)$ on the pump intensity I_L:

$$g(I_L) = G I_{sat} [1 - \exp(-I_L/I_{sat})] , \tag{8.6.15}$$

where I_{sat} is the pump saturation intensity. As is seen from (8.6.15), we have a linear dependence $g(I_L) \approx G I_L$ at $I_L \ll I_{sat}$, whereas at $I_L \gg I_{sat}$ the density of excited molecules is practically at a maximum everywhere and the gain is independent of coordinates, $g \approx G I_{sat}$. The gain of the wave with the profile $E(R, \omega)$ can be calculated from the relation

$$\frac{d}{dz} \int |E(r,z)|^2 d^2r = \int g(r,z) |E(r,z)|^2 d^2r \equiv g_{eff} \int |E(r,z)|^2 d^2r , \tag{8.6.16}$$

which identically follows from the parabolic wave equation. If many speckle inhomogeneities are present in the interaction volume, the integration in (8.6.16) can be changed to averaging over the ensemble of the speckle fields. For configurations uncorrelated with the pump the two cofactors under the integral sign in (8.6.16) can be averaged independently. The probability distribution for the speckle field $E_L(R)$ is given by (3.1.13). If we denote the space-averaged pump intensity by I_0, $\langle I_L(R)\rangle = I_0$, the gain of uncorrelated configurations is given by

$$g_{\text{unc}} = GI_0(1+I_0/I_{\text{sat}})^{-1}.\tag{8.6.17}$$

If $E(\omega,R) = A(z)E_L^*(\omega_L,R)$, then we have

$$g_{\text{con}} = 2GI_0(1+I_0/2I_{\text{sat}}) \cdot (1+I_0/I_{\text{sat}})^{-2}.\tag{8.6.18}$$

Thus, the ratio of these gain coefficients which determines the degree of discrimination is

$$\frac{g_{\text{con}}}{g_{\text{unc}}} = 2\frac{1+I_0/2I_{\text{sat}}}{1+I_0/I_{\text{sat}}}.\tag{8.6.19}$$

At small I_0/I_{sat} this ratio is equal to 2, as in the case of stimulated scattering, whereas at $I_0 \gg I_{\text{sat}}$ the discrimination fails since $g_{\text{con}}/g_{\text{unc}} \to 1$.

OPC at superluminescence was demonstrated in [8.30], where a dye was pumped by the second harmonic of a neodymium laser ($\lambda = 532$ nm) transmitted through a phase plate. A bright central core of diffraction quality was observed in superluminescence radiation ranging from 545 to 565 nm after it had passed through the phase plate. An OPC fraction of the order of 10 to 20% was registered.

8.7 OPC in Forward FWM

Consider a four-wave mixing scheme in which plane reference waves E_1 and E_2 of the same frequency ω copropagate at a small angle α to one another, Fig. 8.6a. In accordance with the results of Sect. 6.2, the permissible propagation directions k_3 of the signal E_3 which satisfy the phase matching condition are positioned, at $\omega_3 = \omega$, on a cone whose vertex angle is also α. The conjugated signal $E_4 \propto \chi^{(3)}E_1E_2E_3^*(R)$ produced in FWM is characterized by the direction k_4 which lies on the same cone and is symmetric to the direction k_3 about the cone axis.

This scheme is similar to that of OPC-TWM in many respects. Back reflection of the complex-conjugated wave by a mirror is needed, a transverse shift of the phase-conjugated wave is also observed, $\Delta x \sim 2\alpha L$, and there exist limitations on the reception angle of the signal wave connected with the phase matching condition. Suppose that the deviation of the signal angular spectrum from the cone is $\Delta\theta_3 \sim \lambda/a_3$, where a_3 is the transverse dimension of the E_3 field inhomogeneity in the corresponding direction. Then the requirement

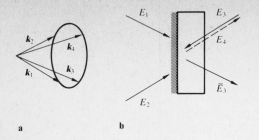

Fig. 8.6. (a) Geometry of wave vectors in copropagating four-wave mixing. **(b)** Scheme of copropagating FWM with a counterpropagating signal; shading denotes a semi-reflecting mirror

that OPC distortions due to the shift of the image be small is: $2\alpha L \leq a_3$ or $\Delta\theta_3 \leq \lambda/2\alpha L$. It is not difficult to check that the latter condition coincides with the limitations placed on $\Delta\theta_3$ following from the phase matching requirement.

It is convenient to direct the original signal $E_3(R)$ in the opposite direction to the reference waves; after passing through a nonlinear medium (without FWM) this signal is reflected from a plane mirror placed normally to the vector k_1+k_2 (Fig. 8.6b). As a result, a wave $\tilde{E}_3(R)$ copropagating with the reference waves is obtained. Then a wave $E_4 \propto E_1 E_2 \tilde{E}_3^*$ will oppose, to within an accuracy of a small shift, the original signal direction and have a complex-conjugated phase, i.e., it will be the phase conjugate of the signal $E_3(R)$ in the proper sense.

For a medium with a local nonlinearity, the equations for the amptides of the copropagating coupled waves E_3 and E_4 under exact phase match conditions, neglecting the depletion of the reference waves E_1 and E_2, are of the form

$$\frac{dE_4^*}{dz} + i\gamma E_1^* E_2^* \tilde{E}_3(z) = 0 ,$$

$$\frac{d\tilde{E}_3}{dz} - i\gamma E_1 E_2 E_4^*(z) = 0 .$$

(8.7.1)

This system of equations differs from (6.5.1) only in the sign reversal in dE_4^*/dz, as in this case both waves are copropagating. It should be noted that equations such as (8.7.1) are also applicable to the three-wave mixing problem in Sect. 8.2. The solution to this system is

$$\tilde{E}_3(z) = \tilde{E}_3(0)\cosh\mu z + e^{i\phi}E_4^*(0)\sinh\mu z ,$$

$$E_4^*(z) = e^{-i\phi}\tilde{E}_3(0)\sinh\mu z + E_4(0)\cosh\mu z ,$$

(8.7.2)

$$\mu = |\gamma E_1 E_2| , \quad e^{i\phi} = i\gamma E_1 E_2 / |\gamma E_1 E_2| .$$

(8.7.3)

Thus, the efficiency with which the input signal $|\tilde{E}_3(0)|^2$ can be converted into the phase-conjugated wave $|E_4(z)|^2$ is $\sinh^2(\mu z)$. In the Born approximation this result is the same as for FWM in the field of two mutually counterpropagating reference waves. However, at $\mu z \gtrsim 1$ the PC reflectivity in the scheme with a copropagating pair of reference waves is high but does not become infinite. In other words, in this case, the instability of the pair of reference waves is not absolute but convective in nature, thus being similar to the instability of a single plane wave, see Sect. 6.7. In the problem considered here, the maximum attainable PC reflectivity should be determined from the condition that the reference waves should not be depleted or distorted by parasitic disturbances in the form of pairs of waves lying on the phase match-ing cone. If the signal power $|E_3(0)|^2$ exceeds the integral power of noise which experiences parametric amplification, the entire power $|E_1|^2$ can be, in prin-ciple, transferred into the phase-conjugated wave if $|E_1|^2 = |E_2|^2$.

The copropagating FWM scheme shown in Fig. 8.6b is advantageous in that it is insensitive to smooth phase distortions introduced by inhomo-geneities of the operating medium. Indeed, if a refractive index inhomo-geneity $\delta n(r, z)$ depends smoothly on r, then the simple change

$$E_{1,2,\tilde{3},4}(r, z) = B_{1,2,\tilde{3},4}(r, z) \exp\left[i\frac{\omega}{c}\int_0^z \delta n(r, z')dz'\right] \tag{8.7.4}$$

will reduce the problem to the previous one, with B_1 and B_2 constant. Thus

$$E_4(r, L) = e^{i\phi}\sinh(\mu L)\exp\left[i\frac{\omega}{c}\int_0^L \delta n(r, z)dz\right]B_3^*(r, z=0),$$

$$e^{i\phi} = i\gamma B_1 B_2 / |\gamma B_1 B_2| = \text{const}, \quad \mu = \text{const}. \tag{8.7.5}$$

If we allow for the fact that on its pass from the input $z = L$ to the mirror $z = 0$ the wave E_3 acquires the phase factor

$$\tilde{B}_3(r, z=0) = E_3(r, z=L)\exp\left[i\frac{\omega}{c}\int_0^L \delta n(r, z)dz\right], \tag{8.7.6}$$

then the final expression for $E_4 \propto E_3^*$ appears in no way affected by distortions due to δn. Hence, in the scheme shown in Fig. 8.6b the medium's inhomo-geneities have no effect either on the interaction efficiency or on the structure of the phase-conjugated field.

Experimental studies of copropagating FWM of a plane-wave signal E_3 in the field of two copropagating plane reference waves E_1 and E_2 are reported in [8.31, 32] and in other publications. *Kucherov* et al. [8.33] realized OPC using the thermal nonlinearity of dye solutions in the scheme shown in Fig. 8.6b.

8.8 Resonators with Phase-Conjugate Mirrors (PCM)

In this section we shall discuss a topic which is slightly separate from the rest of the chapter and which has recently attracted the attention of many investigators, namely, resonators, passive or with amplifying elements, in which one of the mirrors is replaced by an element responsible for ideal OPC. Such resonators possess a number of interesting features. We shall be particularly concerned with the frequency and spatial structures of the fields in such resonators. We will therefore assume that losses inside a resonator are compensated by a single-pass amplification of the radiation.

8.8.1 Structure of OPC-FWM Resonator Modes

Consider first a phase-conjugate mirror (PCM) in a four-wave mixing process with reference waves of frequency $\omega_1 = \omega_2 = \omega_0$. If a wave $E_3 \exp(i k_3 z - i \omega_3 t)$ emerging from inside the resonator is incident on this PCM, the reflected wave will be $E_4 \exp(- i k_4 z - i \omega_4 t)$, where $E_4 = E_3^* \exp(i \phi)$ and $\omega_4 = 2 \omega_0 - \omega_3$, the phase ϕ being specified by the phase of the product $\chi^{(3)} E_1 E_2$. While the wave E_4 propagates from the PCM to the ordinary mirror and back, it acquires the phase factor $\exp(2 i k_4 L)$, where L is the resonator length. In considering the modes of an ordinary resonator, only one round-trip is sufficient, and the field E_4 obtained is equated to the original field E_3. In our particular case, this will not do, since the field E_4 has a differing frequency, $\omega_4 \neq \omega_3$, and we need a second round-trip to complete the equation. The reflection of the wave E_4 from the PCM gives rise to the wave $E_3' = [E_4 \exp(2 i k_4 L)]^* \exp(i \phi)$ of frequency ω_3. Passing from the PCM to the ordinary mirror and back, the wave E_3' acquires a phase shift $\exp(2 i k_3 L)$. Thus, after two complete round-trips we have

$$E_3'' = E_3' \exp(2 i k_3 L) = E_3 \exp[2 i (k_3 - k_4) L] , \qquad (8.8.1)$$

i.e., the field has automatically restored its frequency and acquired a phase shift $2(k_3 - k_4) L = 2(\omega_3 - \omega_4) L/c = 4(\omega_3 - \omega_0) L/c$. This phase shift is independent of the phase of the complex OPC reflectivity.

The mode of a resonator is, by definition, the self-reproducing field. The self-reproduction condition for ordinary resonators specifies the discrete mode frequencies

$$2L \frac{\omega}{c} = 2 \pi m + \Phi(p, q) , \qquad (8.8.2)$$

where $\Phi(p, q)$ is the phase, dependent on the transverse indices p, q, to which quite definite transverse mode configurations $M_{pq}(r, z)$ correspond. Thus, in

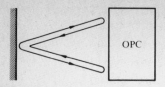

Fig. 8.7. Resonator with one conventional mirror and one phase-conjugate mirror. At sufficiently large transverse mirror dimensions the structure of the transverse modes is arbitrary and is self-reproduced on two trips

an ordinary resonator, the frequency separation of the adjacent modes of a given transverse index (p, q) is $\omega_{m+1} - \omega_m = 2\pi/T$, where $T = 2L/c$ is the complete round-trip time, the absolute frequency position being determined by the resonator length. If the latter is changed, all frequencies become shifted by an amount $\Delta\omega_m = -\omega_m \Delta L/L$.

The self-reproduction condition for a PCM resonator is, according to (8.8.1),

$$4L\frac{\omega_3 - \omega_0}{c} = 2\pi n, \tag{8.8.3}$$

where n is an integer. First of all, there are modes with $n \neq 0$ consisting of two-frequency fields, $\omega_{3n} = \omega_0 + \pi n c/2L$, $\omega_{4n} = \omega_0 - \pi n c/2L$; the difference between modes with $n > 0$ and modes with $n < 0$ is only the choice of which field should be named E_3 and which E_4. The intervals between the adjacent frequencies are $\omega_{n+1} - \omega_n = 2\pi/2T$, where $2T = 4L/c$ is the time of two complete round-trips; these intervals are half as large as those for an ordinary resonator[2]. It is an important fact that the absolute frequency of these modes is determined mainly by the frequency ω_0 of the reference waves and is almost insensitive to small changes in the resonator length: $\Delta\omega_n = (\omega_0 - \omega_n)\Delta L/L$, which is a factor $(\omega_0 - \omega_n)/\omega_0 \sim 10^5$ smaller than for an ordinary resonator.

Suppose that the transverse structures of fields $E_3(r)$ and $E_4(r)$ of the two-frequency mode (with $n \neq 0$) are the same. If such is the case, the time dependence of the intensity at the resonator output dictated by the interference of the fields $E_3(r)$ and $E_4(r)$ is $I(t) \propto [1 + \cos(2\pi nt/T)]$. At $n = 1$, this corresponds to a passage of the intensity maximum from the PCM to the output mirror and back with the velocity of light.

It is remarkable that if both mirrors are sufficiently wide and there are no losses dependent on the transverse structure of the field, the fields E_3 and E_4 are of an absolutely arbitrary form satisfying only the conditions

$$E_{3\,\mathrm{ref}}(r) = E_{4\,\mathrm{inc}}^*(r)\,e^{i\phi}, \qquad E_{4\,\mathrm{ref}}(r) = E_{3\,\mathrm{inc}}^*(r)\,e^{i\phi}, \tag{8.8.4}$$

which relate the incident waves to those reflected from the PCM (Fig. 8.7).

---

2 Since there is no difference between $n > 0$ and $n < 0$, the total number of field oscillators per unit frequency interval is the same as in the case of an ordinary resonator

This statement is also valid in the case of optical inhomogeneities present in the PCM resonator. In the approximation in which the laws of a diffractive change in the transverse structure may be considered identical for different frequencies within the resonator, (8.8.4) will be satisfied for fields of frequencies ω_3 and ω_4 in the whole volume too [3].

Modes with $n = 0$ should be considered separately. All of them, irrespective of their transverse structure and the resonator length L, are of the same frequency ω which coincides with the frequency ω_0 of the reference waves. For simplicity, we shall discuss these modes for resonators with neither gain nor losses and in which the reflectivity of both mirrors (ordinary and PCM) is unity in magnitude for each. The fields $E_3(r, z)$ and $E_4(r, z)$, which were introduced above for $n \neq 0$, are physically identical, $E_3(R) = E_4(R) = E(R)$, in the degenerate case $n = 0$. Then (8.8.4) can be modified as $E(R) = E^*(R) \times \exp[i\phi]$, this giving $E(R) = |E(R)|\exp(i\phi/2)$, where ϕ is specified by the phase of the reference waves. Recalling that in the paraxial approximation

$$E(R) = E_+(r, z) \exp(ikz) + E_-(r, z) \exp(-ikz),$$

we obtain

$$E_+(r, z) = E_-^*(r, z)e^{i\phi}, \qquad E_-(r, z) = E_+^*(r, z)e^{i\phi}. \tag{8.8.5}$$

The fact that the mode field has the same phase within the whole volume, and is a standing wave (with a non-uniform transverse and longitudinal amplitude distribution) is the natural consequence of the definition of the resonator mode. This holds true for the non-degenerate modes of an ordinary resonator lacking losses and active elements. However, in the case of an ordinary resonator the field structure of $E_+(r, z)$ is strictly specified by the set of transverse and longitudinal indices, the phase of the field being arbitrary (because it explicitly depends on the zero time reference). In contrast, in a PCM resonator the transverse structure of the standing wave is arbitrary since the PCM automatically adds the phase-conjugated configuration to any $E_+(r, z)$ configuration. It is important, however, that the absolute phase of light oscillations of this field at the frequency $\omega = \omega_0$ is strictly dictated by the phase ϕ specified by the phase of the reference waves. This fact is conceivable to everybody who can get in phase with the oscillation of a swing as it parametrically sways.

---

3 If the difference in the diffraction laws is essential, i.e., if $|k_3 - k_4|L\theta^2 \gtrsim 1$, where θ is the angle between the rays and the resonator axis, there will exist limitations on the spatial structure of modes with large $|k_3 - k_4| \propto n$. The angular spectrum of such modes must be localized within a ring $|k_3 - k_4|\theta \Delta\theta L \lesssim 1$ round the axis

It should be noted that for both the mode with $n = 0$ in a PCM resonator and the non-degenerate mode of an ordinary resonator the wavefront surface near the ordinary mirror coincides with the mirror surface. This property should not take place for modes in resonators whose losses have a noticeable transverse inhomogeneity (both in ordinary and PCM resonators) and for the modes $n \neq 0$ of ideal PCM resonators.

In the case of OPC-FWM the reflection efficiency decreases as ω_3 is detuned from the frequency ω_0 of the reference waves. As was shown in Sect. 6.2, the spectral interval $|\omega_3 - \omega_0|$ of efficient reflection, which is specified for a fast-responding nonlinear medium by the phase matching condition, is $|\omega_3 - \omega_0| \sim 2\pi c/L'$, where L' is the nonlinear medium's thickness in the direction k_3. Thus, the longitudinal index n of the two-frequency modes of a PCM resonator is limited by the relation $|n| \lesssim 4L/L'$. Also, if the dependence of the phase ϕ of OPC reflection on ω_3 is taken into account, see (6.2.10), the effective resonator length in (8.8.3) can be expressed as $L_{eff} = L + 0.5L'$. As far as the selection of the single longitudinal mode is concerned, the most convenient selection is likely to be the use of a slow-responding FWM nonlinearity with a relaxation time $\tau \gtrsim L_{eff}/c$.

8.8.2 Matrix Method in OPC Resonator Theory

A ray matrix approach is often used in analyzing transverse modes in optical resonators [8.34]. In this approach, the transformation of a ray by an optical system is considered in the paraxial approximation. It is remarkable that an accurate description of the diffraction process can be obtained for parabolic-profile elements in the framework of this approach.

A ray may be characterized by two transverse vector quantities: the distance $r = (x, y)$ from the resonator axis and the slope angle $\theta = (\theta_x, \theta_y)$, see Fig. 8.8. In the paraxial approximation the propagation of the ray through a distance z along the resonator axis corresponds to a transformation $r_2 = r_1 + z\theta_1$, $\theta_2 = \theta_1$. Introducing column vectors composed of (r, θ) we may represent this relation in a matrix form:

$$\begin{pmatrix} r_2 \\ \theta_2 \end{pmatrix} = \begin{pmatrix} A & B \\ C & D \end{pmatrix} \begin{pmatrix} r_1 \\ \theta_1 \end{pmatrix} \tag{8.8.6}$$

with the matrix

$$\begin{pmatrix} A & B \\ C & D \end{pmatrix} = \begin{pmatrix} 1 & z \\ 0 & 1 \end{pmatrix}. \tag{8.8.7}$$

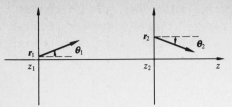

Fig. 8.8. Matrix formalism in paraxial optics

It is not difficult to verify that refraction through a thin lens (or reflection by a mirror) of focal length f can also be treated in terms of (8.8.6) with the matrix

$$\begin{pmatrix} A & B \\ C & D \end{pmatrix} = \begin{pmatrix} 1 & 0 \\ f^{-1} & 1 \end{pmatrix}. \tag{8.8.8}$$

Matrix (8.8.8) with $f = \infty$, i.e., the unit matrix, corresponds to reflection by a plane mirror. The passage of the ray through several optical elements and between them is described by the product of the corresponding matrices taken in the appropriate order, $\hat{M} = \hat{M}_k \dots \hat{M}_2 \hat{M}_1$.

Since the determinant of matrices (8.8.7, 8) is equal to unity, the determinant of any matrix product is also unity; in other words, $\det \hat{M} = AD - BC = 1$ holds true for any transformation of the matrix $\hat{M}$. This relationship represents the theorem of conservation of luminance (Lagrange-Helmholtz theorem).

Let us introduce the matrix

$$\hat{T} = \begin{pmatrix} 1 & 0 \\ 0 & -1 \end{pmatrix}, \quad \hat{T} = \hat{T}^{-1} \tag{8.8.9a}$$

for the backward-going ray tracing effect; it corresponds to the transformation $r_2 = r_1$, $\theta_2 = -\theta_1$. Then the matrix $\hat{M}_{\text{back}}$ corresponding to retracing the elements $(1, 2, \dots, k)$ is of the form

$$\hat{M}_{\text{back}} = \hat{T}\hat{M}^{-1}\hat{T} \equiv \hat{T}\hat{M}_1^{-1}\hat{M}_2^{-1}\dots\hat{M}_k^{-1}\hat{T}. \tag{8.8.9b}$$

With diffraction taken into account, the solution of the parabolic wave equation turns out to be a Gaussian beam:

$$E(r) \propto \exp\left(i\frac{kr^2}{2q}\right) \tag{8.8.10}$$

where $1/q$ is the complex curvature of the wavefront. The imaginary part $\text{Im}\{1/q\}$ corresponds to a Gaussian fall in the amplitude along the radial coordinate to a level e^{-2} of the maximum intensity: the beam radius w is given by $w^2 = (\text{HWe}^{-2}\text{M})^2 = 2(k\,\text{Im}\{1/q\})^{-1}$. The real part $\text{Re}\{1/q\}$ corresponds to the wavefront curvature R in its proper sense: $q^{-1} = R^{-1} + 2i/kw^2$. It can be shown [8.34] that the transformation of a Gaussian beam by the matrix-$\hat{M}$ optical system, with due regard for diffraction, again yields a Gaussian beam, and that

$$q_2 = \frac{Aq_1 + B}{Cq_1 + D}, \tag{8.8.11}$$

where A, B, C, D are the elements of the matrix $\hat{M}$.

To determine the complex curvature q^{-1} of the Gaussian mode in the conventional resonator at the output mirror, it is necessary to calculate the round-trip matrix $\hat{M}$ with the elements A, B, C, D. The mode self-reproduction condition will take the form of (8.8.11) with $q_1 = q_2$, from which

$$q = \frac{A - D \pm \sqrt{(A+D)^2 - 4}}{2C}. \tag{8.8.12}$$

For the mode to be localized in the vicinity of the resonator axis, it is necessary that $|A + D| < 2$; in this case one of the two complex-conjugate roots of (8.8.12) contains the positive imaginary part $\mathrm{Im}\{1/q\} > 0$, responsible for a Gaussian intensity drop along the radius, Such resonators are called stable. In a stable resonator, in the geometrical-optics approximation, a ray with arbitrary initial values r_0, θ_0 will never deviate far from the axis either in r space or in angle even after an infinite number of passes. In this sense, a PCM resonator is always stable (Fig. 8.7).

The presence of a diaphragm with an amplitude transmittance $\propto \exp(-r^2/w_D^2)$ in the resonator can be described by the matrix

$$\hat{M}_D = \begin{pmatrix} 1 & 0 \\ 2i/kw_D^2 & 1 \end{pmatrix}, \tag{8.8.13}$$

whose determinant is also unity. Equation (8.8.9b) also holds in the case when amplitude diaphragms satisfying (8.8.13) are present in the sequence of elements with matrices $\hat{M}_i$.

We now proceed to a matrix description of the PCM resonator. The action of a PCM on the rays can be described by the matrix $\hat{T}$ from (8.8.9a) along with the complex conjugation operator K:

$$\hat{M}(\mathrm{OPC}) = \hat{T}K. \tag{8.8.14}$$

The validity of (8.8.14) for rays is quite understandable since the operation K for them is reduced to the identity transformation. The necessity for the additional operation K is easily seen from (8.8.11) which gives the complex curvature. If K were replaced by I, from (8.8.14) we would obtain the relation

$q_2 = -q_1$, corresponding to the transition from a limited Gaussian beam to a beam whose intensity exponentially rises along the radial coordinate; whereas using K in (8.8.14) we obtain the true relation $q_2 = -q_1^*$.

Let the output (conventional) mirror be plane and serve as the reference plane. The operation $\hat{M}$ for a round-trip in the resonator, see Fig. 8.7, $\hat{M} = \hat{M}_- \hat{T} K \hat{M}_+$, where $\hat{M}_+$ is the matrix for the passage through the optical elements from the conventional mirror to the PCM and $\hat{M}_- = \hat{T} \hat{M}_+^{-1} \hat{T}$ is the matrix corresponding to the passage via the same elements in the reverse order, see (8.8.9). Finally, we have $\hat{M} = \hat{T} \hat{M}_+^{-1} K \hat{M}_+$. If there are no amplitude diaphragms in the resonator, then $\hat{M}_+$ is the real matrix $K \hat{M}_+ = \hat{M}_+ K$, and thus $\hat{M} = \hat{T} K$.

Therefore, if the resonator has no transverse limiters (diaphragms, etc.), the operation of two round trips $\hat{M}^2 = \hat{T} K \hat{T} K$ is always unity, $\hat{M}^2 = 1$. Thus, we have justified the conclusion that in a lossless PCM resonator any field automatically recovers after two round-trips.

Self-reproduction in a round-trip is the condition necessary for single-frequency modes of a PCM resonator. This gives for q:

$$q = (Aq^* + B)/(Cq^* + D) = -q^*,$$

from which it follows that $\text{Re}\{q\} = 0$, $\text{Im}\{q\}$ is arbitrary. The condition $\text{Re}\{q\} = 0$ means that the wavefront of the output radiation is planar, i.e., its shape coincides with that of the output (plane) mirror. If the output mirror is spherical, it is easy to show that the wavefront curvature of the single-frequency mode of a PCM resonator without diaphragms coincides with the mirror curvature, as is the case for the conventional resonator without transverse and longitudinal losses.

8.8.3 Selection of Mode with Lowest Transverse Index

If a resonator (either a conventional one or with a PCM) has no limiting diaphragms and the mirrors are not transversely limited, many modes with different transverse indices are generated, and the divergence of the radiation may be arbitrary. It is of practical interest to make the laser generate the only Gaussian mode with zero transverse index, corresponding to the diffraction-directivity radiation. In conventional resonator this problem can be solved in two ways [8.35]:

1) by using unstable resonators in which losses are high for the zero-index mode and rise especially quickly for higher-index transverse modes; a high total gain during one pass is required to exceed the generation threshold.

2) By using a stable resonator with a diaphragm which transmits the zero mode well but causes high losses in higher transverse modes whose intensities are localized within larger transverse dimensions; only such systems are

suitable for laser media with a low one-pass gain. The required diaphragm size w_D is governed by the fact that the Fresnel number $w_D^2/\lambda L_{eff}$ calculated with the effective resonator length L_{eff} should be of the order of 1. Unfortunately, if the resonator elements possess optical inhomogeneities, the diaphragm must be reduced in size, decreasing the quality factor and power takeoff.

In the presence of phase (non-absorbing) optical inhomogeneities a PCM of large transverse dimensions and, therefore, capable of conjugating all the incident radiation operates as ideally as if inhomogeneities were absent. In other words, from the viewpoint of the field transverse structure, the PCM together with the inhomogeneous medium can be replaced by just the PCM. Though this statement holds true for arbitrary (including irregular) distributed inhomogeneities of the refractive index, we shall illustrate it using, as an example, parabolic inhomogeneities to which the matrix approach applies. The passage through a system with the matrix $\hat{M}_+$, the OPC operation $\hat{T}K$, and the reverse passage through the same system (matrix $\hat{M}_-$) give the operation $\hat{R} = \hat{M}_-\hat{T}K\hat{M}_+$. In the absence of amplitude diaphragms $\hat{M}_+ = \hat{M}_+^*$, i.e., $K\hat{M}_+ = \hat{M}_+K$, and by virtue of (8.8.9), $\hat{M}_- = \hat{T}\hat{M}_+^{-1}\hat{T}$, so that $\hat{R} = \hat{T}K$, i.e., the properties of the matrix $\hat{M}_+$ do not influence the result.

Taking this into account, we can consider the resonator shown in Fig. 8.9. We assume that the radius of curvature of the output mirror is R_0 and that there are no optical inhomogeneities between the diaphragm and the mirror. We limit ourselves to modes with frequency $\omega = \omega_0$, i.e., with the longitudinal index $n = 0$; the reasoning for two-frequency modes, $n \neq 0$, is similar.

If we take as a reference a plane in the vicinity of the output mirror and calculate the field component on it, the round-trip operation can be represented as

$$\hat{M} = \begin{pmatrix} 1 & L \\ 0 & 1 \end{pmatrix} \begin{pmatrix} 1 & 0 \\ 2i/kw_D^2 & 1 \end{pmatrix} \begin{pmatrix} 1 & 0 \\ 0 & -1 \end{pmatrix}$$

$$\times K \begin{pmatrix} 1 & 0 \\ 2i/kw_D^2 & 1 \end{pmatrix} \begin{pmatrix} 1 & L \\ 0 & 1 \end{pmatrix} \begin{pmatrix} 1 & 0 \\ -2/R_0 & 1 \end{pmatrix}$$

$$= \begin{pmatrix} 1+2ibL(1-2L/R_0) & 2ibL^2 \\ 2/R_0+2ib(1-2L/R_0) & -1+2ibL \end{pmatrix} K. \tag{8.8.15}$$

Here R_0 is the radius of curvature of the output mirror, w_D is the transverse dimension of the Gaussian diaphragm, and $b = 2/kw_D^2$. Applying (8.8.11) taking into account the complex conjugation operation K, and equating the curvature q^{-1} after the round-trip to the original curvature we obtain the equation $q = (Aq^*+B)/(Cq^*+D)$, where A, B, C, D are the elements of the matrix $\hat{M}$ in (8.8.15). The explicit expressions for the real radius of curvature

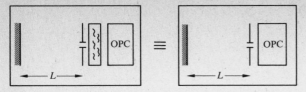

Fig. 8.9. Resonator with a phase inhomogeneous medium between the diaphragm and PCM is equivalent, in the sense of the transverse mode structure, to a resonator with a PCM near the diaphragm

R of the mode wavefront and for the mode waist dimension w on the output mirror is

$$R = L\left[\alpha + \frac{(1-\alpha)|1-\alpha|}{\sqrt{N^2+(1-\alpha)^2}}\right]^{-1}, \quad w = \left[\frac{L\lambda\sqrt{N^2+(1-\alpha)^2}}{\pi N|1-\alpha|}\right]^{1/2},$$

$$N = \frac{1}{2bL} = \frac{\pi w_D^2}{2L\lambda}, \quad \alpha = L/R_0. \tag{8.8.16}$$

Let us discuss these results for a plane output mirror, $R_0 \to \infty$, $\alpha \to 0$. In this case $R = L\sqrt{1+N^2}$, $w = (\lambda L\sqrt{N^2+1}/\pi N)^{1/2}$, where N is the Fresnel number, w and w_D are the radii of the beam on the output mirror and of the diaphragm, respectively, at the $\exp(-2)$ level of intensity. We fix the distance L from the output mirror to the diaphragm and vary w_D or, which is the same, the Fresnel number N.

At $N \ll 1$, i.e., at small values of the radius w_D of the diaphragm, we have $R \approx +L$ and $w \approx (2/\pi^2)^{1/2}L\lambda/w_D$. This means that a diverging Gaussian beam emerges from the resonator with the waist in the vicinity of the diaphragm. When this wave reaches the diaphragm after its reflection from the plane mirror, the radius of curvature of the wavefront is $R' \approx 2L$ and the beam radius $w' \approx 2w$. At $N \ll 1$, the resonator's quality factor is low: most of the beam is cut off by the diaphragm since $w' \gg w_D$. For the lowest mode, the energy fraction transmitted through the diaphragm on the round-trip amounts to $[1+2w'^2/w_0^2]^{-1}$; for $N \lesssim 1$ this fraction is about $N^2/4 \ll 1$. Hermitian-Gaussian modes with transverse indices s, p experience still larger losses; they are localized on the diaphragm with a radius of the order of $w'_{s,p} \sim w'(s+p+1)^{1/2}$.

If the Fresnel number is large, $N \gg 1$, we obtain $R \approx LN \approx \pi w_D^2/2\lambda$, $w \approx (\lambda L/\pi)^{1/2}$. In this case, the lowest mode corresponds to a Gaussian beam with a nearly planar wavefront, and the beam structure is practically the same in the pass from the diaphragm to the output mirror. In this case losses are not high: the energy fraction passing through the diaphragm on the round-trip, for the Hermitian-Gaussian mode with indices s, p, is

$$1 - 2(s+p+1)L\lambda/\pi w_{\mathrm{D}}^2 = 1 - (s+p+1)/N. \tag{8.8.17}$$

Unfortunately, at $N \to \infty$, the selection of higher transverse modes weakens.

The structure of the modes, their quality factors, and which modes are selected in a resonator with a diaphragm and PCM at different values of N appear to be, in general, the same as those for a resonator employing a conventional mirror and a diaphragm at the same value of N, but only if there are no optical inhomogeneities in the conventional resonator. The advantage of the PCM resonator is its complete insensitivity to those phase inhomogeneities which lie between the selecting diaphragm and the PCM. At $N \sim 2$ to 3 and for a not-very-large excess over the threshold, higher transverse modes do not participate in generation either in conventional or in PCM resonators. In this case the zeroth mode in the conventional resonator is strongly distorted and, in addition, possesses a low quality factor due to the loss on the diaphragm of those rays deflected by inhomogeneities. As opposed to this, the lowest mode in the PCM resonator is of diffraction quality at the same N, and possesses a good quality factor even if optical inhomogeneities are present. Therefore, a considerably larger diaphragm may be used in a PCM resonator generating the lowest mode; in this case, however, the output mirror should be placed at a larger distance from it to keep the Fresnel number N unchanged. Thus, we can decrease the divergence of the radiation and increase the efficiency of energy extraction from the whole volume of the operating medium.

8.8.4 Resonators with an SBS Cell

The application of an SBS cell to modulation of the quality factor of a pulsed laser is a matter of long-standing interest [8.36, 37]. In recent years, considerable interest in such resonators has been aroused in connection with the discovery of self-phase conjugation in SBS. A typical laser scheme employing an SBS cell is illustrated in Fig. 8.10, see [8.38]. Unfortunately, to date there has not been any quantitative theory or even an understanding of all processes essential for the operation of such a laser. Here we shall confine ourselves to a brief description of some experimental results.

A typical frequency shift in SBS is $\Omega_{\mathrm{B}}/2\pi c \sim 0.1 \ \mathrm{cm}^{-1}$ (for acetone at $\lambda \approx 1 \ \mu\mathrm{m}$). For a resonator of length $L = 2 \ \mathrm{m}$ the spacing between modes with adjacent transverse indices is $\Delta \Omega_{n,\,n-1}/2\pi c = (2L)^{-1} \approx 2.5 \times 10^{-3} \ \mathrm{cm}^{-1}$, i.e., about a factor of 40 lower. Therefore, there are practically always resonator modes whose frequency difference coincides with Ω_{B} to an accuracy of the scattering linewidth Γ. Because of this, the resonator length has little effect on the laser operation. However, as experiments show, this type of operation is rather sensitive to the position of the scattering region inside the resonator [8.39]; the position of an SBS mirror is determined by the localization of the

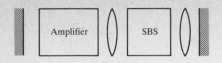

Fig. 8.10. Typical scheme of an SBS-cell resonator

telescope focal waist inside an SS-active medium (Fig. 8.10). Different types of operation can be realized in such a laser. First, it is the free running mode in which SBS fails to develop. Second, with the resonator turned appropriately, different frequency components of the counterpropagating waves generated by the laser are matched in phase within the waist in such a way that a hypersonic wave is efficiently excited. If, in addition, the reflectivity of the right mirror in Fig. 8.10 is considerably lower than 1, the quality factor of the laser sharply increases and a gigantic pulse is generated, provided the sound reflection hologram in the waist is sufficiently effective.

With the initiation of SBS, the radiation spectrum may consist of several equidistant components separated by the shift Ω_B [8.39]. *Lesnik* et al. in [8.40] observed a gradual shift of the mean generation frequency towards the Stokes region, and the frequency was shifted by an amount Ω_B during the round-trip time.

A low selectivity, in the sense of the radiation transverse structure, was observed for resonators with an SBS cell [8.39, 40], which conforms with the general properties of PCM resonators. In this case, the divergence of the radiation generated was far larger than the diffraction limit. Nevertheless, the transverse (spatial) coherence of the radiation measured by Young's method appeared to be high; this was connected, according to *Leschev* et al. [8.39], with the sluggishness of the hypersound buildup.

Rather interesting results of investigations into the spatial and temporal structure of radiation from a laser in which the quality factor was modulated by SBS or STS have been obtained [8.41 – 43].

Il'ichev et al. [8.44] used mirror configurations which corresponded to an unstable resonator in the absence of SBS. They demonstrated that as SBS was switched on, a gigantic pulse was generated whose output beam had a transverse structure of diffration quality.

8.9 Literature

OPC by a reflecting surface (OPC-S) was suggested in [8.17] and realized in [8.1]. It was subsequently realized in [8.2 – 7] for optical radiation and in [8.16, 45] for sound waves. The application of optically controllable liquid-crystal masks to OPC-S of weak signals was suggested in [8.46] and realized in [8.6]. The idea of using three-wave mixing (TWM) to obtain a complex-conjugated wave was suggested in [8.47]; for the realization of the TWM method see [8.12 – 15]. Bragg TWM for OPC was considered theoretically in [8.48]. Parametric interaction and generation of pairs of counterpropagating waves by modulating a medium's parameter at twice the signal frequency was studied theoretically and experimentally in [8.49 – 50] for the acoustic region. The application of these processes to OPC of acoustic waves was suggested in [8.52]; for the realization see [8.18]. OPC by a surface twinkling at double frequency was suggested in [8.53] and independently realized in [8.19]. A review on OPC of acoustic waves is given in [8.54]. The first publications on the application of retroreflectors to compensation of laser medium distortions are [8.22, 55]. Compensation of inhomogeneities using a retroreflector is considered theoretically in [8.56]. A retroreflector-resonator laser was investigated in [8.23, 56]. The phenomenon of photon and spin echo is discussed in [8.24, 25, 57]. Experiments on the phase-conjugated echo wave are discussed in [8.26, 27, 58, 59]. A theoretical consideration of OPC with a discrimination mechanism in a saturable amplifying medium was carried out in [8.60, 61]; for an experiment on OPC at superluminescence see [8.30]. OPC by forward FWM is discussed in [8.33, 62, 63].

The first experiment with a PCM laser is reported in [8.64], see also [8.38]. The theory of PCM resonators was reported in [8.40, 65 – 70], and many subsequent publications reviewed in [8.20, 21] in which experimental results are also discussed. For resonators with an SBS cell or STS cell see [8.36 – 44].

References

Chapter 1

1.1 B. Ya. Zel'dovich, N. F. Pilipetsky, V. V. Shkunov: Optical phase conjugation in stimulated scattering, Usp. Fiz. Nauk **133**, 249 (1982) [English transl.: Sov. Phys. − Usp. **25**, 713 (1982)]

1.2 D. M. Pepper: Nonlinear optical phase conjugation. Opt. Eng. **21**, 156 (1982)

1.3 R. W. Hellwarth: Optical beam phase conjugation by stimulated backscattering. Opt. Eng. **21**, 257 (1982)

1.4 A. Yariv: Phase conjugate optics and real-time holography. IEEE J. Quant. Electron. **14**, 650 (1978)

1.5 R. Fisher (ed.): *Optical Phase Conjugation* (Academic, New York 1983)

1.6a) V. I. Bespalov (ed.): *Optical Phase Conjugation in Nonlinear Media* (IPF AN SSSR, Gorky, USSR 1979) (in Russian)

   b) V. I. Bespalov (ed.): *Phase Conjugation in Nonlinear Media* (IPF AN SSSR, Gorky, USSR 1982) (in Russian)

1.7 D. Gabor: Nature **161**, 777 (1948); Proc. Roy. Soc. Ser. A **197**, 454 (1949)

1.8 W. L. Bragg: Microscopy on the base of wave front reconstruction. Nature **166**, 399 (1950)

1.9 D. Gabor: Patent No. 2.770.166 (USA), Application July 6, 1951

1.10 Yu. N. Denisyuk: On mapping optical properties of an object in the scattered field. Opt. Spektrosk. **15**, 522 (1963) [English transl.: Opt. Spectrosc. **15**, 279 (1962)]

1.11 H. W. Kogelnik: Bell Syst. Techn. J. **44**, 2451 (1965)

1.12 E. N. Leith, J. Upatnieks: Holographic imaging through diffusing media. J. Opt. Soc. Am. **56**, 523 (1966)

1.13 H. Kogelnik: Controlled transmission of waves through inhomogeneous media, Patent No. 3.449.577 (USA), Application October 23, 1965

1.14 G. Sincerbox: Optical data recording, Patent No. 1.218.331 (GB), Application April 9, 1968

1.15 W. P. Cathey: Holographic simulation of compensation for atmospheric wavefront distortions. Proc. IEEE **56**, 340 (1968)

1.16 H. J. Gerritsen, E. G. Ramberg, S. Freeman: In *Proc. Symposium on Modern Optics* (Polytechnic Press, New York 1967) p.109

1.17 V. I. Stepanov, E. I. Ivakin, A. S. Rubanov: On recording plane and volume dynamic holograms in saturable absorbers. Dokl. Akad. Nauk SSSR **196**, 567 (1971) [English transl.: Sov. Phys. − Dokl. **16**, 46 (1971)]

1.18 J. P. Woerdman: Formation of a transient free carrier hologram in Si. Opt. Comm. **2**, 212 (1970)

1.19 Hellwarth, R. W.: Generation of time-reversed wavefront by nonlinear reflection. J. Opt. Soc. Am. **67**, 1 (1977)

1.20 D. Bloom, G. C. Bjorklund: Conjugate wavefront generation and image reconstruction by four-wave mixing. Appl. Phys. Lett. **31**, 592 (1977)

1.21 A. Yariv, D. M. Pepper: Amplified reflection, phase conjugation, and oscillation in degenerate four-wave mixing. Opt. Lett. **1**, 16 (1977)

1.22 B. Ya. Zel'dovich, N. F. Pilipetsky, A. N. Sudarkin, V. V. Shkunov: Phase conjugation by a surface. Dok. Akad. Nauk SSSR **252**, 92 (1980) [English transl.: Sov. Phys. − Dokl. **25**, 377 (1980)]

1.23 O. L. Kulikov, N. F. Pilipetsky, A. N. Sudarkin, V. V. Shkunov: Realization of phase conjugation by a surface. Pis'ma Zh. Eksp. Teor. Fiz. **31**, 377 (1980) [English transl.: JETP Lett. **31**, 345 (1980)]

1.24 B. Ya. Zel'dovich, V. I. Popovichev, V. V. Ragulsky, F. S. Faizullov: On relation between wavefronts of reflected and exciting radiation in stimulated Brillouin scattering. Pis'ma Zh. Eksp. Teor. Fiz. **15**, 160 (1972) [English transl.: JETP Lett. **15**, 109 (1972)]

1.25 O. Yu. Nosach, V. I. Popovichev, V. V. Ragulsky, F. S. Faizullov: Compensation of phase distortions in an amplifying medium by a "Brillouin mirror". Pis'ma Zh. Eksp. Teor. Fiz. **16**, 617 (1972) [English transl.: JETP Lett. **16**, 435 (1972)]

1.26 V. N. Blaschuk, B. Ya. Zel'dovich, N. A. Mel'nikov, N. F. Pilipetsky, V. I. Popovichev, V. V. Ragulsky: Phase conjugation by stimulated scattering of focused beams. Pis'ma Zh. Tekhn. Fiz. **3**, 211 (1977) [English transl.: Sov. Phys. – Tech. Phys. Lett. **3**, 83 (1977)]

1.27 V. I. Bespalov, A. A. Betin, G. A. Pasmanik: On reconstruction effects in stimulated scattering. Pis'ma Zh. Tekhn. Fiz. **3**, 215 (1977) [English transl.: Sov. Phys. – Tech. Phys. Lett. **3**, 85 (1977)]

1.28 V. Wang, G. R. Giuliano: Corrections of phase aberrations via stimulated Brillouin scattering. Opt. Lett. **2**, 4 (1978)

1.29 V. G. Sidorovich: To the theory of a "Brillouin mirror". Zh. Tekhn. Fiz. **46**, 2168 (1976) [English transl.: Sov. Phys. – Tech. Phys. **21**, 1270 (1976)]

1.30 I. M. Bel'dyugin, M. G. Galushkin, E. M. Zemskov, V. I. Mandrosov: On complex field conjugation in SBS, Kvantovaya Elektron. **3**, 2467 (1976) [English transl.: Sov. J. Quant. Electron. **6**, 1349 (1976)]

1.31 B. Ya. Zel'dovich, V. V. Shkunov: On wavefront reproduction in SRS. Kvantovaya Elektron. **4**, 1090 (1977) [English transl.: Sov. J. Quant. Electron. **7**, 610 (1977)]

1.32 B. Ya. Zel'dovich, N. A. Mel'nikov, N. F. Pilipetsky, V. V. Ragulsky: Demonstration of optical phase conjugation in stimulated Raman scattering. Pis'ma Zh. Eksp. Teor. Fiz. **25**, 41 (1977) [English transl.: JETP Lett. **25**, 36 (1977)]

1.33 A. I. Sokolovskaya, G. L. Brekhovskykh, A. D. Kudryavtseva: Wavefront reconstruction of light beams in stimulated Raman scattering. Dokl. Akad. Nauk SSSR **233**, 356 (1977) [English transl.: Sov. Phys. – Dokl. **22**, 156 (1977)]

1.34 A. Yariv: On transmission and recovery of three-dimensional image information in optical waveguides. J. Opt. Soc. Am. **66**, 301 (1976)

1.35 P. V. Avizonis, F. A. Hopf, W. D. Bomberger, S. F. Jacobs, A. Tomita, K. H. Womack: Optical phase conjugation in lithium formate crystal. Appl. Phys. Lett. **31**, 435 (1977)

1.36 S. I. Shostko, Ya. G. Podoba, Yu. A. Anan'ev, V. D. Volosov, A. V. Gorlanov: On a possibility of compensating for optical inhomogeneities in laser devices. Pis'ma Zh. Tekhn. Fiz. **5**, 29 (1979) [English transl.: Sov. Phys. – Tech. Phys. Lett. **5**, 11 (1979)]

1.37 A. A. Chaban: Instability of elastic oscillations in piezo-electrics in alternating electric field. Pis'ma Zh. Eksp. Teor. Fiz. **6**, 381 (1967) [English transl.: JETP Lett. **6**, 381 (1967)]

1.38 A. Billman, C. Frenois, J. Joffrin, A. Levelut, S. Ziolkiewich: Les Echos de Phonons. J. Phys. (Paris) **34**, 453 (1973)

1.39 F. V. Bunkin, D. V. Vlasov, Yu. A. Kravtsov: On sound phase conjugation with amplification of phase-conjugated wave. Kvantovaya Elektron. **8**, 1144 (1981) [English transl.: Sov. J. Quant. Electron. **11**, 687 (1981)]

1.40 A. P. Brysev, F. V. Bunkin, D. V. Vlasov, Yu. E. Kozarov: Experimental realization of a NbLi parametric phase-conjugate sound amplifier. Pis'ma Zh. Tekhn. Fiz. **8**, 546 (1982)

1.41 A. P. Brysev, F. V. Bunkin, D. V. Vlasov, L. L. Gervits: A phase-conjugate parametric plane mirror. Pis'ma Zh. Tekhn. Fiz. **8**, 554 (1982)

1.42 N. F. Pilipetsky, A. N. Sudarkin, V. V. Shkunov: Wave front reversal by a flickering surface. Kvantovaya Elektron. **9**, 835 (1982) [English transl.: Sov. J. Quant. Electron. **12**, 528 (1982)]

1.43 B. Ya. Zel'dovich, N. F. Pilipetsky, V. V. Ragulsky, V. V. Shkunov: Phase conjugation by nonlinear optics methods. Kvantovaya Elektron. **5**, 1800 (1978) [English transl.: Sov. J. Quant. Electron. **8**, 1021 (1978)]

1.44 V. Wang: Nonlinear optical phase conjugation for laser systems. Opt. Eng. **17**, 267 (1978)

1.45 D. M. Pepper (ed.): Nonlinear optical phase conjugation. Special issue of Opt. Eng. **21**, 155 (1982)

1.46 N. Basov, I. Zubarev: Powerful laser systems with phase conjugation by SMBS mirror. Appl. Phys. **20**, 61 (1979)

1.47 C. R. Giuliano: Applications of optical phase conjugation. Phys. Today, April 1981, p. 27

1.48 Yu. V. Dolgopolov, V. A. Komarevsky, S. B. Kormer, G. G. Kochemasov, S. M. Kulikov, V. M. Murugov, V. D. Nikolaev, S. A. Sukharev: Experimental investigations of applicability of optical phase conjugation by stimulated Brillouin scattering. Zh. Eksp. Teor. Fiz. **76**, 908 (1979) [English transl.: Sov. Phys. − JETP **49**, 458 (1979)]

1.49 S. A. Lesnik, M. S. Soskin, A. I. Khizhnyak: An SBS complex-conjugate mirror laser. Zh. Tekhn. Fiz. **49**, 2257 (1979) [English transl.: Sov. Phys. − Tech. Phys. **24**, 1249 (1979)]

1.50 I. M. Bel'dyugin, M. G. Galushkin, E. M. Zemskov: On properties of PCM resonators. Kvantovaya Elektron. **6**, 38 (1979) [English transl.: Sov. J. Quant. Electron. **9**, 20 (1979)]

1.51 J. Au Yeung, D. Fekete, D. M. Pepper, A. Yariv: A theoretical and experimental investigation of the modes of resonators with phase-conjugate mirrors. IEEE J. Quant. Electron. **15**, 1180 (1979)

1.52 G. J. Dunning, R. C. Lind: Demonstration of image transmission through fibers by optical phase conjugation. Opt. Lett. **7**, 558 (1982)

1.53 A. Yariv, D. Fekete, D. M. Pepper: Compensation for channel dispersion by nonlinear optical phase conjugation. Opt. Lett. **4**, 52 (1979)

1.54 Yu. M. Kruzhilin: A self-adjusting laser-target system for laser thermonuclear fusion. Kvantovaya Elektron. **5**, 625 (1978) [English transl.: Sov. J. Quant. Electron. **8**, 359 (1978)]

1.55 N. F. Pilipetsky, V. I. Popovichev, V. V. Ragulsky: Light concentration by optical phase conjugation. Pis'ma Zh. Eksp. Teor. Fiz. **27**, 619 (1978) [English transl.: JETP Lett. **27**, 595 (1978)]

1.56 A. A. Ilyukhin, G. v. Peregudov, M. E. Plotkin, E. N. Ragozin, V. A. Chirkov: Laser radiation focusing on a target by optical phase conjugation in SBS. Pis'ma Zh. Eksp. Teor. Fiz. **29**, 364 (1979) [English transl.: JETP Lett. **29**, 328 (1979)]

1.57 B. Ya. Zel'dovich, V. V. Shkunov: "Spatial-polarization optical phase conjugation by four-photon interaction", in *Optical Phase Conjugation in Nonlinear Media*, ed. by V. I. Bespalov (IPF AN SSSR, Gorky, USSR 1979) p. 23 (in Russian)

1.58 V. N. Blaschuk, A. V. Mamaev, N. F. Pilipetsky, V. V. Shkunov, B. Ya. Zel'dovich: Wavefront reversal with angular tilting-theory and experiment for the four-wave mixing. Opt. Comm. **31**, 383 (1979)

1.59 J. H. Marburger: Optical pulse integration and chirp reversal in degenerate four-wave mixing. Appl. Phys. Lett. **32**, 372 (1978)

1.60 D. M. Pepper, J. Au Yeung, D. Fekete, A. Yariv: Spatial convolution and correlation of optical fields via degenerate four-wave mixing. Opt. Lett. **3**, 7 (1978)

1.61 G. P. Agrawal, C. Flytzanis: Bistability and hysteresis in phase-conjugated reflectivity. IEEE J. Quant. Electron. **17**, 374 (1981)

1.62 A. A. Borsch, M. S. Brodin, V. I. Volkov, N. V. Kukhtarev: Bistability and hysteresis in phase-conjugated wave in degenerate six-photon interaction in cadmium sulphide. Kvantovaya Elektron. **8**, 1304 (1981) [English transl.: Sov. J. Quant. Electron. **11**, 777 (1981)]

1.63 D. M. Pepper, R. L. Abrams: Narrow optical bandpass filter via nearly degenerate four-wave mixing. Opt. Lett. **3**, 212 (1978)

1.64 V. N. Blaschuk, N. F. Pilipetsky, V. V. Shkunov: Four-wave interaction as a controllable frequency filter. Dokl. Akad. Nauk SSSR **251**, 70 (1980) [English transl.: Sov. Phys. − Dokl. **25**, 185 (1980)]

1.65 N. F. Pilipetsky, V. V. Shkunov: Narrowband four-wave reflecting filter with frequency and angular tuning. Opt. Comm. **37**, 217 (1981)

1.66 A. G. Gyulamiryan, A. V. Mamaev, N. F. Pilipetsky, V. V. Shkunov: Tunable nonlinear four-wave filter. Opt. Spektrosk. **52**, 387 (1982)

1.67 J. Nilsen, N. S. Gluck, A. Yariv: Narrow-band optical filter through phase conjugation by nondegenerate four-wave mixing in sodium vapor. Opt. Lett. **6**, 370 (1981)

1.68 V. N. Blaschuk, V. N. Krasheninnikov, N. A. Melnikov, N. F. Pilipetsky, V. V. Ragulsky, V. V. Shkunov, B. Ya. Zel'dovich: SBS wavefront reversal for the depolarized light – theory and experiment. Opt. Comm. **27**, 137 (1978)

1.69 V. I. Bespalov, V. G. Manishin, G. A. Pasmanik: Nonlinear selection of optical radiation by its reflection from an SBS mirror. Zh. Eksp. Teor. Fiz. **77**, 1756 (1979) [English transl.: Sov. Phys. – JETP **50**, 879 (1979)]

1.70 J. F. Lam: Doppler-free laser spectroscopy via degenerate four-wave mixing. Opt. Eng. **21**, 219 (1982)

1.71 V. V. Ragulsky: The possibility of registration of low light absorption by phase conjugation. Pis'ma Zh. Tekhn. Fiz. **6**, 687 (1980) [English transl.: Sov. Phys. – Tech. Phys. Lett. **6**, 297 (1980)]

1.72 N. G. Basov, I. G. Zubarev, A. V. Mironov, S. I. Mikhailov, A. Yu. Okulov: Laser interferometer with phase-conjugate mirrors. Zh. Eksp. Teor. Fiz. **79**, 1678 (1980) [English transl.: Sov. Phys. – JETP **52**, 847 (1980)]

1.73 A. E. Siegman: Dynamic interferometry and differential holography of irregular phase objects using phase conjugate reflection. Opt. Comm. **31**, 257 (1979)

Chapter 2

2.1 L. D. Landau, E. M. Lifshitz: *Electrodynamics of Continuous Media,* 3rd ed. (Pergamon, Oxford 1978)

2.2 L. D. Landau, E. H. Lifshitz: *Mechanics of Continuous Media* (Pergamon, Oxford 1959)

2.3 I. L. Fabelinskii: *Molecular Scattering of Light* (Plenum, New York 1968)

2.4 V. S. Starunov, I. L. Fabelinskii: Stimulated Mandelstam-Brillouin scattering and stimulated entropian (temperature) scattering of light. Usp. Fiz. Nauk **98**, 441 (1969) [English transl.: Sov. Phys. – Usp. **12**, 463 (1970)]

2.5 V. V. Ragulsky: Stimulated Brillouin scattering lasers. Trudy FIAN **85**, 3 (1976) [English transl.: Lebedev Trans. **85**, 1 (1976)]

2.6 Yu. V. Dolgopolov, Yu. F. Kiryianov, S. B. Kormer, G. G. Kochemasov, S. M. Kulikov, V. D. Nikolaev, S. A. Sukharev: "Study of Phase Conjugation in SBS and Its Application to Laser Thermonuclear Fusion Set-up", in *Optical Phase Conjugation in Nonlinear Media,* ed. by V. I. Bespalov (IPF AN SSSR, Gorky, USSR 1979) p. 117 (in Russian)

2.7 B. Ya. Zel'dovich, V. V. Shkunov: Influence of a wave's group velocity mismatch on reproduction of pump spectrum under stimulated scattering. Kvantovaya Elektron. **5**, 2659 (1978) [English transl.: Sov. J. Quant. Electron. **8**, 1505 (1978)]

2.8 I. M. Bel'dyugin, I. G. Zubarev, S. I. Michailov: Analysis of conditions for stimulated Raman scattering of multimode pump radiation in dispersive media. Kvantovaya Elektron. **7**, 1471 (1980) [English transl.: Sov. J. Quant. Electron. **10**, 847 (1980)]

2.9 G. Plachek: *Rayleigh Scattering and Raman Effect* (GNTIU, Khar'kov, USSR 1935) (in Russian); *Handbuch der Radiologie,* Vol. 6, ed. by E. Marx (Akademische Verlagsges., Leipzig 1934)

2.10 N. Bloembergen: The stimulated Raman effect. Amer. J. Phys. **35**, 989 (1967)

2.11 B. Ya. Zel'dovich, I. I. Sobel'man: Stimulated light scattering induced by absorption. Usp. Fiz. Nauk **101**, 3 (1970) [English transl.: Sov. Phys. – Usp. **13**, 307 (1970)]

2.12 I. L. Fabelinskii: "Stimulated Mandelstam-Brillouin Process", in *Quantum Electronics: a Treatise,* Vol. 1, ed. by H. Rabin and C. L. Tang (Academic, New York 1975) p. 364

2.13 C. S. Wang: "The Stimulated Raman Process", in *Quantum Electronics: a Treatise,* Vol. 1, ed. by H. Rabin und C. L. Tang (Academic, New York 1975) p. 447

2.14 E. I. Woodbury, W. K. Ng: Proc. IRE **50**, 2367 (1962)

2.15 G. Eckhardt, R. W. Hellwarth, F. J. McClung, S. E. Schwarz, D. Weiner, E. I. Woodbury: Stimulated Raman scattering from organic liquids. Phys. Rev. Lett. **9**, 455 (1962)

2.16 E. Garmire, F. Pandarese, C. H. Townes: Coherently driven molecular vibrations and light modulation. Phys. Rev. Lett. **11**, 160 (1963)

2.17 R. Y. Chiao, C. H. Townes, B. P. Stoicheff: Stimulated Brillouin scattering and coherent generation of intense hypersonic waves. Phys. Rev. Lett. **12**, 592 (1964)

2.18 R. Y. Mash, V. V. Morosov, V. S. Starunov, I. L. Fabelinskii: Rayleigh-wing stimulated light scattering. Pis'ma Zh. Eksp. Teor. Fiz. **2**, 41 (1965) [English transl.: JETP Lett. **2**, 41 (1965)]

2.19 R. L. Carman, R. Y. Chiao, P. L. Kelley: Phys. Rev. Lett. **17**, 1281 (1966)

2.20 R. M. Hermann, M. A. Gray: Theoretical prediction of the stimulated Rayleigh scattering in liquids. Phys. Rev. Lett. **19**, 824 (1967)

2.21 D. H. Rank, C. W. Cho, N. D. Foltz, T. A. Wiggins: Stimulated thermal Rayleigh scattering. Phys. Rev. Lett. **19**, 828 (1967)

2.22 G. I. Zaitsev, Yu. I. Kysylasov, V. S. Starunov, I. L. Fabelinskii: Stimulated temperature scattering of light in liquids. Pis'ma Zh. Eksp. Teor. Fiz. **6**, 802 (1967) [English transl.: JETP Lett. **5**, 255 (1967)]

2.23 I. L. Fabelinskii, D. I. Mash, V. V. Morozov, V. S. Starunov: Stimulated scattering of light in hydrogen gas at low pressures. Phys. Lett. **27A**, 253 (1968)

2.24 R. W. Hellwarth: Theory of stimulated Raman scattering. Phys. Rev. **130**, 1850 (1963)

2.25 C. L. Tang: Saturation and spectral characteristics of the Stokes emission in the stimulated Brillouin process. J. Appl. Phys. **37**, 2445 (1966)

2.26 N. M. Kroll: J. Appl. Phys. **36**, 34 (1965)

2.27 N. M. Kroll, P. L. Kelley: Temporal and spatial gain in stimulated light scattering. Phys. Rev. **4A**, 763 (1971)

2.28 B. Ya. Zel'dovich: Time of establishment of stationary regime of stimulated light scattering. Pis'ma Zh. Eksp. Teor. Fiz. **15**, 226 (1972) [English transl.: JETP Lett. **15**, 158 (1972)]

2.29 Yu. E. Dyakov: The influence of nonmonochromatics on the spectrum shape of stimulated Brillouin scattering. Pis'ma Zh. Eksp. Teor. Fiz. **9**, 489 (1969) [English transl.: JETP Lett. **10**, 347 (1969)]

2.30 R. L. German, F. Shimizu, C. S. Wang, N. Bloembergen: Theory of Stokes pulse shapes in transient stimulated Raman scattering. Phys. Rev. **2A**, 60 (1970)

2.31 T. I. Kusnetsova: Phasing of the spectrum and short light pulses in stimulated Raman scattering. Pis'ma Zh. Eksp. Teor. Fiz. **10**, 153 (1969) [English transl.: JETP Lett. **10**, 98 (1969)]

2.32 S. A. Akhmanov, Yu. E. Dyakov, A. S. Chirkin: *Introduction to Statistical Radiophysics and Optics*. Springer Ser. Inf. Sci., Vol. 16 (Springer, Berlin, Heidelberg, to be published)

2.33 G. P. Dzhotyan, Yu. E. Dyakov, I. G. Zubarev, A. B. Mironov, S. I. Mikhailov: Amplification in SRS in a nonmonochromatic pump field. Zh. Eksp. Teor. Fiz. **73**, 822 (1977) [English transl.: Sov. Phys. − JETP **46**, 431 (1977)]

2.34 I. G. Zubarev, S. I. Mikhailov: Influence of parametric effects on the stimulated scattering of nonmonochromatic pump radiation. Kvantovaya Elektron. **5**, 2383 (1978) [English transl.: Sov. J. Quant. Electron. **8**, 1338 (1978)]

2.35 V. G. Sidorovich: Reproduction of the pump spectrum in SRS. Kvantovaya Electron. **5**, 1356 (1978) [English transl.: Sov. J. Quant. Electron. **8**, 784 (1978)]

2.36 B. Ya. Zel'dovich, V. V. Shkunov: Theory of phase-locking in transient stimulated scattering. Kvantovaya Elektron. **6**, 1926 (1979) [English transl.: Sov. J. Quant. Electron. **9**, 1173 (1979)]

2.37 S. A. Akhmanov, R. V. Khokhlov: *Problems in Nonlinear Optics (Electromagnetic Waves in Nonlinear Dispersive Media)* (INI AN SSSR, Moscow 1964) (in Russian)

2.38 N. Bloembergen: *Nonlinear Optics* (Benjamin, New York 1965)

2.39 V. N. Lugovoy: *Introduction to the Theory of Stimulated Raman Scattering* (Nauka, Moscow 1968) (in Russian)

2.40 B. Ya. Zel'dovich, V. V. Shkunov: Specific features of stimulated scattering in counter-propagating pump beams. Kvantovaya Elektron. **9**, 393 (1982) [English transl.: Sov. J. Quant. Electron. **12**, 223 (1982)]

2.41 N. B. Baranova, A. A. Golubtsov, B. Y. Zel'dovich, N. A. Mel'nikov, N. F. Pilipetsky, A. N. Rusetsky: Stimulated light backscattering in the media with induced anisotropy. Kvantovaya Elektron. **4**, 844 (1977) [English transl.: Sov. J. Quant. Electron. **7**, 469 (1977)]

Chapter 3

3.1 N. B. Baranova, B. Ya. Zel'dovich: Wavefront dislocations and zeros of the amplitude. Zh. Eksp. Teor. Fiz. **80**, 1789 (1981) [English transl.: Sov. Phys. − JETP **54**, 925 (1981)]

3.2 N. B. Baranova, B. Ya. Zel'dovich, A. V. Mamaev, N. F. Pilipetsky, V. V. Shkunov: Dislocations of the wavefront of a speckle-inhomogeneous field (theory and experiment). Pis'ma Zh. Eksp. Teor. Fiz. **33**, 206 (1981) [English transl.: JETP Lett. **33**, 195 (1981)]

3.3a) N. B. Baranova, B. Ya. Zel'dovich, A. V. Mamaev, N. F. Pilipetsky, V. V. Shkunov: Experimental investigation of density of wavefront dislocations of speckle-inhomogeneous light fields. Zh. Eksp. Teor. Fiz. **83**, 1702 (1982) [English transl.: Sov. Phys. − JETP **56**, 983 (1982)]

   b) N. B. Baranova, B. Ya. Zel'dovich, A. V. Mamaev, N. F. Pilipetsky, V. V. Shkunov: Wave-front dislocations: topological limitations for adaptive systems with phase conjugation. J. Opt. Soc. Am. **73**, 525 (1983)

3.4a) S. M. Rytov: *Introduction to Statistical Radiophysics, Part I, Random Processes* (Nauka, Moscow 1976) (in Russian)

   b) S. M. Rytov, Yu. A. Kravtsov, V. I. Tatarsky: *Introduction to Statistical Radiophysics, Part II, Random Fields* (Nauka, Moscow 1978) (in Russian)

3.5 S. M. Rytov, Y. A. Kravtsov, V. I. Tatarsky: *Principles of Statistical Radiophysics*, Vols. 1−4, Springer Ser. Electrophys. (Springer, Berlin, Heidelberg, to be published)

3.6 S. A. Akhmanov, Yu. E. D'yakov, A. S. Chirkin: *Introduction to Statistical Radiophysics and Optics*. Springer Ser. Inf. Sci., Vol. 16 (Springer, Berlin, Heidelberg, to be published)

3.7 J. Klauder, E. Sudarshan: *Fundamentals of Quantum Optics* (Benjamin, New York 1968)

3.8 M. Born, E. Wolf: *Principles of Optics*, 6th ed. (Pergamon, Oxford 1980)

3.9 E. Wolf, L. Mandel: Rev. Mod. Phys. **37**, 231 (1965)

3.10 J. C. Dainty (ed.): *Laser Speckle and Related Phenomena*, Topics Appl. Phys., Vol. 9, 2nd ed. (Springer, Berlin, Heidelberg 1984)

3.11 M. Franson: *La granularité laser (speckle) et ses applications en optique* (Masson, Paris 1978)

3.12 G. A. Pasmanik: Raman interaction in the field of an incoherent pump beam. Izv. Vyssh. Uchebn. Zaved. Radiofiz. **17**, 970 (1974)

3.13 N. B. Baranova, B. Ya. Zel'dovich, V. V. Shkunov: Phase conjugation by stimulated light scattering in a focused spatially inhomogeneous pump beam. Kvantovaya Elektron. **5**, 973 (1978) [English transl.: Sov. J. Quant. Electron. **8**, 559 (1978)]

3.14 J. F. Nye, M. V. Berry: Proc. Roy. Soc. London **A336**, 165 (1974)

3.15 F. J. Wright: "Wavefront dislocations and their analysis using catastrophe theory", in *Structural Stability in Physics,* ed. by W. Güttinger, H. Eikemeier, Springer Ser. Syn., Vol. 4 (Springer, Berlin, Heidelberg 1979) p.141

Chapter 4

4.1 B. Ya. Zel'dovich, V. I. Popovichev, V. V. Ragulsky, F. S. Faizullov: On the relationship between wavefronts of reflected and exciting light in stimulated Brillouin scattering. Pis'ma Zh. Eksp. Teor. Fiz. **15**, 160 (1972) [English transl.: JETP Lett. **15**, 109 (1972)]

4.2 O. Yu. Nosach, V. I. Popovichev, V. V. Ragulsky, F. S. Faizullov: Compensation of phase distortions in a gain medium by a "Brillouin mirror". Pis'ma Zh. Eksp. Teor. Fiz. **16**, 617 (1972) [English transl.: JETP Lett. **16**, 435 (1972)]

4.3 N. F. Pilipetsky, V. I. Popovichev, V. V. Ragulsky: On reproduction accuracy of a light field at stimulated scattering. Dokl. Akad. Nauk SSSR **248**, 1097 (1979) [English transl.: Sov. Phys. − Dokl. **24**, 845 (1979)]

4.4 N. F. Pilipetsky, V. I. Popovichev, V. V. Ragulsky: The reproduction of weak components of a light field at stimulated scattering. Opt. Comm. **31**, 97 (1979)

4.5 N. F. Pilipetsky, V. I. Popovichev, V. V. Ragulsky: "The precision of the reproduction of a light field at stimulated scattering", in Proc. Intern. Conf. Lasers 79, Orlando, Florida, USA, ed. by V. J. Corcoran, p. 673

4.6   V. I. Bespalov, A. A. Betin, G. A. Pasmanik: Experimental investigation of the SS thresh-
      old of multimode light beams and the pump reproduction degree in scattered radiation. Izv.
      Vyssh. Uchebn. Zaved. Radiofiz. **20**, 791 (1977) [English transl.: Sov. Phys. − Radiophys.
      Quant. Electron. **20**, 544 (1977)]

4.7   V. N. Belousov, L. A. Bol'shov, H. G. Koval'sky, Yu. K. Nizienko: "On the fine structure
      of SBS spectra during optical phase conjugation", in *Phase Conjugation in Nonlinear
      Media,* ed. by V. I. Bespalov (IPF AN SSSR, Gorky, USSR 1982) p. 176 (in Russian)

4.8   N. F. Pilipetsky, V. I. Popovichev, V. V. Ragulsky: On the relation between gain coeffi-
      cients of phase-conjugated and non-conjugated waves in stimulated scattering. Dokl. Akad.
      Nauk SSSR **257**, 1116 (1981); Opt. Commun. **40**, 73 (1981)

4.9   V. F. Efimkov, I. G. Zubarev, A. V. Kotov, A. B. Mironov, S. I. Mikhailov: On gain incre-
      ments of Stokes fields in stimulated scattering of spatially inhomogeneous radiation. Kvan-
      tovaya Electron. **8**, 891 (1981) [English transl.: Sov. J. Quant. Elektron. **11**, 584 (1981)]

4.10 N. B. Baranova, B. Ya. Zel'dovich, V. V. Shkunov: Optical phase conjugation at stimulat-
      ed scattering in a focused spatially inhomogeneous pump beam. Kvantovaya Elektron. **5**,
      973 (1978) [English transl.: Sov. J. Quant. Electron. **8**, 559 (1978)]

4.11 V. E. Yashin, V. I. Kryzhanovsky, V. A. Serebryakov: Optical phase conjugation of nano-
      and subnanosecond pulses in SBS. Kvantovaya Elektron. **9**, 1695 (1982) [English transl.:
      Sov. J. Quant. Electron. **12**, 1086 (1982)]

4.12 K. V. Gratsianov, V. I. Kryzhanovsky, V. V. Lyubimov, A. A. Mak, V. G. Pankov, V. A.
      Serebryakov, A. I. Stepanov, V. E. Yashin: "Investigation of optical phase conjugation ac-
      curacy in SBS", in *Phase Conjugation in Nonlinear Media,* ed. by V. I. Bespalov (IPF AN
      SSSR, Gorky, USSR 1982) p. 143 (in Russian)

4.13 V. V. Ragulsky: Stimulated Brillouin scattering lasers. Trudy FIAN **85**, 3 (1976) [English
      transl.: Levedev Trans. **85**, 1 (1976)]

4.14 R. G. Brewer: Phys. Rev. Ser. A **140**, 800 (1965)

4.15 D. H. Rank, C. W. Cho, N. D. Foltz, T. A. Wiggins: Stimulated thermal Rayleigh scatter-
      ing. Phys. Rev. Lett. **19**, 828 (1967)

4.16 V. I. Bespalov, A. M. Kubarev: "Some of the observation results on the propagation and
      stimulated scattering of intense light in hinolin", in Trans. 2nd Symposium on Nonlinear
      Optics. Novosibirsk, 1968, p. 247

4.17 A. D. Kudryavtseva, A. I. Sokolovskaya, M. M. Suschinsky: Stimulated Raman scattering
      and self-focusing of light in liquid nitrogen. Zh. Eksp. Teor. Fiz. **59**, 1556 (1970) [English
      transl.: Sov. Phys. − JETP **32**, 849 (1971)]

4.18 A. A. Betin, G. A. Pasmanik: On the spatial structure of Stokes radiation in backward SBS
      of light beams. Kvantovaya Elektron. **3**, 2215 (1976) [English transl.: Sov. J. Quant. Elec-
      tron. **6**, 1204 (1976)]

4.19 G. G. Kochemasov, V. D. Nikolaev: On the reproduction of spatial distributions of the
      pump beam amplitude and phase in SBS. Kvantovaya Elektron. **4**, 115 (1977) [English
      transl.: Sov. J. Quant. Electron. **7**, 60 (1977)]

4.20 V. G. Sidorovich: To the theory of the "Brillouin mirror". Zh. Tekhn. Fiz. **46**, 2168 (1976)
      [English transl.: Sov. Phys. − Tech. Phys. **21**, 1270 (1980)]

4.21 I. M. Bel'dyugin, M. G. Galushkin, E. M. Zemskov, V. I. Mandrosov: On complex conju-
      gation of fields in SBS. Kvantavaya Elektron. **3**, 2467 (1976) [English transl.: Sov. J. Quant.
      Electron. **6**, 1349 (1976)]

4.22 B. Ya. Zel'dovich, V. V. Shkunov: On wavefront reproduction in SRS. Kvantovaya Elek-
      tron. **4**, 1090 (1977) [English transl.: Sov. J. Quant. Electron. **7**, 610 (1977)]

4.23 B. Ya. Zel'dovich, V. V. Shkunov: On optical phase conjugation limits in stimulated scat-
      tering. Kvantovaya Elektron. **5**, 36 (1978) [English transl.: Sov. J. Quant. Electron. **8**, 15
      (1978)]

4.24 B. Ya. Zel'dovich, T. B. Yakovleva: Small-scale distortions in wave-front reversal of a
      beam with incomplete spatial modulation (backward SBS, theory). Kvantovaya Elektron. **7**,
      316 (1980) [English transl.: Sov. J. Quant. Electron. **10**, 181 (1980)]

4.25 B. Ya. Zel'dovich, N. F. Pilipetsky, V. V. Shkunov: Phase Conjugation in stimulated scat-
      tering. Usp. Fiz. Nauk **138**, 249 (1982) [English transl.: Sov. Phys. − Usp. **25**, 713 (1982)]

4.26 V. N. Blaschuk, B. Ya. Zel'dovich, N. A. Melnikov, N. F. Pilipetsky, V. I. Popovichev, V. V. Ragulsky: Optical phase conjugation in stimulated scattering of focused beams. Pis'ma Zh. Tekhn. Fiz. **3**, 211 (1977) [English transl.: Sov. Phys. − Tech. Phys. Lett. **3**, 83 (1977)]

4.27 V. I. Bespalov, A. A. Betin, G. A. Pasmanik: On reproduction effects in stimulated scattering. Pis'ma Zh. Techn. Fiz. **3**, 215 (1977) [English transl.: Sov. Phys. − Tech. Phys. Lett. **3**, 85 (1977)]

4.28 V. Wang, G. R. Ginliano: Correction of phase aberrations via stimulated Brillouin scattering. Opt. Lett. **2**, 4 (1978)

4.29 V. I. Kryzhanovsky, V. A. Serebryakov, V. E. Yashin: Experimental investigation of a neodymium glass double-pass laser amplifier with a $\lambda/4$ decoupling and SBS mirror. Zh. Tekhn. Fiz. **52**, 1356 (1982)

4.30 V. I. Bespalov, A. A. Betin, G. A. Pasmanik: Reproduction of pump wave in stimulated scattering. Izv. Vyssh. Uchebn. Zaved. Radiofiz. **21**, 961 (1978) [English transl.: Sov. Phys. − Radiophys. Quant. Electron. **21**, 675 (1979)]

4.31 N. B. Baranova, B. Ya. Zel'dovich: Optical phase conjugation of focused beams (backward SBS, theory). Kvantovaya Elektron. **7**, 973 (1980) [English transl.: Sov. J. Quant. Electron. **10**, 555 (1980)]

4.32 V. I. Bespalov, A. A. Betin, V. G. Manishin, G. A. Pasmanik: "The amplification of a Stokes wave reproducing the pump in stimulated scattering of multimode beams", in *Nonlinear Waves,* ed. by R. V. Khokhlov (Nauka, Moscow 1979) p. 239 (in Russian)

Chapter 5

5.1 B. Ya. Zel'dovich, V. V. Shkunov: On the limits of optical phase conjugation by stimulated scattering. Kvantovaya Elektron. **5**, 36 (1978) [English transl.: Sov. J. Quant. Electron. **8**, 15 (1978)]

5.2 V. N. Blaschuk, B. Ya. Zel'dovich, V. N. Krasheninnikov, N. A. Mel'nikov, N. F. Pilipetsky, V. V. Ragulsky, V. V. Shkunov: Stimulated scattering of depolarized radiation. Dokl. Akad. Nauk SSSR **241**, 1322 (1978) [English transl.: Sov. Phys. − Dokl. **23**, 588 (1978)]

5.3 V. N. Blaschuk, V. N. Krasheninnikov, N. A. Mel'nikov, N. F. Pilipetsky, V. V. Ragulsky, V. V. Shkunov, B. Ya. Zel'dovich: SBS wavefront reversal for the depolarized light (theory and experiment). Opt. Commun. **28**, 137 (1978)

5.4 N. G. Basov, V. F. Efimkov, I. G. Zubarev, A. V. Kotov, A. B. Mironov, S. I. Mikhailov, M. G. Smirnov: The effect of some of the radiation parameters on phase conjugation of pump in a "Brillouin mirror". Kvantovaya Elektron. **6**, 765 (1979) [English transl.: Sov. J. Quant. Electron. **9**, 455 (1979)]

5.5 V. I. Bespalov, A. A. Betin, G. A. Pasmanik, A. A. Shilov: Observation of temporal field oscillations in the SBS-scattered radiation. Pis'ma Zh. Eksp. Teor. Fiz. **31**, 668 (1980) [English transl.: JETP Lett. **31**, 630 (1980)]

5.6 M. V. Vasil'ev, A. L. Gyulamiryan, A. V. Mamaev, V. V. Ragulsky, P. M. Semenov, V. G. Sidorovich: Registration of phase fluctuations of stimulated-scattered light. Pis'ma Zh. Eksp. Teor. Fiz. **31**, 668 (1980) [English transl.: JETP Lett. **31**, 634 (1980)]

5.7 N. G. Basov, I. G. Zubarev, A. B. Mironov, S. I. Mikhailov, A. Yu. Okulov: On Stokeswave phase fluctuations in stimulated scattering. Pis'ma Zh. Eksp. Teor. Fiz. **31**, 685 (1980) [English transl.: JETP Lett. **31**, 645 (1980)]

5.8 N. G. Basov, V. F. Efimkov, I. G. Zubarev, A. V. Kotov, S. I. Mikhailov, M. G. Smirnov: Optical phase conjugation in SBS of depolarized pump. Pis'ma Zh. Eksp. Teor. Fiz. **28**, 215 (1978) [English transl.: JETP Lett. **28**, 197 (1978)]

5.9 E. L. Bubis, M. V. Vasil'ev, A. A. Leschev, G. A. Pasmanik, V. G. Sidorovich, A. A. Shilov: Optical phase conjugation of incoherent radiation via stimulated Brillouin scattering. Opt. Spektrosk. **53**, 921 (1982)

5.10 V. G. Sidorovich: "Optical phase conjugation by stimulated scattering of incoherent radiation", in *Phase Conjugation in Nonlinear Media,* ed. by V. I. Bespalov (IPF AN SSSR, Gorky, USSR 1982) p. 160 (in Russian)

5.11 A. I. Sokolovskaya, G. L. Brekhovskikh, A. D. Kudryavtseva: Wavefront reconstruction of light beams in stimulated Raman scattering. Dokl. Akad. Nauk SSSR **233**, 356 (1977) [English transl.: Sov. Phys. − Dokl. **22**, 156 (1977)]

5.12 G. L. Brekhovskikh, N. V. Okladnikov, A. I. Sokolovskaya: Experimental investigation of the effect of gain saturation at wavefront reconstruction in SRS. Zh. Prikl. Spektrosk. **32**, 24 (1980)

5.13 A. D. Kudriavtseva, A. I. Sokolovskaia, J. Gazengel, N. Phu Xuan, G. Rivoire: Reconstruction of the laser wave-front by stimulated scattering in the pico-second range. Opt. Commun. **28**, 446 (1978)

5.14 V. F. Efimkov, I. G. Zubarev, A. V. Kotov, S. I. Mikhailov, A. B. Mironov, G. A. Pasmanik, A. A. Shilov: Inertia of stimulated Brillouin scattering and nonthreshold reflection of short pulses with phase conjugation. Zh. Eksp. Teor. Fiz. **77**, 526 (1979) [English transl.: Sov. Phys. − JETP **50**, 267 (1979)]

5.15 A. L. Gyulamiryan, A. V. Mamaev, N. F. Pilipetsky, V. V. Shkunov: Optical phase conjugation of weak signals shifted in frequency in SBS. Opt. Spektrosk. **51**, 204 (1981) [English transl.: Opt. Spectrosc. **51**, 113 (1981)]

5.16 N. G. Basov, V. F. Efimkov, I. G. Zubarev, A. V. Kotov, S. I. Mikhailov: Control of characteristics of phase-conjugate mirrors in the amplification regime. Kvantovaya Elektron. **8**, 2191 (1981) [English transl.: Sov. J. Quant. Electron. **11**, 1335 (1981)]

5.17 B. Ya. Zel'dovich, N. A. Mel'nikov, N. F. Pilipetsky, V. V. Ragulsky: Demonstration of optical phase conjugation by stimulated Raman scattering. Pis'ma Zh. Eksp. Teor. Fiz. **25**, 41 (1977) [English transl.: JETP Lett. **25**, 36 (1977)]

5.18 N. F. Andreev, V. I. Bespalov, A. N. Kiselev, G. A. Pasmanik: Experimental investigation of spatial structure of the first Stokes component in SRS. Kvantovaya Elektron. **6**, 996 (1979) [English transl.: Sov. J. Quant. Electron. **9**, 585 (1979)]

5.19 G. V. Krivoschekov, S. G. Struts, M. F. Stupak: Spectral features of stimulated temperature scattering in optical phase conjugation. Pis'ma Zh. Tekhn. Fiz. **6**, 428 (1980) [English transl.: Sov. Phys. − Tech. Phys. Lett. **6**, 184 (1980)]

5.20 V. N. Belousov, L. A. Bel'shov, N. G. Kovalsky, Yu. K. Nizienko: Experimental investigation of optical phase conjugation by stimulated temperature and Brillouin scattering in liquids. Zh. Eksp. Teor. Fiz. **79**, 2119 (1980) [English transl.: Sov. Phys. − JETP **52**, 1071 (1980)]

5.21 V. I. Kislenko, V. L. Strizhevsky: Optical phase conjugation by stimulated temperature scattering and its application. Izv. Akad. Nauk SSSR **45**, 976 (1981)

5.22 A. A. Betin, G. A. Pasmanik: Conservation of spatial coherence of Stokes beams at their amplification in the field of a multimode pump. Pis'ma Zh. Eksp. Teor. Fiz. **23**, 577 (1976) [English transl.: JETP Lett. **23**, 528 (1976)]

5.23 V. G. Sidorovich: To the theory of a "Brillouin mirror". Zh. Tekhn. Fiz. **46**, 2168 (1976) [English transl.: Sov. Phys. − Tech. Phys. **21**, 1270 (1976)]

5.24 I. M. Belgyugin, M.G. Galushkin, E. M. Zemskov, V. I. Mandrosov: On complex conjugation of fields in SBS. Kvantolaya Elektron. **3**, 2467 (1976) [English transl.: Sov. J. Quant. Electron. **6**, 1349 (1976)]

5.25 B. Ya. Zel'dovich, T. V. Yakovleva: Small-structure distortions in phase conjugation of a beam with incomplete spatial modulation (backward SBS, theory). Kvantovaya Elektron. **7**, 316 (1980) [English transl.: Sov. J. Quant. Electron. **10**, 181 (1980)]

5.26 B. Ya. Zel'dovich, T. V. Yakovleva: Distortions of small-structure wave picture due to self-focusing nonlinearity. Kvantovaya Elektron. **7**, 1325 (1980) [English transl.: Sov. J. Quant. Electron. **10**, 757 (1980)]

5.27 B. Ya. Zel'dovich, T. V. Yakovleva: Fine-structure distortions in OPC-SBS under transient conditions. Kvantovaya Elektron. **7**, 2243 (1980) [English transl.: Sov. J. Quant. Electron. **10**, 1306 (1980)]

5.28 B. Ya. Zel'dovich, V. V. Shkunov: On wavefront reproduction in SRS. Kvantovaya Elektron. **4**, 1090 (1977) [English transl.: Sov. J. Quant. Electron. **7**, 610 (1977)]

5.29 V. G. Sidorovich, V. V. Shkunov: On spectral selectivity of a three-dimensional hologram. Opt. Spektrosk. **44**, 1001 (1978) [English transl.: Opt. Spectrosc. **44**, 586 (1978)]

5.30 N. B. Baranova, B. Ya. Zel'dovich, V. V. Shkunov, T. V. Yakovleva: "The theory of field reconstruction by volume holograms and spectral-and-angular distortions", in *Holography and Optical Processing of Information: Methods and Equipment,* ed. by G. V. Skrotsky (Leningrad Inst. Nucl. Phys. Editions, Leningrad 1980) p. 3 (in Russian)

5.31 N. B. Baranova, B. Ya. Zel'dovich: Transverse coherentization of scattered field in optical phase conjugation. Kvantovaya Elektron. **7**, 299 (1980) [English transl.: Sov. J. Quant. Electron. **10**, 172 (1980)]

5.32 V. I. Bespalov, A. A. Betin, G. A. Pasmanik: Experimental investigation of the SS threshold of multimode beams and the degree of pump reproduction in scattered radiation. Izv. Vyssh. Uchebn. Zaved. Radiofiz. **20**, 791 (1977) [English transl.: Sov. Phys. – Radiophys. Quant. Electron. **20**, 544 (1977)]

5.33 N. B. Baranova, B. Ya. Zel'dovich, V. V. Shkunov: Optical phase conjugation by stimulated scattering in a focused spatially inhomogeneous pump beam. Kvantovaya Elektron. **5**, 973 (1981) [English transl.: Sov. J. Quant. Electron. **8**, 559 (1978)]

5.34 A. A. Betin, V. I. Bespalov, G. A. Pasmanik: Reproduction of the pump wave in stimulated scattered radiation. Izv. Vyssh. Uchebn. Zaved. Radiofiz. **21**, 961 (1978) [English transl.: Sov. Phys. – Radiophys. Quant. Electron. **21**, 675 (1979)]

5.35 N. B. Baranova, B. Ya. Zel'dovich: Optical phase conjugation of focused beams (backward SBS, theory). Kvantovaya Elektron. **7**, 973 (1980) [English transl.: Sov. J. Quant. Electron. **10**, 555 (1980)]

5.36 B. Ya. Zel'dovich, V. V. Shkunov: Optical phase conjugation by a depolarized pump. Zh. Eksp. Teor. Fiz. **75**, 428 (1978) [English transl.: Sov. Phys. – JETP **48**, 214 (1978)]

5.37 G. A. Pasmanik: Wavefront reproduction of complicated signals in backward stimulated scattering. Pis'ma Zh. Tekhn. Fiz. **4**, 504 (1978) [English transl.: Sov. Phys. – Tech. Phys. Lett. **4**, 201 (1978)]

5.38 V. I. Bespalov, V. G. Manishin, G. A. Pasmanik: Nonlinear selection of optical radiation through its reflection by an SBS mirror. Pis'ma Zh. Eksp. Teor. Fiz. **77**, 1756 (1979) [English transl.: JETP Lett. **50**, 879 (1979)]

5.39 I. M. Bel'dyugin, M. G. Galushkin, E. M. Zemskov: On stimulated scattering of nonmonochromatic spatially inhomogeneous radiation. Kvantovaya Elektron. **6**, 587 (1979) [English transl.: Sov. J. Quant. Electron. **9**, 348 (1979)]

5.40 O. M. Vokhnik, V. I. Odintsov: Calculation of stimulated radiation intensity at different shapes of broad-band pump spectrum. Opt. Spektrosk. **49**, 371 (1980)

5.41 I. M. Bel'dyugin, E. M. Zemskov: On the effect of pump field variations on the kind of amplified signal field in stimulated scattering. Kvantovaya Elektron. **5**, 2055 (1978) [English transl.: Sov. J. Quant. Electron. **8**, 1163 (1978)]

5.42 G. G. Kochemasov, V. D. Nikolaev: Investigation of spatial characteristics of Stokes radiation in stimulated scattering under saturation. Kvantovaya Elektron. **6**, 1960 (1979) [English transl.: Sov. J. Quant. Electron. **9**, 1155 (1979)

5.43 V. I. Bespalov, A. A. Betin, S. N. Kulagina, A. Z. Matveev, G. A. Pasmanik, A. A. Shilov: "Optical phase conjugation of weak signals", in *Optical Phase Conjugation in Nonlinear Media,* ed. by V. I. Bespalov (IPF AN SSSR, Gorky, USSR 1979) p. 44

5.44 B. Ya. Zel'dovich, V. V. Shkunov: The effect of spatial interference on the gain in stimulated scattering. Kvantovaya Elektron. **4**, 2353 (1977) [English transl.: Sov. J. Quant. Electron. **7**, 1345 (1977)]

5.45 A. I. Sokolovskaya, G. L. Brekhovskikh: Dynamic holograms in stimulated scattering. Dokl. Akad. Nauk SSSR **243**, 630 (1978)

5.46 A. I. Sokolovskaya, G. L. Brekhovskikh, A. D. Kudryavtseva: Light beam wave-front reconstruction and real volume image reconstruction of the object at the stimulated Raman scattering. Opt. Commun. **24**, 74 (1978)

5.47 N. V. Okladnikov, G. L. Brekhovskikh, A. I. Sokolovskaya, A. A. Garmonov: Optical phase conjugation (wave-front reproduction) and diffractive efficiency of dynamic holograms in stimulated scattering. Pis'ma Zh. Tekhn. Fiz. **7**, 373 (1981)

5.48 N. G. Basov, I. G. Zubarev, A. V. Kotov, S. I. Mikhailov, M. G. Smirnov: Optical phase conjugation of weak signals by thresholdless reflection from an SBS mirror. Kvantovaya Elektron. **6**, 394 (1979) [English transl.: Sov. J. Quant. Electron. **9**, 237 (1979)]

5.49 V. V. Ragulsky: Optical phase conjugation of weak beams in stimulated scattering. Pis'ma Zh. Tekhn. Fiz. **5**, 251 (1979) [English transl.: Sov. Phys. – Tech. Phys. Lett. **6**, 297 (1980)]

5.50 N. F. Pilipetsky, V. I. Popovichev, V. V. Ragulsky: The reproduction of weak components of a light field at stimulated scattering. Opt. Comm. **31**, 97 (1979)

5.51 B. Ya. Zel'dovich, T. V. Yakovleva: Spatial-polarization phase conjugation in stimulated Rayleigh-wing scattering. Kvantovaya Elektron. **7**, 880 (1980) [English transl.: Sov. J. Quant. Electron. **10**, 501 (1980)]

5.52 R. H. Lehmberg: Numerical study of phase conjugation in stimulated Brillouin scattering from an optical waveguide. J. Opt. Soc. Am. **73**, 558 (1983)

Chapter 6

6.1 D. Gabor: A new microscopic principle. Nature **161**, 777 (1948)

6.2 D. Gabor: Microscopy by reconstructed wavefront. Proc. Roy. Soc. Ser. A **197**, 454 (1949)

6.3 W. L. Bragg: Microscopy on the base of wavefront reconstruction. Nature **166**, 399 (1950)

6.4 S. A. Akhmanov, V. V. Khokhlov: *Problems of Nonlinear Optics* (VINITI, Moscow 1964) (in Russian)

6.5 N. Bloembergen: *Nonlinear Optics* (Benjamin, New York 1965)

6.6 A. L. Gyulamiryan, A. V. Mamaev, N. F. Pilipetsky, V. V. Ragulsky, V. V. Shkunov: Investigation of the efficiency of nondegenerate four-wave interaction. Kvantovaya Elektron. **8**, 196 (1981) [English transl.: Sov. J. Quant. Electron. **11**, 115 (1981)]

6.7 V. N. Blaschuk, A. V. Mamaev, N. F. Pilipetsky, V. V. Shkunov, B. Ya. Zel'dovich: Wavefront reversal with angular tilting theory and experiment for the four-wave mixing. Opt. Commun. **31**, 383 (1979)

6.8 V. N. Blaschuk, B. Ya. Zel'dovich, A. V. Mamaev, N. F. Pilipetsky, V. V. Shkunov: Complete phase conjugation of depolarized radiation by degenerate four-photon interaction (theory and experiment). Kvantovaya Elektron. **7**, 627 (1980) [English transl.: Sov. J. Quant. Electron. **10**, 356 (1980)]

6.9 G. Martin, L. K. Lam, R. W. Hellwarth: Generation of a time-reversed replica of a nonuniformly polarized image-bearing optical beam. Opt. Lett. **5**, 185 (1980)

6.10 R. L. Abrams, R. C. Lind: Degenerate four-wave mixing in absorbing media. Opt. Lett. **2**, 94 (1978); ibid, **3**, 205

6.11 E. M. Lifshitz, L. P. Pitaevsky: *Physical Kinetics* (Pergamon, Oxford 1981)

6.12 S. N. Vlasov, V. I. Talanov: "On some features of scattering of a signal wave by counter-propagating pump beams in degenerate four-photon interaction", in *Optical Phase Conjugation in Nonlinear Media,* ed. by V. I. Bespalov (IPF AN SSSR, Gorky, USSR 1979) p. 85

6.13 H. Kogelnik: Bell Syst. Tech. J. **44**, 2451 (1965)

6.14 E. N. Leit, J. Upatnieks: Holographic imaging through diffusing media. J. Opt. Soc. Am. **56**, 523 (1966)

6.15 H. W. Kogelnik: Controlled transmission of waves through inhomogeneous media, Patent No. 3.449.577 (USA), Application October 23, 1965

6.16 G. T. Sincerbox: Optical data recording, Patent No. 1.218.331 (GB), Application April 9, 1968

6.17 W. T. Cathey: Holographic simulation of compensation for atmospheric wavefront distortions. Proc. IEEE **56**, 340 (1968)

6.18 H. J. Gerritsen, E. G. Ramberg, S. Freeman: In *Proc. Symposium on Modern Optics* (Polytechnic Press, New York 1967) p. 109

6.19 B. I. Stepanov, E. V. Ivakin, A. S. Rubanov: On recording plane and volume dynamic holograms in saturable absorbers. Dokl. Akad. Nauk SSSR **196**, 567 (1971) [English transl.: Sov. Phys. – Dokl. **16**, 46 (1971)]

6.20 J. P. Woerdman: Formation of a transient free carrier hologram in Si. Opt. Commun. **2**, 212 (1970)

6.21 R. W. Hellwarth: Generation of time-reversed wavefront by nonlinear reflection. J. Opt. Soc. Am. **67**, 1 (1977)

6.22 D. Bloom, G. C. Bjorklund: Conjugate wavefront generation and image reconstruction by four-wave mixing. Appl. Phys. Lett. **31**, 592 (1977)

6.23 S. M. Jensen, R. W. Hellwarth: Observation of the time-reversed replica of a monochromatic optical wave. Appl. Phys. Lett. **32**, 166 (1978)

6.24 B. Ya. Zel'dovich, V. V. Shkunov: "Spatial-polarization phase conjugation by four-photon interaction", in *Optical Phase Conjugation in Nonlinear Media,* ed. by V. I. Bespalov (IPF AN SSSR, Gorky, USSR 1979) p. 23

6.25 V. N. Blaschuk, A. V. Mamaev, N. F. Pilipetsky, V. V. Shkunov, B. Ya. Zel'dovich: Wavefront reversal with angular tilting — theory and experiment for four wave mixing. Opt. Commun. **31**, 383 (1979)

6.26 D. M. Pepper, R. L. Abrams: Narrow optical bandpass filter via nearly degenerate four-wave mixing. Opt. Lett. **3**, 212 (1978)

6.27 J. Nilsen, A. Yariv: Nearly degenerate four-wave mixing applied to optical filters. Appl. Opt. **18**, 143 (1979)

6.28 Fu Tao-yi, M. Sargent III: Effects of signal detuning on phase conjugation. Opt. Lett. **4**, 366 (1979)

6.29 V. N. Blaschuk, N. F. Pilipetsky, V. V. Shkunov: Four-wave interaction as a controllable frequency filter. Dokl. Akad. Nauk SSSR **251**, 70 (1980) [English transl.: Sov. Phys. — Dokl. **25**, 185 (1980)]

6.30 N. F. Pilipetsky, V. V. Shkunov: Narrowband four-wave reflecting filter with frequency and angular tuning. Opt. Commun. **37**, 217 (1981)

6.31 A. L. Gyulamiryan, A. V. Mamaev, N. F. Pilipetsky, V. V. Shkunov: Tunable nonlinear four-wave filter. Opt. Spektrosk. **52**, 387 (1982)

6.32 S. Saikan, H. Waketa: Configuration dependence of optical filtering characteristics in backward nearly degenerate four-wave mixing. Opt. Lett. **6**, 281 (1981)

6.33 J. Nilsen, N. S. Gluck, A. Yariv: Narrow-band optical filter through phase conjugation by nondegenerate four-wave mixing in sodium vapor. Opt. Lett. **6**, 380 (1981)

6.34 R. C. Lind, D. G. Steel, G. J. Dunning: Phase conjugation by resonantly enhanced degenerate four-wave mixing. Opt. Eng. **21**, 190 (1982)

6.35 J. F. Lam: Doppler-free laser spectroscopy via degenerate four-wave mixing. Opt. Eng. **21**, 219 (1982)

6.36 A. L. Gyulamiryan, B. Ya. Zel'dovich, A. V. Mamaev, N. F. Pilipetsky, V. V. Shkunov: "Four-wave phase conjugation with selection and control of signal", in *Phase Conjugation in Nonlinear Media*, ed. by V. I. Bespalov (IPF AN SSSR, Gorky, USSR 1982) p. 91

6.37 B. Ya. Zel'dovich, V. V. Shkunov: Spatial-polarization phase conjugation by four-photon interaction. Kvantovaya Elektron. **6**, 629 (1979) [English transl.: Sov. J. Quant. Electron. **9**, 379 (1979)]

6.38 J. Feinberg, R. W. Hellwarth: CW phase conjugation of a nonuniformly polarized optical beam. J. Opt. Soc. Am. **70**, 602 (1980)

6.39 S. Saikan, M. Kiguchi: Generation of phase-conjugated vector wavefronts in atomic vapors. Opt. Lett. **7**, 555 (1982)

6.40 J. O. Tocho, W. Sibbett, D. J. Bradley: Picosecond phase-conjugate reflection from organic dye saturable absorber. Opt. Commun. **34**, 122 (1980)

6.41 B. Ya. Zel'dovich, T. V. Yakovleva: The effect of linear absorption and reflection on characteristics of four-wave OPC. Kvantovaya Elektron. **8**, 1891 (1981) [English transl.: Sov. J. Quant. Electron. **11**, 1144 (1981)]

6.42 R. G. Caro, M. Gower: Phase conjugation by degenerate four-wave mixing in absorbing media. IEEE J. Quant. Electron. **18**, 1376 (1982)

6.43 L. A. Vasil'ev, M. G. Galushkin, A. M. Sergeev, N. V. Cheburkin: Phase conjugation by four-wave interaction in a medium with thermal nonlinearity. Kvantovaya Elektron. **9**, 1571 (1982) [English transl.: Sov. J. Quant. Electron. **12**, 1007 (1982)]

6.44 A. Yariv, D. M. Pepper: Amplified reflection, phase conjugation and oscillation in degenerate four-wave mixing. Opt. Lett. **1**, 16 (1977)

6.45 J. H. Marburger, J. F. Lam: Nonlinear theory of degenerate four-wave mixing. Appl. Phys. Lett. **34**, 389 (1979)

6.46 H. G. Winful, J. H. Marburger: Hysteresis and optical bistability in degenerate four-wave mixing. Appl. Phys. Lett. **36**, 613 (1980)

6.47 A. A. Borsch, M. S. Brodin, V. I. Volkov, N. V. Kukhtarev: Optical bistability and hysteresis in a phase-conjugated wave in degenerate six-photon interaction in cadmium sulphide. Kvantovaya Elektron. **8**, 1304 (1981) [English transl.: Sov. J. Quant. Electron. **11**, 777 (1981)]

6.48 G. P. Agrawal, C. Flitzanis, R. Frey, F. Pradere: Bistable reflectivity of phase-conjugated signal through intracavity degenerate four-wave mixing. Appl. Phys. Lett. **38**, 492 (1981)

6.49 J. H. Marburger: Optical pulse integration and chirp reversal in degenerate four-wave mixing. Appl. Phys. Lett. **32**, 372 (1978)

6.50 B. Ya. Zel'dovich, M. A. Orlova, V. V. Shkunov: Nonstationary theory and calculation of four-wave OPC build-up time. Dokl. Akad. Nauk SSSR **252**, 592 (1980) [English transl.: Sov. Phys. − Dokl. **25**, 390 (1980)]

6.51 W. W. Rigrod, R. A. Fisher, B. J. Feldman: Transient analysis of nearly degenerate four-wave mixing. Opt. Lett. **5**, 105 (1980)

6.52 T. R. O'Meara, A. Yariv: Time-domain signal processing via four-wave mixing in nonlinear delay lines. Opt. Eng. **21**, 237 (1982)

6.53 B. R. Suydam, R. A. Fisher: Transient response of Kerr-like phase conjugators: a review. Opt. Eng. **21**, 184 (1982)

6.54 A. L. Gyulamiryan, A. V. Mamaev, N. F. Pilipetsky, V. V. Shkunov: Phase conjugation of weak frequency-shifted signals in SBS. Opt. Spektrosc. **51**, 204 (1981) [English transl.: Opt. Spectrosc. **51**, 113 (1981)]

6.55 D. G. Steel, R. C. Lind: Multiresonant behavior in nearly degenerate four-wave mixing: the ac Stark effect. Opt. Lett. **6**, 587 (1981)

6.56 G. A. Askar'yan: Zh. Eksp. Teor. Fiz. **42**, 1567 (1962)

6.57 R. Y. Chiao, E. Garmire, C. H. Townes: Self-trapping of optical beams. Phys. Rev. Lett. **13**, 479 (1964)

6.58 N. F. Pilipetsky, A. R. Rustamov: Observation of light self-focusing in liquid. Pis'ma Zh. Eksp. Teor. Fiz. **2**, 88 (1965) [English transl.: JETP Lett. **2**, 55 (1965)]

6.59 V. I. Bespalov, V. T. Talanov: Filamentary structure of light beams in nonlinear liquids. Pis'm Zh. Eksp. Teor. Fiz. **3**, 471 (1966) [English transl.: JETP Lett. **3**, 307 (1966)]

6.60 D. M. Pepper, D. Fekete, A. Yariv: Observation of amplified phase conjugate reflection and optical parametric oscillation by four-wave mixing in a transparent medium. Appl. Phys. Lett. **3**, 44 (1978)

6.61 L. A. Bolshov, D. V. Vlasov, A. M. Dykhne, V. V. Korobkin, Kh. Sh. Saidov, A. N. Starostin: On the possibility of complete compensation for nonlinear distortions of a light beam by means of phase conjugation. Pis'ma Zh. Eksp. Teor. Fiz. **31**, 311 (1980) [English transl.: JETP Lett. **31**, 286 (1980)]

6.62 V. I. Kryzhanovsky, A. A. Mak, V. A. Serebryakov, V. E. Yashin: Application of phase conjugation to suppression of small-scale self-focusing. Pis'ma Zh. Tekhn. Fiz. **7**, 400 (1981) [English transl.: Sov. Phys. − Tech. Phys. Lett. **7**, 170 (1981)]

6.63 V. N. Blaschuk, B. Ya. Zel'dovich, V. V. Shkunov: Four-wave phase conjugation in the field of coded reference waves. Kvantovaya Elektron. **7**, 2559 (1980) [English transl.: Sov. J. Quant. Electron. **10**, 1494 (1980)]

6.64 A. M. Lazaruk: On compensation for small-scale nonlinear distortions of light beams by phase conjugation. Kvantovaya Elektron. **8**, 2461 (1981) [English transl.: Sov. J. Quant. Electron. **11**, 1502 (1981)]

6.65 D. M. Pepper: Nonlinear optical phase conjugation. Opt. Eng. **21**, 155 (1982)

6.66 A. Yariv: Phase conjugate optics and real time holography. IEEE J. Quant. Electron. **14**, 650 (1978)

6.67 R. Fisher (ed.): *Optical Phase Conjugation* (Academic, New York 1983)

Chapter 7

7.1   D. Bloom, G. E. Bjorklund: Conjugate wave-front generation and image reconstruction by four-wave mixing. Appl. Phys. Lett. **31**, 592 (1977)

7.2   S. M. Jensen, R. W. Hellwarth: Observation of the time-reversed replica of a monochromatic optical wave. Appl. Phys. Lett. **32**, 166 (1978)

7.3   D. M. Pepper, D. Fekete, A. Yariv: Observation of amplified phase conjugate reflection and optical parametric oscillation by four-wave mixing in a transparent medium. Appl. Phys. Lett. **33**, 41 (1978)

7.4   S. M. Jensen, R. W. Hellwarth: Generation of time-reversed waves by nonlinear refraction in a waveguide. Appl. Phys. Lett. **33**, 404 (1978)

7.5   V. N. Blaschuk, B. Ya. Zel'dovich, A. V. Mamaev, N. F. Pilipetsky, V. V. Shkunov: Complete phase conjugation of depolarized radiation via degenerate four-photon interaction (theory and experiment). Kvantovaya Elektron. **7**, 627 (1980) [English transl.: Sov. J. Quant. Electron. **10**, 356 (1980)]

7.6   D. M. Bloom, P. F. Liao, N. P. Economou: Observation of amplified reflection by degenerate four-wave mixing in atomic sodium vapor. Opt. Lett. **2**, 58 (1978)

7.7   R. L. Abrams, R. C. Lind: Degenerate four-wave mixing in absorbing media. Opt. Lett. **2**, 94 (1978); ibid, **3**, 205 (1978)

7.8   R. C. Lind, D. G. Steel, G. J. Dunning: Phase conjugation by resonantly enhanced degenerate four-wave mixing. Opt. Eng. **21**, 190 (1982)

7.9   S. Saikan, H. Waketa: Configuration dependence of optical filtering characteristics in backward nearly degenerate four-wave mixing. Opt. Lett. **6**, 281 (1981)

7.10 J. Nilsen, N. S. Gluck, A. Yariv: Narrow-band optical filter through phase conjugation by nondegenerate four-wave mixing in sodium vapor. Opt. Lett. **6**, 380 (1981)

7.11 J. F. Lam: Doppler-free laser spectroscopy via degenerate four-wave mixing. Opt. Eng. **21**, 219 (1982)

7.12 B. I. Stepanov, E. V. Ivakin, A. S. Rubanov: On recording plane and volume dynamic holograms in saturable absorbers. Dokl. Akad. Nauk SSSR **196**, 567 (1971) [English transl.: Sov. Phys. − Dokl. **16**, 46 (1971)]

7.13 V. N. Blaschuk, A. V. Mamaev, N. F. Pilipetsky, V. V. Shkunov, B. Ya. Zel'dovich: Wavefront reversal with angular tilting − theory and experiment for the four-wave mixing. Opt. Commun. **31**, 383 (1979)

7.14 R. A. Fisher, B. J. Feldman: On resonant phase-conjugate reflection and amplification at 10.6 μm in inverted CO_2. Opt. Lett. **4**, 140 (1979)

7.15 A. Tomita: Phase conjugation using gain saturation of a Nd: YAG laser. Appl. Phys. Lett. **34**, 463 (1979)

7.16 F. V. Bunkin, V. V. Savransky, G. A. Shafeev: Resonant phase conjugation in an active medium with copper vapour. Kvantovaya Elektron. **8**, 1346 (1981) [English transl.: Sov. J. Quant. Electron. **11**, 810 (1981)]

7.17 Yu. F. Kir'yanov, G. G. Kochemasov, S. M. Martynova, V. D. Nikolaev: Four-wave mixing in resonantly amplifying media under inversion saturation. Kvantovaya Elektron. **8**, 1734 (1981) [English transl.: Sov. J. Quant. Electron. **11**, 1047 (1981)]

7.18 R. K. Jain: Degenerate four-wave mixing in semiconductors: application to phase conjugation and to picosecond-resolved studies of transient carrier dynamics. Opt. Eng. **21**, 199 (1982)

7.19 E. E. Bergmann, I. J. Bigio, B. J. Feldman, R. A. Fisher: High-efficiency pulsed 10.6 μm phase-conjugate reflection via degenerate 4-wave mixing. Opt. Lett. **3**, 82 (1978)

7.20 D. E. Watkins, C. R. Phipps, Jr., S. J. Thomas: Observation of amplified reflection through degenerate four-wave mixing at CO_2 laser wavelength in germanium. Opt. Lett. **6**, 76 (1981)

7.21 I. J. Bigio, B. J. Feldman, R. A. Fisher, E. E. Bergmann: High-efficient phase conjugation in germanium and inverted CO_2 (review). Kvantovaya Elektron. **6**, 2318 (1979) [English transl.: Sov. J. Quant. Electron. **9**, 1365 (1979)]

7.22   N. G. Basov, B. Ya. Zel'dovich, V. I. Kovalev, F. S. Faizullov, V. B. Fedorov: Reflection of a multifrequency signal in four-wave interaction in germanium at 10.6 μm. Kvantovaya Elektron. **8**, 860 (1981) [English transl.: Sov. J. Quant. Electron. **11**, 514 (1981)]

7.23   E. O. Kane: J. Phys. Chem. Solids **1**, 249 (1957)

7.24a) M. A. Khan, P. W. Kruse, J. F. Ready: Optical phase conjugation in $Hg_{1-x}Cd_xTe$. Opt. Lett. **5**, 261 (1980)

   b) R. K. Jain, D. G. Steel: Degenerate four-wave mixing of 10.6 μm radiation in $Hg_{1-x}Cd_xTe$. Appl. Phys. Lett. **37**, 1 (1980)

7.25   S. Y. Yuen: Four-wave mixing via optically generated free carriers in $Hg_{1-x}Cd_xTe$. Appl. Phys. Lett. **41**, 590 (1982)

7.26   R. K. Jain, M. B. Klein: Degenerate four-wave mixing near the band gap of semiconductors. Appl. Phys. Lett. **35**, 454 (1979)

7.27   R. K. Jain, M. B. Klein, R. C. Lind: High-efficiency degenerate four-wave mixing of 1.06 μm radiation in silicon. Opt. Lett. **4**, 328 (1979)

7.28   A. Borshch, M. Brodin, V. Volkov, N. Kukhtarev: Phase conjugation by the degenerate six-photon mixing in semiconductors. Opt. Commun. **35**, 287 (1980)

7.29   M. A. Khan, R. L. Bennet, P. W. Kruse: Bandgap – resonant optical phase conjugation in n-type $Hg_{1-x}Cd_xTe$ at 10.6 μm. Opt. Lett. **6**, 560 (1981)

7.30   V. L. Vinetsky, N. V. Kukhtarev, S. G. Odulov, M. S. Soskin: Dynamic self-diffraction of coherent light beams. Usp. Fiz. Nauk **129**, 113 (1979)

7.31   N. V. Kukhtarev, S. G. Odulov: Optical phase conjugation via four-wave interaction in media with nonlocal nonlinearity. Pis'ma Zh. Eksp. Teor. Fiz. **30**, 6 (1979) [English transl.: JETP Lett. **30**, 4 (1979)]

7.32   V. L. Vinetsky, N. V. Kukhtarev, V. B. Markov, S. G. Odulov, M. S. Soskin: Amplification of coherent beams by dynamic holograms in ferroelectric crystals. Izv. Akad. Nauk SSSR **41**, 811 (1977)

7.33   V. P. Kondilenko, S. G. Odulov, M. S. Soskin: Amplified reflection of phase-conjugate waves in crystals with the linear electrooptical effect in an external electrical field. Izv. Akad. Nauk SSSR, Ser. Fiz. **45**, 958 (1981)

7.34   D. S. Hamilton, D. Heiman, J. Feinberg, R. W. Hellwarth: Spatial-diffusion measurements in impurity-doped solids by degenerate four-wave mixing. Opt. Lett. **4**, 124 (1979)

7.35   J. Feinberg: Self-pumped, continuous-wave phase conjugator using internal reflection. Opt. Lett. **7**, 486 (1982)

7.36   J. Feinberg, R. W. Hellwarth: Phase conjugating mirror with continuous-wave gain. Opt. Lett. **5**, 519 (1980); ibid, **6**, 257 (1981)

7.37   B. Fischer, M. Cronin-Colomb, J. O. White, A. Yariv: Amplified reflection, transmission and self-oscillation in real-time holography. Opt. Lett. **6**, 519 (1981)

7.38   V. I. Bespalov, A. A. Betin, G. A. Pasmanik, A. A. Shilov: Optical phase conjugation by Raman transformation of Stokes wave in the field of counterpropagating pump beams. Pis'ma Zh. Tekhn. Fiz. **5**, 242 (1979) [English transl.: Sov. Phys. – Tech. Phys. Lett. **5**, 97 (1979)]

7.39   N. F. Andreev, V. I. Bespalov, A. M. Kiselev, A. Z. Matveev, G. A. Pasmanik, A. A. Shilov: Phase conjugation of weak optical signals with a large reflection coefficient. Pis'ma Zh. Eksp. Teor. Fiz. **32**, 639 (1980) [English transl.: JETP Lett. **32**, 625 (1980)]

7.40   B. Ya. Zel'dovich, V. V. Shkunov: Specific features of stimulated scattering in counterpropagating pump beams. Kvantovaya Elektron. **9**, 393 (1982) [English transl.: Sov. J. Quant. Electron. **12**, 223 (1982)]

7.41   N. F. Andreev, V. I. Bespalov, A. M. Kiselev, G. A. Pasmanik, A. A. Shilov: Raman interaction in the field of counterpropagating optical waves. Zh. Eksp. Teor. Fiz. **82**, 1047 (1982) [English transl.: Sov. Phys. – JETP **55**, 537 (1982)]

7.42   S. N. Vlasov, V. I. Talanov: "On some features of signal wave scattering by counterpropagating pump beams in degenerate four-photon interaction", in *Optical Phase Conjugation in Nonlinear Media,* ed. by V. I. Bespalov (IPF AN SSSR, Gorky, USSR 1979) p. 85

7.43   J. P. Huignard, J. P. Herrian, P. Auborg, E. Spitz: Phase-conjugate wavefront generation via real-time holography in $Bi_{12}SiO_{20}$ crystals. Opt. Lett. **4**, 21 (1979)

7.44 J. O. Tocho, W. Sibbett, D. J. Bradley: Thermal effects in phase conjugation in saturable absorbers with picosecond pulses. Opt. Commun. **37**, 67 (1981)

7.45 R. G. Caro, M. C. Gower: Phase conjugation of KrF laser radiation. Opt. Lett. **6**, 557 (1981)

7.46 Yu. I. Kucherov, S. A. Lesnik, M. S. Soskin, A. I. Khizhnyak: "Copropagating four-beam interaction in slowly-responding media", in *Phase Conjugation in Nonlinear Media,* ed. by V. I. Bespalov (IPF AN SSSR, Gorky, USSR 1982) p. 111

7.47 Materials of the 1st Soviet-Japanese Symposium on Ferroelectricity. Izv. AN SSSR, Ser. Fiz., 3 and 4, 1977

7.48 J. J. Amodei: RCA Rev. **32**, 185 (1971)

7.49 V. I. Bespalov, A. A. Betin, S. N. Kulagina, G. A. Pasmanik, A. A. Shilov: Phase conjugation of radiation with a spatially inhomogeneous polarization state in four-wave Raman interaction. Pis'ma Zh. Tekhn. Fiz. **6**, 1288 (1980)

7.50 V. I. Bespalov, A. A. Betin, A. I. Dyatlov, S. N. Kulagina, V. G. Manishin, G. A. Pasmanik, A. A. Shilov: Optical phase conjugation by four-photon processes at two-quantum resonance. Zh. Eksp. Teor. Fiz. **79**, 378 (1980) [English transl.: Sov. Phys. − JETP **52**, 190 (1980)]

Chapter 8

8.1 O. L. Kulikov, N. F. Pilipetsky, A. N. Sudarkin, V. V. Shkunov: Realization of phase conjugation by a surface. Pis'ma Zh. Eksp. Teor. Fiz. **31**, 377 (1980) [English transl.: JETP Lett. **31**, 345 (1980)]

8.2 N. F. Pilipetsky, A. N. Sudarkin, V. V. Shkunov, V. V. Yakimenko: Experimental observation of beam amplification by surface dynamic holograms. Kvantovaya Elektron. **10**, 353 (1983) [English transl.: Sov. J. Quant. Electron. **13**, 265 (1983)]

8.3 A. A. Golubtsov, N. F. Pilipetsky, A. N. Sudarkin, V. V. Shkunov: Optical phase conjugation by light-induced profiling of absorber surface. Kvantovaya Elektron. **8**, 663 (1981) [English transl.: Sov. J. Quant. Electron. **11**, 402 (1981)]

8.4 G. B. Altshuler, K. I. Krylov, V. A. Romanov, L. M. Studenkin, V. Yu. Khramov: Optical phase conjugation by a semiconductor surface. Pis'ma Zh. Tekhn. Fiz. **7**, 1453 (1981)

8.5 A. V. Mamaev, N. A. Melnikov, N. F. Pilipetsky, A. N. Sudarkin, V. V. Shkunov: Optical phase conjugation by a semiconductor surface via plasma reflection. Zh. Eksp. Teor. Fiz. **86**, 232 (1984); J. Opt. Soc. Am. (1984)

8.6 O. V. Garibyan, I. N. Companetz, A. V. Parfyonov, N. F. Pilipetsky, V. V. Shkunov, A. N. Sudarkin, A. V. Sukhov, N. V. Tabiryan, A. A. Vasiliev, B. Ya. Zel'dovich: Optical phase conjugation by microwatt power of reference waves via liquid crystal light valve. Opt. Commun. **38**, 67 (1981)

8.7 T. E. Zaporozhets, S. G. Odulov: Optical phase conjugation at recording dynamic holograms on the surface of a mercury mirror. Pis'ma Zh. Techn. Fiz. **6**, 1391 (1980)

8.8 A. Akhmanov, R. V. Khokhlov: *Problems of Nonlinear Optics* (VINITI, Moscow 1964) (in Russian)

8.9 N. Bloembergen: *Nonlinear Optics* (Benjamin, New York 1965)

8.10 A. Yariv: *Quantum Electronics* (Wiley, New York 1975)

8.11 L. D. Landau, E. M. Lifshitz: *Electrodynamics of Continuous Media* (Pergamon, Oxford 1978)

8.12 P. V. Avizonis, F. A. Hopf, W. D. Bomberger, S. F. Jacobs, A. Tomita, K. H. Womack: Optical phase conjugation in a lithium formate crystal. Appl. Phys. Lett. **31**, 435 (1977)

8.13 S. N. Shostko, Ya. G. Podoba, Yu. A. Anan'ev, V. D. Volosov, A. V. Gorlanov: On a possibility of compensation of optical inhomogeneities in laser devices. Pis'ma Zh. Tekhn. Fiz. **5**, 29 (1979) [English transl.: Sov. Phys. − Tech. Phys. Lett. **5**, 11 (1979)]

8.14 F. A. Hopf, A. Tomita, T. Liepman: Quality of phase conjugation in silicon. Opt. Commun. **37**, 72 (1981)

8.15 E. S. Voronin, V. V. Ivakhnik, V. M. Petnikova, V. S. Solomatin, V. V. Shuvalov: Compensation of phase distortions in three-frequency parametric interaction. Kvantovaya Elektron. **6**, 1306 (1979) [English transl.: Sov. J. Quant. Electron. **9**, 785 (1979)]

8.16 N. P. Andreeva, F. V. Bunkin, D. V. Vlasov, K. Karshiev: Demonstration of phase conjugation of sound on the surface of a liquid. Pis'ma Zh. Tekhn. Fiz. **8**, 104 (1982)

8.17 B. Ya. Zel'dovich, N. F. Pilipetsky, A. N. Sudarkin, V. V. Shkunov: Phase conjugation by a surface. Dokl. Akad. Nauk SSSR **252**, 92 (1980) [English transl.: Sov. Phys.-Dokl. **25**, 377 (1980)]

8.18 A. P. Brysev, F. V. Bunkin, D. V. Vlasov, Yu. E. Kazarov: Model realization of a NbLi parametric phase-conjugating sound amplifier. Pis'ma Zh. Tokhn. Fiz. **8**, 546 (1982)

8.19 A. P. Brysev, N. F. Bunkin, D. V. Vlasov, L. L. Gervits: Parametric phase conjugating plane mirror. Pis'ma Zh. Tekhn. Fiz. **8**, 554 (1982)

8.20 T. R. O'Meara: Wavefront compensation with pseudoconjugation. Opt. Eng. **21**, 271 (1982)

8.21 S. F. Jacobs: Experiments with retrodirective arrays. Opt. Eng. **21**, 281 (1982)

8.22 V. K. Orlov, Yu. Z. Virnik, S. P. Vorotilin, V. B. Gerasimov, Yu. A. Kalinin, A. Yu. Sagalovich: Corner reflectors for dynamic compensation of optical inhomogeneities. Kvantovaya Elektron. **5**, 1389 (1978) [English transl.: Sov. J. Quant. Electron. **8**, 799 (1978)]

8.23 Z. E. Badgasarov, Ya. Z. Virnik, S. P. Vorotilin, V. B. Gerasimov, V. M. Zaika, M. V. Zakharov, V. M. Kazansky, Yu. A. Kalinin, V. K. Orlov, A. K. Piskunov, A. Ya. Sagalovich, A. F. Suchkov, N. D. Ustinov: Investigation of peculiarities of divergence shaping of laser radiation in resonators with retroreflectors. Kvantovaya Elektron. **8**, 2397 (1981) [English transl.: Sov. J. Quant. Electron. **11**, 1465 (1981)]

8.24 I. D. Abella, N. A. Kurnit, S. R. Hartmann: Photon echoes. Phys. Rev. **141**, 391 (1966)

8.25 A. Allen, J. Eberly: *Optical Resonance and Two-Level Atoms* (Wiley, New York 1975)

8.26 E. I. Shtyrkov, V. S. Lovkov, N. G. Yarmukhametov: Gratings induced in Rb by interference of atomic states. Pis'ma Zh. Eksp. Teor. Fiz. **27**, 685 (1978) [English transl.: JETP Lett. **27**, 648 (1978)]

8.27 V. A. Zuikov, V. V. Samartsev, R. G. Usmanov: Reversed light echo in ruby. Pis'ma Zh. Eksp. Teor. Fiz. **31**, 654 (1980) [English transl.: JETP Lett. **31**, 617 (1980)]

8.28 S. L. McCall, E. L. Hahn: Self-induced transparency by pulsed coherent light. Phys. Rev. Lett. **18**, 908 (1967)

8.29 S. A. Shakir: Zero-area optical pulse processing by degenerate four-wave mixing. Opt. Commun. **40**, 151 (1981)

8.30 V. G. Koptev, A. M. Lazaruk, I. P. Petrovich, A. S. Rubanov: Optical phase conjugation at superluminescence. Pis'ma Zh. Eksp. Teor. Fiz. **28**, 468 (1978) [English transl.: JETP Lett. **28**, 434 (1979)]

8.31 A. R. Bogdan, Y. Prior, N. Bloembergen: Opt. Lett. **6**, 82 (1981)

8.32 L. A. Bol'shov, D. V. Vlasov, R. A. Garaev: On spatial resonance at four-photon interaction of copropagating waves in a third-order nonlinear medium. Kvantovaya Elektron. **9**, 83 (1982) [English transl.: Sov. J. Quant. Electron. **12**, 52 (1982)]

8.33 Yu. I. Kucherov, S. A. Lesnik, M. S. Soskin, A. I. Khizhnyak: "Copropagating four-beam interaction in slowly-responding media", in *Phase Conjugation in Nonlinear Media,* ed. by V. I. Bespalov (IPF AN SSSR, Gorky, USSR 1982) p. 111

8.34 A. Gerrard, J. M. Burch: *Introduction to Matrix Methods in Optics* (Wiley, New York 1975)

8.35 Yu. A. Anan'ev: *Optical resonators and the problem of laser radiation divergence* (Nauka, Moscow 1979)

8.36 E. A. Tikhonov, M. T. Shpak: Passive Q-switching of solid-state laser based on stimulated Brillouin scattering. Pis'ma Zh. Eksp. Teor. Fiz. **8**, 282 (1968) [English transl.: JETP Lett. **8**, 173 (1968)]

8.37 A. Z. Grasyuk, V. V. Ragulsky, F.S. Faizullov: Generation of powerful nanosecond pulses by Brillouin scattering and stimulated Raman scattering. Pis'ma zh. Eksp. Teor. Fiz. **9**, 11 (1969) [English transl.: JETP Lett. **9**, 6 (1969)]

8.38 S. A. Lesnik, M. S. Soskin, A. I. Khizhnyak: Laser with an SBS phase-conjugate mirror. Zh. Tekhn. Fiz. **49**, 2257 (1979) [English transl.: Sov. Phys. − Tech. Phys. **24**, 1249 (1979)]

8.39 A. A. Leschev, P. M. Semenov, V. G. Sidorovich: "On the effect of an SBS mirror on laser radiation parameters", in *Optical Phase Conjugation in Nonlinear Media,* ed. by V. I. Bespalov (IPF AN SSSR, Gorky, USSR 1979) p. 135

8.40 S. A. Lesnik, M. G. Reznikov, M. S. Soskin, A. I. Khizhnyak: "Structure of radiation of a phase-conjugate mirror laser", in *Optical Phase Conjugation in Nonlinear Media*, ed. by V. I. Bespalov (IPF AN SSSR, Gorky, USSR 1979) p. 146

8.41 V. I. Bezrodny, F. I. Ibragimov, V. I. Kislenko, R. A. Petrenko, V. L. Strizhevsky, E. A. Tikhonov: Mechanism of Laser Q-switching by intracavity stimulated scattering. Kvantovaya Elektron. **7**, 664 (1980) [English transl.: Sov. J. Quant. Electron. **10**, 382 (1980)]

8.42 V. I. Kislenko, V. L. Strizhevsky: Optical phase conjugation by stimulated temperature scattering and its application. Izv. Akad. Nauk SSSR, Ser. Fiz. **45**, 976 (1981)

8.43 S. G. Karpenko, F. N. Marchevsky, V. L. Strizhevsky: Passive modulation of Q-factor of solid-state lasers on the basis of stimulated Brillouin scattering. Kvantovaya Elektron. **21**, 12 (1981)

8.44 N. N. Il'ichev, A. A. Malyutin, P. P. Pashinin: Laser with diffraction-limited divergence and Q-switching by SBS. Kvantovaya Elektron. **9**, 1803 (1982) [English transl.: Sov. J. Quant. Electron. **12**, 1164 (1982)]

8.45 F. V. Bunkin, D. V. Vlasov, Yu. A. Kravtsov: Phase conjugation and self-focusing of sound through nonlinear interaction with the surface of a liquid. Pis'ma Zh. Tekhn. Fiz. **7**, 325 (1981) [English transl.: Sov. Phys. – Tech. Phys. Lett. **7**, 138 (1981)]

8.46 B. Ya. Zel'dovich, N. V. Tabiryan: On the possibility of phase conjugation by liquid-crystal masks. Kvantovaya Elektron. **8**, 421 (1981) [English transl.: Sov. J. Quant. Electron. **11**, 257 (1981)]

8.47 A. Yariv: On transmission and recovery of three-dimensional image information in optical waveguides. J. Opt. Soc. Am. **66**, 301 (1976)

8.48 N. B. Baranova, B. Ya. Zel'dovich: Bragg three-wave mixing for optical phase conjugation. Dokl. Akad. Nauk SSSR **263**, 325 (1982)

8.49 A. A. Chaban: Instability of elastic oscillations in piezoelectrics in an alternating electric field. Pis'ma Zh. Eksp. Teor. Fiz. **6**, 967 (1967) [English transl.: JETP Lett. **6**, 381 (1967)]

8.50 R. B. Thompson, C. F. Quate: Nonlinear interaction of microwave electric fields and sound in LiNbO$_3$. J. Appl. Phys. **42**, 907 (1971)

8.51 M. Luukkala, J. Surakka: Acoustic convolution and correlation and the associated nonlinear parameters in LiNbO$_3$. J. Appl. Phys. **43**, 2510 (1972)

8.52 F. V. Bunkin, D. V. Vlasov, Yu. A. Kravtsov: To the problem of phase conjugation of sound with amplification of phase-conjugated wave. Kvantovaya Elektron. **8**, 1144 (1981) [English transl.: Sov. J. Quant. Electron. **11**, 687 (1981)]

8.53 N. F. Pilipetsky, A. N. Sudarkin, V. V. Shkunov: Phase conjugation by a twinkling surface. Kvantovaya Elektron. **9**, 835 (1982)

8.54 F. V. Bunkin, D. V. Vlasov, Yu. A. Kravtsov: "Phase conjugation in acoustics", in *Phase Conjugation in Nonlinear Media,* ed. by V. I. Bespalov (IPF AN SSSR, Gorky, USSR 1982) p. 63 (in Russian)

8.55 H. H. Barrett, S. F. Jacobs: Retroreflective array as approximate phase conjugation. Opt. Lett. **4**, 190 (1979)

8.56 I. M. Bel'dyugin: On the properties of resonators with mirrors formed of a set of inverting elements. Kvantovaya Elektron. **8**, 2345 (1981)

8.57 A. N. Oraevsky: Radiation echo. Usp. Fiz. Nauk **91**, 181 (1967)

8.58 N. C. Griffen, C. V. Heer: Focusing and phase conjugation of photon echoes in Na vapor. Appl. Phys. Lett. **33**, 865 (1978)

8.59 M. Fujita, M. Natatsuka, H. Nakanishi, M. Matsouka: Backward echo in two-level system. Phys. Rev. Lett. **42**, 974 (1979)

8.60 A. N. Oraevsky: On the possibility of application of resonantly excited media for optical phase conjugation. Kvantovaya Elektron. **6**, 218 (1979) [English transl.: Sov. J. Quant. Electron. **9**, 119 (1979)]

8.61 A. M. Lazaruk: Optical phase conjugation in amplifying dynamic holograms in dye solution. Kvantovaya Elektron. **6**, 1770 (1979) [English transl.: Sov. J. Quant. Electron. **9**, 1041 (1979)]

8.62 Yu. A. Anan'ev, V. D. Solov'ev: Opt. Spektrosk. **54**, 212 (1983)

8.63 V. D. Solov'ev, A. I. Khizhnyak: Copropagating four-beam interaction. Opt. Spektrosk. **53**, 723 (1982)

8.64 S. A. Lesnik, M. S. Soskin, A. I. Khizhnyak: Proc. of the All-Union Conference: "Problems of control of laser radiation parameters" (Tashkent, USSR 1978) p. 715

8.65 I. M. Bel'dyugin, E. M. Zemskov: Calculation of the field in a laser resonator with a wave-front-reversing mirror. Kvantovaya Elektron. **7**, 1334 (1980) [English transl.: Sov. J. Quant. Electron. **10**, 764 (1980)]

8.66 M. G. Reznikov, A. I. Khizhnyak: On the properties of resonators with a phase-conjugate mirror. Kvantovaya Elektron. **7**, 1105 (1980) [English transl.: Sov. J. Quant. Electron. **10**, 633 (1980)]

8.67 I. M. Bel'dyugin, M. G. Galushkin, E. M. Zemskov: On the properties of resonators with phase-conjugate mirrors. Kvantovaya Elektron. **6**, 38 (1979) [English transl.: Sov. J. Quant. Electron. **9**, 20 (1979)]

8.68 J. Au Yeung, D. Fekete, D. M. Pepper, A. Yariv: A theoretical and experimental investigation of the modes of optical resonators with phase-conjugate mirrors. IEEE J. Quant. Electron. **15**, 1180 (1979)

8.69 P. A. Belanger, A. Hardy, A. E. Siegman: Resonant modes of optical cavities with phase conjugate mirrors. Appl. Opt. **19**, 479; ibid, 602 (1980)

8.70 J. F. Lam, W. P. Brown: Optical resonators with phase-conjugate mirrors. Opt. Lett. **5**, 61 (1980)

Subject Index

This Subject Index covers only those subjects which are not included in the Table of Contents

Aberrator (phase plate) 85, 105, 128
Adaptive optics 83
Amplifying hologram 142
Atmospheric inhomogeneities 17

Collision broadening 176
Complex curvature 222

Depolarizer 128
Diffusion of carriers
  in photorefractive crystals 187
  in semiconductors 186
Diffusion of envelope 118
Discrimination in focused beam 99
Discrimination in superluminescence 213
Dispersion in waveguides 16
Doppler broadening 176
"Doubly-Gaussian" speckle beam 77

Electrostriction force 25
Energy transfer in SS 27
Extinction coefficient
  for spontaneous scattering 33
  of specklon into distortions 91, 111

Fraction of conjugation
  definition 115
  experimental determination 87, 106
Frequency-shifted specklon 113
Fresnel number 224

Gain coefficient for SS 28

Gain for SBS
  dependence on scattering angle 30
  dependence on wavelength 28
  experimental determination 52, 97
  numerical values 31
Geometries of FWM 172
Group velocity, in SS-active media 47, 57

Helmholtz wave equation 72
High-temperature plasma 18
Hydrodynamic equations 26

Instability absolute, convective 165
Interference gratings
  in FWM 6, 145
  in stimulated scattering 10, 27
Irreversibility of mechanical motion 2

Laser fusion 20
Liquid-crystal valve 198

Manley-Row relation 42, 156
Mirror wedge 105
Mask, amplitude or phase 136
Matrix for diaphragm 222

Nonlinear distortions in FWM 168

Orientation of molecules by light 37, 171
Overlap integral in OPC-SS 12

Parametric generation
  in FWM   155
  via SBS   189
Phase matching
  ellipsoid   148
  for anti-Stokes Raman
    components   58
  for FWM   147
  for three-wave mixing   200
Phase plate (aberrator)   85, 105,
  128
Photo-lithography   18
Polarization conjugation
  by FWM   151
  by SS   130
Polarization specklons   126

Reproduction of time behavior
  in SS   54, 131
Reversal of wavefront (optical phase
  conjugation)   3
Reversibility in optics   2

Scalar nature of SBS   63
Semiconductors, for surface OPC
  197
Sound pressure   204

Spatial resonance   113
Speckle field
  correlation functions   71, 74
  "cucumbers" and "serpents"
    76
  moments   70
Spontaneous emission, Einstein
  relation   33
Stationary-phase method   49
"Stick" model of non-mono-
  chromatic pump   55
Stimulated emission   31
Stimulated scattering, Einstein-
  Hellwarth relation to spontaneous
  scattering   34

Thermal defocusing, compensation
  for   19
Thermal gratings   181
Threshold for temperature SS   39
Threshold of SBS, influence of
  parametric interaction   62
Threshold of SS   29
Time contrast of pulse   20
Turned specklon   116

Van Zittert-Zernike theorem   75

Lasers

D.C.Brown

High-Peak-Power Nd: Glass Laser Systems

1981. 135 figures. IX, 276 pages. (Springer Series in Optical Sciences, Volume 25) ISBN 3-540-10516-6

Contents: Glass Laser Physics. – Optical and Physical Properties of Laser Glasses. – Optical-Pump Sources for Nd: Glass Lasers. – Amplified Spontaneous Emission and Parasitic Oscillations in Nd: Glass Amplifiers. – Amplifiers for High-Peak-Power ND: Glass Laser Systems. – Damage Effects in High-Peak-Power Nd: Glass Laser Systems. – Nonlinear Effects in High-Peak-Power Nd: Glass Laser Systems. – The Design of High-Peak-Power Nd: Glass Laser Systems. – Acronyms. – References. – Subject Index.

S.A.Losev

Gasdynamic Laser

1981. 100 figures. X, 297 pages. (Springer Series in Chemical Physics, Volume 12) ISBN 3-540-10503-4

Contents: Introduction. – Basic Concepts of Quantum Electronics. – Physico-Chemical Gas Kinetics. – Relaxation in Nozzle Gas Flow. – Infrared CO_2 Gasdynamic Laser. – Gasdynamic Lasers with Other Active Medium. – Appendixes. – List of the Most Used Symbols. – References. – Subject Index.

Excimer Lasers

Editor: **C.K.Rhodes**
2nd revised and enlarged edition. 1984. 100 figures. XII, 271 pages. (Topics in Applied Physics, Volume 30) ISBN 3-540-13013-6

Contents: *P. W. Hoff, C. K. Rhodes:* Introduction. – *M. Krauss, F. H. Mies:* Electronic Structure and Radiative Transitions of Excimer Systems. – *M. V. McCusker:* The Rare Gas Excimers. – *C. A. Brau:* Rare Gas Halogen Excimers. – *A. Gallagher:* Metal Vapor Excimers. – *D. L. Huestis, G. Marowsky, F. K. Tittel:* Triatomic Rare-Gas-Halide Excimers. – *H. Pummer, H. Egger, C. K. Rhodes:* High-Spectral-Brightness Excimer Systems. – *K. Hohla, H. Pummer, C. K. Rhodes:* Applications of Excimer Systems. – List of Figures. – List of Tables. – Subject Index.

W.Koechner

Solid-State Laser Engineering

1976. 287 figures, 38 tables. XI, 620 pages. (Springer Series in Optical Sciences, Volume 1). ISBN 3-540-90167-1

Contents: Optical amplification. – Properties of solid-state laser materials. – Laser oscillator. – Laser amplifier. – Optical resonantor. – Optical pump systems. – Heat removal. – Q-switches and external switching devices. – Mode locking. – Nonlinear devices. – Design of lasers relevant to their application. – Damage of optical elements. – Appendices: Laser safety. Conversion factors and constants. – Subject Index.

G.Brederlow, E.Fill, K.J.Witte

The High-Power Iodine Laser

1983. 46 figures. IX, 182 pages. (Springer Series in Optical Sciences, Volume 34) ISBN 3-540-11792-X

Contents: Introduction. – Basic Features. – Principles of High-Power Operation. – Beam Quality and Losses. – Design and Layout of an Iodine Laser System. – The ASTERIX III System. – Scalability and Prospect of the Iodine Laser. – Conclusion. – References. – Subject Index.

Springer-Verlag
Berlin
Heidelberg
New York
Tokyo

P.S. Theocaris, E.E. Gdoutos

Matrix Theory of Photoelasticity

1979. 93 figures, 6 tables. XIII, 352 pages
(Springer Series in Optical Sciences,
Volume 11)
ISBN 3-540-08899-7

Contents: Introduction. – Electromagnetic Theory of Light. – Description of Polarized Light. – Passage of Polarized Light Through Optical Elements. – Measurement of Elliptically Polarized Light. – The Photoelastic Phenomenon. – Two-Dimensional Photoelasticity. – Three-Dimensional Photoelasticity. – Scattered-Light Photoelasticity. – Interferometric Photoelasticity. – Holographic Photo elasticity. – The Method of Birefringent Coatings. – Graphical and Numerical Methods in Polarization Optics, Based on the Poincaré Sphere and the Jones Calculus.

W. Schuman, M. Dubas

Holographic Interferometry

From the Scope of Deformation Analysis of Opaque Bodies

1979. 73 figures, 3 tables. X, 194 pages
(Springer Series in Optical Sciences,
Volume 16)
ISBN 3-540-09371-0

From the reviews:
"... Let me say that I have been immensely delighted and stimulated by this book. It is very carefully written, extremely logical, and excellently printed. Its 194 pages cover the work of these men in a thorough and methodical manner that is as enjoyable as it is instructive. ... It is may personal feeling that this book would be wonderful in the hands of a gifted teacher, and that I should very much enjoy taking such a course ..."

J. Opt. Soc. Am.

Y.I. Ostrovsky, M.M. Butusov, G.V. Ostrovskaya

Interferometry by Holography

1980. 184 figures, 4 tables. X, 330 pages
(Springer Series in Optical Sciences,
Volume 20)
ISBN 3-540-09886-0)

This book is concerned with holographic interferometry, i.e. the interferometric applications of holography. It starts with the design of a holographic set-up and the evaluation of information extracted from a hologram. The physical principles as well as practical details are treated thoroughly.

The book will thus assist newcomers entering the field to utilize interferometry via holograms. It is suitable as background material for courses on holography.

Holographic Recording Materials

Editor: H.M. Smith

1977. 96 figures, 17 tables. XIII, 252 pages
(Topics in Applied Physics, Volume 20)
ISBN 3-540-08293-X

Contents: Basic Holographic Principles. – Silver Halide Photographic Materials. – Dichromated Gelatin. – Ferroelectric Crystals. – Inorganic Photochromic Materials. – Thermoplastic Hologram Recording. – Photoresists. – Other Materials and Devices.

Springer-Verlag
Berlin
Heidelberg
New York
Tokyo